Die Grundlehren der mathematischen Wissenschaften

in Einzeldarstellungen
mit besonderer Berücksichtigung
der Anwendungsgebiete

Band 195

Sterling K. Berberian

Baer *-Rings

Springer-Verlag New York Heidelberg Berlin 1972

Sterling K. Berberian

Professor of Mathematics
The University of Texas at Austin

AMS Subject Classifications (1970)
Primary 16 A 34
Secondary 46 L 10, 06 A 30, 16 A 28, 16 A 30

ISBN 0-387-05751-X Springer-Verlag New York Heidelberg Berlin
ISBN 3-540-05751-X Springer-Verlag Berlin Heidelberg New York

 Library of Congress Catalog Card Number 72-189105. Printed in Germany. Monophoto typesetting and offset printing: Zechnersche Buchdruckerei, Speyer. Bookbinding: Konrad Triltsch, Würzburg.

To Kap

Preface

This book is an elaboration of ideas of Irving Kaplansky introduced in his book *Rings of operators* ([**52**], [**54**]).

The subject of Baer $*$-rings has its roots in von Neumann's theory of 'rings of operators' (now called von Neumann algebras), that is, $*$-algebras of operators on a Hilbert space, containing the identity operator, that are closed in the weak operator topology (hence also the name W^*-algebra). Von Neumann algebras are blessed with an excess of structure—algebraic, geometric, topological—so much, that one can easily obscure, through proof by overkill, what makes a particular theorem work.

The urge to axiomatize at least portions of the theory of von Neumann algebras surfaced early, notably in work of S. W. P. Steen [**84**], I. M. Gel'fand and M. A. Naĭmark [**30**], C. E. Rickart [**74**], and von Neumann himself [**53**]. A culmination was reached in Kaplansky's AW^*-algebras [**47**], proposed as a largely algebraic setting for the intrinsic (nonspatial) theory of von Neumann algebras (i. e., the parts of the theory that do not refer to the action of the elements of the algebra on the vectors of a Hilbert space).

Other, more algebraic developments had occurred in lattice theory and ring theory. Von Neumann's study of the projection lattices of certain operator algebras led him to introduce continuous geometries (a kind of lattice) and regular rings (which he used to 'coordinatize' certain continuous geometries, in a manner analogous to the introduction of division ring coordinates in projective geometry).

Kaplansky observed [**47**] that the projection lattice of every 'finite' AW^*-algebra is a continuous geometry. Subsequently [**51**], he showed that certain abstract lattices were also continuous geometries, employing 'complete $*$-regular rings' as a basic tool. A similar style of ring theory—emphasizing $*$-rings, idempotents and projections, and annihilating ideals—underlies both enterprises.

Baer $*$-rings, introduced by Kaplansky in 1955 lecture notes [**52**], are a common generalization of AW^*-algebras and complete $*$-regular rings. The definition is simple: A Baer $*$-ring is a ring with involution in which the right annihilator of every subset is a principal right ideal generated by a projection. The AW^*-algebras are precisely the Baer

∗-rings that happen to be C^*-algebras; the complete ∗-regular rings are the Baer ∗-rings that happen to be regular in the sense of von Neumann.

Although Baer ∗-rings provided a common setting for the study of (1) certain parts of the algebraic theory of von Neumann algebras, and (2) certain lattices, the two themes were not yet fully merged. In AW^*-algebras, one is interested in '∗-equivalence' of projections; in complete ∗-regular rings, 'algebraic equivalence'. The finishing touch of unification came in the revised edition of Kaplansky's notes [**54**]: one considers Baer ∗-rings with a postulated equivalence relation (thereby covering ∗-equivalence and algebraic equivalence simultaneously).

"Operator algebra" would have been a conceivable subtitle for the present book, alluding to the roots of the subject in the theory of operator algebras and to the fact that the subject is a style of argument as well as a coherent body of theorems; the book falls short of earning the subtitle because large areas of the algebraic theory of operator algebras are omitted (for example, general linear groups and unitary groups, module theory, derivations and automorphisms, projection lattice isomorphisms) and because the theory elaborated here—∗-equivalence in Baer ∗-rings—does not develop Kaplansky's theory in its full generality. My reason for limiting the scope of the book to ∗-equivalence in Baer ∗-rings is that the reduced subject is more fully developed and is more attuned to the present state of the theory of Hilbert space operator algebras; the more general theories (as far as they go) are beautifully exposed in Kaplansky's book, and need no re-exposition here.

Perhaps the most important thing to be explained in the Preface is the status of functional analysis in the exposition that follows. The subject of Baer ∗-rings is essentially pure algebra, with historic roots in operator algebras and lattice theory. Accordingly, the exposition is written with two principles in mind: (1) if all the functional analysis is stripped away (by hands more brutal than mine), what remains should stand firmly as a substantial piece of algebra, completely accessible through algebraic avenues; (2) it is not very likely that the typical reader of this book will be unacquainted with, or uninterested in, Banach algebras.

Interspersed with the main development are examples and applications pertaining to C^*-algebras, AW^*-algebras and von Neumann algebras. In principle, the reader can skip over all such matters. One possible exception is the theory of commutative AW^*-algebras (Section 7). The situation is as follows. Associated with every Baer ∗-ring there is a complete Boolean algebra (the set of central projections in the ring); the Stone representation space of a complete Boolean algebra is an extremally disconnected, compact topological space (briefly, a Stonian space); Stonian spaces are precisely the compact spaces $\mathscr{X}$ for which the algebra $C(\mathscr{X})$ of continuous, complex-valued functions on $\mathscr{X}$ is a commutative AW^*-algebra. These algebras play an important role in the dimension theory and reduction theory of finite rings (Chapters 6 and 7). They can be approached either through the theory of commutative Banach algebras (as in the text) or from general topology. The choice is mainly one of order of development; give or take some terminology, commutative AW^*-algebras are essentially a topic in general topology.

The reader can avoid topological considerations altogether by restricting attention to factors, i.e., rings in which 0 and 1 are the only central projections (this amounts to restricting $\mathscr{X}$ to be a singleton). However, the chapter on reduction theory (Chapter 7) then disappears, the objects under study (finite factors) being already irreducible. There is ample precedent for limiting attention to the factorial case the first time through; this is in fact how von Neumann wrote out the theory of continuous geometries [**71**], and the factorial case dominates the early literature of rings of operators.

Baer $*$-rings are a compromise between operator algebras and lattice theory. Both the operator-theorist ("but this is too general!") and the lattice-theorist ("but this can be generalized!") will be unhappy with the compromise, since neither has any need to feel that the middle ground makes his own subject easier to understand; but uncommitted algebraists may find them enjoyable. I personally believe that Baer $*$-rings have the didactic virtue just mentioned, but the issue is really marginal; the test that counts is the test of intrinsic appeal. The subject will flourish if and only if students find its achievements exciting and its problems provocative.

Exercises are graded A–D according to the following mnemonics:

A ("Above"): can be solved using preceding material.

B ("Below"): can be solved using subsequent material.

C ("Complements"): can be solved using outside references.

D ("Discovery"): open questions.

I am indebted to the University of Texas at Austin, and Indiana University at Bloomington, for making possible the research leave at Indiana University in 1970–71 during which this work took form.

Austin, Texas
October, 1971

Sterling K. Berberian

Interdependence of Chapters

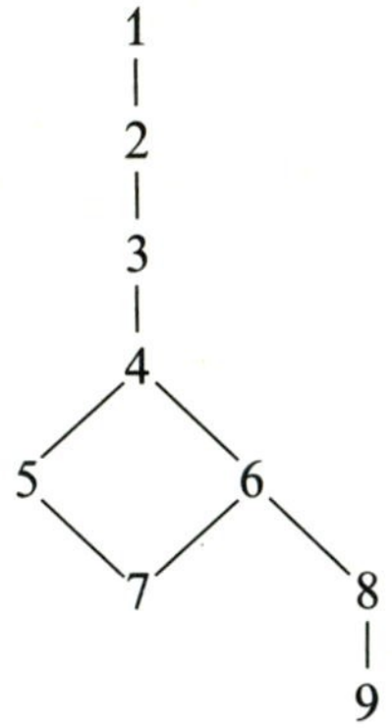

Contents

Part 1: General Theory

Part 2: Structure Theory

Part 3: Finite Rings

Part 1: General Theory

Chapter 1

Rickart $*$-Rings, Baer $*$-Rings, AW^*-Algebras: Generalities and Examples

§ 1. $*$-Rings

All rings considered in this book are associative, and, except in a few of the excercises, they are equipped with an involution in the sense of the following definition:

Definition 1. A *$*$-ring* (or *involutive ring*, or *ring with involution*) is a ring with an involution $x \mapsto x^*$:

$$(x^*)^* = x, \qquad (x+y)^* = x^* + y^*, \qquad (xy)^* = y^* x^* .$$

When A is also an algebra, over a field with involution $\lambda \mapsto \lambda^*$ (the identity involution is allowed), we assume further that

$$(\lambda x)^* = \lambda^* x^*$$

and call A a *$*$-algebra.* {The complex $*$-algebras are especially important special cases, but the main emphasis of the book is actually on $*$-rings.}

The decision to limit attention to $*$-rings is crucial; it shapes the entire enterprise. {For example, functional-analysts contemplating the voyage are advised to leave their Banach spaces behind; the subject of this book is attuned to Hilbert space (the involution alludes to the adjoint operation for Hilbert space operators).} From the algebraic point of view, the intrinsic advantage of $*$-rings over rings is that projections are vastly easier to work with than idempotents.

For the rest of the section, A denotes a $*$-ring.

Definition 2. An element $e \in A$ is called a *projection* if it is self-adjoint ($e^* = e$) and idempotent ($e^2 = e$). We write $\tilde{A}$ for the set of all projections in A; more generally, if S is any subset of A we write $\tilde{S} = S \cap \tilde{A}$.

If x and y are self-adjoint, then $(xy)^* = yx$ shows that xy is self-adjoint if and only if x and y commute ($xy = yx$). It follows that if e and f are projections, then ef is a projection iff e and f commute.

A central feature of the theory is the ordering of projections:

Definition 3. For projections e, f, we write $e \leq f$ in case $e = ef$ (therefore $ef = fe = e$).

Proposition 1. (1) *The relation* $e \leq f$ *is a partial ordering of projections.*

(2) $e \leq f$ *iff* $eA \subset fA$ *iff* $Ae \subset Af$.

(3) $e = f$ *iff* $eA = fA$ *iff* $Ae = Af$.

Proof. (2) If $e \leq f$ then $e = ef \in Af$, hence $Ae \subset Af$. Conversely, if $Ae \subset Af$ then $e = ee \in Ae \subset Af$, say $e = xf$; then $ef = xff = xf = e$, thus $e \leq f$.

(3) In view of (2), $eA = fA$ means $e = ef$ $(= fe)$ and $f = fe$, thus $e = f$.

(1) Immediate from (2) and (3). ∎

Definition 4. Projections e, f are called *orthogonal* if $ef = 0$ (equivalently, $fe = 0$).

Proposition 2. (1) *If* e, f *are orthogonal projections, then* $e + f$ *is a projection.*

(2) *If* e, f *are projections with* $e \leq f$, *then* $f - e$ *is a projection orthogonal to e and* $\leq f$.

Proof. Trivial. {See Exercises 1 and 2 for partial converses.} ∎

In general, extra conditions on A are needed to make $\tilde{A}$ a lattice (such assumptions are invoked from Section 3 onward); a drastic condition that works is commutativity:

Proposition 3. *If* $e, f \in \tilde{A}$ *commute, then* $e \cap f$ *and* $e \cup f$ *exist and are given by the formulas* $e \cap f = ef$ *and* $e \cup f = e + f - ef$.

Proof. Set $g = ef$, $h = e + f - ef$. The proof that g and h have the properties required of $\inf\{e, f\}$ and $\sup\{e, f\}$ is routine. ∎

To a remarkable degree (see Section 15), certain $*$-rings may be classified through their projection-sets; this classification entails the following relation in the set of projections:

Definition 5. Projections e, f in A are said to be *equivalent* (relative to A), written $e \sim f$, in case there exists $w \in A$ such that $w^* w = e$ and $w w^* = f$.

Proposition 4. *With notation as in Definition 5, one can suppose, without loss of generality, that* $w \in fAe$.

Proof. Set $v = w w^* w = we = fw$. Then $v \in fAe$ and $v^* v = (e w^*)(we) = e^3 = e$, $v v^* = (fw)(w^* f) = f^3 = f$. ∎

Definition 6. An element $w \in A$ such that $ww^*w = w$ is called a *partial isometry*.

Proposition 5. *An element $w \in A$ is a partial isometry if and only if $e = w^*w$ is a projection such that $we = w$. Then $f = ww^*$ is also a projection (thus $e \sim f$) and $fw = w$. Moreover, e is the smallest projection such that $we = w$, and f the smallest such that $fw = w$.*

Proof. If w has the indicated property, then $ww^*w = we = w$, thus w is a partial isometry.

Conversely, if $ww^*w = w$ then, setting $e = w^*w$, we have $we = w$ and $e^2 = (w^*w)(w^*w) = w^*(ww^*w) = w^*w = e = e^*$. It follows from $w^*ww^* = w^*$ that w^* is also a partial isometry, and, setting $f = (w^*)^*w^* = ww^*$, we have $w^*f = w^*$, $fw = w$. If g is any projection such that $wg = w$, then $w^*wg = w^*w$, $eg = e$, thus $e \leq g$. Similarly, f is minimal in the property $fw = w$. ∎

Definition 7. With notation as in Proposition 5, e is called the *initial projection* and f the *final projection* of the partial isometry w. The equivalence $e \sim f$ is said to be *implemented* by w.

Proposition 6. *Let e, f be projections in A. Then $e \sim f$ if and only if there exists a partial isometry with initial projection e and final projection f.*

Proof. The "if" part is noted in Proposition 5. Conversely, suppose $e \sim f$. By Proposition 4, there exists $w \in fAe$ with $w^*w = e$ and $ww^* = f$; since $w = we = ww^*w$, w is a partial isometry. ∎

The term "equivalence" is justified by the following proposition:

Proposition 7. *The relation $e \sim f$ is an equivalence relation in $\tilde{A}$:*

(1) $e \sim e$,

(2) $e \sim f$ *implies* $f \sim e$,

(3) $e \sim f$ *and* $f \sim g$ *imply* $e \sim g$.

Moreover,

(4) $e \sim 0$ *if and only if* $e = 0$,

(5) $e \sim f$ *implies* $he \sim hf$ *for every central projection* h.

Proof. (1) $e^*e = ee^* = e^2 = e$.

(2) This is clear from Definition 5.

(3) By Proposition 6, there exist partial isometries w, v such that $w^*w = e$, $ww^* = f$ and $v^*v = f$, $vv^* = g$. Setting $u = vw$, it results from $fw = w$ and $vf = v$ that $u^*u = e$ and $uu^* = g$.

(4) If $e \sim 0$ then, by Proposition 4, there exists $w \in 0Ae = \{0\}$ with $w^*w = e$, thus $e = 0$.

(5) If $w^*w=e$, $ww^*=f$ and h is a projection in the center of A, then $(wh)^*(wh)=hw^*w=he$ and $(wh)(wh)^*=hf$. ∎

The next proposition shows that equivalence is finitely additive; to a large extent, the first four chapters are a struggle to extend this result to families of arbitrary cardinality:

Proposition 8. *If* $e_1,\dots,e_n$ *are orthogonal projections, and* $f_1,\dots,f_n$ *are orthogonal projections such that* $e_i\sim f_i$ *for* $i=1,\dots,n$, *then*

$$e_1+\cdots+e_n\sim f_1+\cdots+f_n.$$

Proof. Let w_i be a partial isometry with $w_i^*w_i=e_i$, $w_iw_i^*=f_i$, and set $w=w_1+\cdots+w_n$. Since $w_ie_i=w_i=f_iw_i$, it is routine to check that w is a partial isometry implementing the desired equivalence. {Incidentally, $we_i=w_i=f_iw$ for all i.} ∎

If two projections are equivalent, what happens 'under' one of them is reflected in what happens under the other:

Proposition 9. *If* $e\sim f$ *via the partial isometry* w, *then the formula*

$$\varphi(x)=wxw^* \qquad (x\in eAe)$$

defines a $*$*-isomorphism* $\varphi\colon eAe\to fAf$. *In particular,* φ *is an order-preserving bijection of the set of projections* $\le e$ *onto the set of projections* $\le f$; φ *preserves orthogonality and equivalence; for every projection* $g\le e$, *one has* $g\sim\varphi(g)$.

Proof. Since $w\in fAe$, $\varphi(x)\in fAf$ for all $x\in eAe$. Obviously φ is additive: $\varphi(x+y)=\varphi(x)+\varphi(y)$.

φ is multiplicative: if $x,y\in eAe$ then $\varphi(xy)=wxyw^*=wxeyw^*$ $=wxw^*wyw^*=\varphi(x)\varphi(y)$.

φ is injective: if $x\in eAe$ and $wxw^*=0$, then $0=w^*(wxw^*)w=exe=x$.

φ is surjective: if $y\in fAf$, then $w^*yw\in eAe$ and $\varphi(w^*yw)=ww^*yww^*$ $=fyf=y$.

For all $x\in eAe$, $\varphi(x^*)=wx^*w^*=(wxw^*)^*=(\varphi(x))^*$. Thus φ is a $*$-isomorphism.

Note that the projections in eAe are precisely the projections $g\in A$ with $g\le e$. If $g\le e$, then $g\sim\varphi(g)$ is implemented by the partial isometry wg. It is clear from the definitions that φ preserves order, orthogonality, and equivalence. ∎

The classification theory requires an ordering of projections more subtle than $e\le f$:

Definition 8. For projections e, f in A, we write $e\precsim f$, and say that e is *dominated* by f, in case $e\sim g\le f$, that is, e is equivalent to a subprojection of f. {This means (Proposition 6) that there exists a partial isometry w with $w^*w=e$ and $ww^*\leq f$.}

Proposition 10. *The relation* $e\precsim f$ *has the following properties:*

(1) $e\leq f$ *implies* $e\precsim f$,

(2) $e\sim f$ *implies* $e\precsim f$,

(3) $e\precsim f$ *and* $f\precsim g$ *imply* $e\precsim g$.

Proof. (3) Choose partial isometries w and v such that $w^*w=e$, $ww^*=f'\leq f$ and $v^*v=f$, $vv^*=g'\leq g$. Set $u=vw$. Then $u^*u=e$ and $ue=u$; by Proposition 5, u is a partial isometry with initial projection e, and $gu=u$ shows that $uu^*\leq g$. ∎

In the presence of lattice completeness, there is a theorem of Schröder-Bernstein type (see also Section 12):

Theorem 1. *If A is a $*$-ring such that $\tilde{A}$ is conditionally complete with respect to the ordering* $e\leq f$, *then* $e\precsim f$ *and* $f\precsim e$ *imply* $e\sim f$.

The proof is based on a general lattice-theoretic result:

Lemma. *If L is a complete lattice and* $\varphi\colon L\to L$ *is an order-preserving mapping, then φ has at least one fixed point.*

Proof. The hypothesis on φ is that $x\leq y$ implies $\varphi(x)\leq\varphi(y)$. Write $0=\inf L$. Let $S=\{x\in L\colon \varphi(x)\geq x\}$ (e.g., $0\in S$) and define $a=\sup S$; we show that $\varphi(a)=a$. For all $x\in S$, we have $a\geq x$, therefore $\varphi(a)\geq\varphi(x)\geq x$; thus $\varphi(a)\geq\sup S=a$. Then $\varphi(\varphi(a))\geq\varphi(a)$, thus $\varphi(a)\in S$, hence $\varphi(a)\leq\sup S=a$. ∎

Proof of Theorem 1. For any projection e, write $[0,e]$ for the set of all projections $g\leq e$, that is, $[0,e]=(eAe)^{\sim}$. The hypothesis on A is that $[0,e]$ is a complete lattice for each projection e.

Suppose $e\precsim f$ and $f\precsim e$. Let w and v be partial isometries with $w^*w=e$, $ww^*=f'\leq f$ and $v^*v=f$, $vv^*=e'\leq e$. The plan of the proof is to construct an order-preserving mapping $\varphi\colon[0,f]\to[0,f]$, to which the Lemma is applied; the mapping φ is taken to be the composite of four mappings $\varphi_1,\dots,\varphi_4$, defined as follows. Define

$$\varphi_1\colon[0,f]\to[0,e]$$

by the formula $\varphi_1(g)=vgv^*$; by Proposition 9, φ_1 is an order-preserving mapping of $[0,f]$ onto $[0,e']\subset[0,e]$. Define

$$\varphi_2\colon[0,e]\to[0,e]$$

by $\varphi_2(g)=e-g$; thus φ_2 is order-reversing. Similarly, define

$$\varphi_3\colon[0,e]\to[0,f]$$

by $\varphi_3(g)=wgw^*$; and

$$\varphi_4: [0, f] \to [0, f]$$

by $\varphi_4(g)=f-g$. Finally, define

$$\varphi: [0, f] \to [0, f]$$

to be the composite $\varphi=\varphi_4\circ\varphi_3\circ\varphi_2\circ\varphi_1$ (thus φ is order-preserving); explicitly,

$$\varphi(g)=f-w(e-vgv^*)w^*$$

for all $g\le f$. Since $[0, f]$ is complete, the Lemma yields a projection $g_0\le f$ such that $\varphi(g_0)=g_0$, thus

$$w(e-vg_0v^*)w^*=f-g_0;$$

setting $x=w(e-vg_0v^*)$, this reads $xx^*=f-g_0$; since $w^*w=e$, one has $x^*x=e-vg_0v^*$, thus

$$(*) \qquad f-g_0\sim e-vg_0v^*.$$

Also, setting $y=vg_0$, one calculates $y^*y=g_0$ and $yy^*=vg_0v^*$, thus

$$(**) \qquad g_0\sim vg_0v^*.$$

Combining $(*)$ and $(**)$, $f\sim e$ by Proposition 8. ∎

Recalling the classical set-theoretic result, one expects that countable lattice operations should suffice for a theorem of Schröder-Bernstein type; a result of this sort is proved in Section 12.

Exercises

1A. Let A be a $*$-ring in which $2x=0$ implies $x=0$, and let e, f be projections in A. (i) If $f-e$ is a projection, then $e\le f$. (ii) If $e+f$ is a projection, then $ef=0$.

2A. Let A be a $*$-ring in which $x^*x+y^*y=0$ implies $x=y=0$, and let e, f be projections in A. Then (i) $e\le f$ iff $f-e=x^*x$ for some $x\in A$. Also (ii) $e\le f$ iff $f-e$ is a projection, and (iii) $e+f$ is a projection iff $ef=0$.

3A. If $e_1,\dots,e_n$ are orthogonal projections, and $f_1,\dots,f_n$ are orthogonal projections such that $e_i\precsim f_i$ $(i=1,\dots,n)$, then $e_1+\dots+e_n\precsim f_1+\dots+f_n$.

4A. Let A be a $*$-ring, let e, f be projections in A such that $e\sim f$, and suppose $e_1,\dots,e_n$ are orthogonal projections with $e=e_1+\dots+e_n$. Then there exist orthogonal projections $f_1,\dots,f_n$ with $f=f_1+\dots+f_n$ and $e_i\sim f_i$ $(i=1,\dots,n)$.

5A. If e, f are projections in a $*$-ring A such that $e\sim f$, then Ae and Af are isomorphic left A-modules. (See Exercise 8 for a converse.)

6A. Pursuing Exercise 5, let A be any ring and let e, f be idempotents in A. The following conditions are equivalent: (a) Ae and Af are isomorphic left A-modules; (b) there exist $x\in fAe$, $y\in eAf$ such that $yx=e$, $xy=f$; (c) there exist $x, y \in A$ such that $yx = e$, $xy = f$. (Such idempotents are sometimes called *algebraically equivalent*.)

7C. If A is a *symmetric* $*$-ring and f is any idempotent in A, then $fA = eA$ for a suitable projection e. {When A has a unity element, symmetry means that $1 + a^*a$ is invertible for every $a \in A$; when A is unitless, symmetry means that $-a^*a$ is quasiregular for every $a \in A$ ($x \in A$ is quasiregular if there exists $y \in A$ with $x + y - xy = 0$).}

8A. If e, f are projections in a $*$-ring A, then algebraic equivalence in the sense of Exercise 6 implies $e \sim f$ in the sense of Definition 5, provided A satisfies the following condition (called the *weak square-root axiom*): for each $x \in A$, there exists $r \in \{x^*x\}''$ (the bicommutant of x^*x [§ 3, Def. 5]) such that $x^*x = r^*r (= rr^*)$.

9A. If A is a ring with unity, and e, f are idempotents in A such that $Ae = Af$, then e and f are similar (that is, $e = xfx^{-1}$ for a suitable invertible element x).

10A. If, in a ring A, e and f are algebraically equivalent idempotents (in the sense of Exercise 6), then the subrings eAe and fAf are isomorphic.

11B. Let A be a Rickart $*$-ring [§ 3, Def. 2] and suppose e, f are projections in A that are algebraically equivalent (in the sense of Exercise 6). Then the $*$-subrings eAe and fAf have isomorphic projection lattices.

12A. Let A be a $*$-ring, e a projection in A, $x \in eAe$, and suppose x is invertible in eAe; say $y \in eAe$, $xy = yx = e$. Then $y \in \{x, x^*\}''$ (the bicommutant of the set $\{x, x^*\}$ [§ 3, Def. 5]).

13A. Let $(A_\iota)_{\iota \in I}$ be a family of $*$-rings and let $A = \prod_{\iota \in I} A_\iota$ be their *complete direct product* (i.e., A is the Cartesian product of the A_ι, endowed with the coordinatewise $*$-ring operations). Then (i) A has a unity element if and only if every A_ι has one; (ii) an element $x = (x_\iota)_{\iota \in I}$ of A is self-adjoint (idempotent, partially isometric, unitary, a projection, etc.) if and only if every x_ι is self-adjoint (idempotent, etc.); (iii) if $e = (e_\iota)_{\iota \in I}$ and $f = (f_\iota)_{\iota \in I}$ are projections in A, then $e \sim f$ iff $e_\iota \sim f_\iota$ for all $\iota \in I$.

14B. Let A be a complex $*$-algebra and let M be a $*$-subset of A (that is, $x \in M$ implies $x^* \in M$). The following conditions on M are equivalent: (a) M is maximal among commutative $*$-subsets of A; (b) M is maximal among commutative $*$-subalgebras of A; (c) $M' = M$; (d) M is maximal among commutative subsets of A. (Here M' denotes the commutant of M in A [§ 3, Def. 5].) Such an M is called a *masa* ('maximal abelian self-adjoint' subalgebra). Every commutative $*$-subset of A can be enlarged to a masa; in particular, if $x \in A$ is *normal* (i.e., $x^*x = xx^*$), then x belongs to some masa.

15A. If $e \precsim h$, where h is a central projection, then $e \le h$.

16A. If $(A_\iota)_{\iota \in I}$ is a family of $*$-rings [$*$-algebras over the same involutive field K], we define their *P*-sum* A as follows: let $B = \prod_{\iota \in I} A_\iota$ be the complete direct product of the A_ι (Exercise 13), write B_0 for the $*$-ideal of all $x = (a_\iota)_{\iota \in I}$ in B such that $a_\iota = 0$ for all but finitely many ι (thus, B_0 is the 'weak direct product' of the A_ι), and define A to be the $*$-subring [$*$-subalgebra] of B generated by B_0 and the set of all projections in B. Thus, if P is the subring [subalgebra] of B generated by the projections of B, then $A = B_0 + P$.

17A. Let A be the $*$-ring of all 2×2 matrices over the field of three elements, with transpose as involution. The set of all projections in A is $\{0, 1, e, 1-e, f, 1-f\}$, where

$$e=\begin{pmatrix}1 & 0\\ 0 & 0\end{pmatrix}, \quad f=\begin{pmatrix}2 & 2\\ 2 & 2\end{pmatrix}.$$

The only equivalences (other than the trivial equivalences $g\sim g$) are $e\sim 1-e$ and $f\sim 1-f$.

18A. The projections e, f of Exercise 17 are algebraically equivalent, but not equivalent.

19A. With notation as in Exercise 17, eAe and fAf are $*$-isomorphic, although e and f are not equivalent.

20A. Let A be a $*$-ring with unity and let A_2 be the $*$-ring of all 2×2 matrices over A (with $*$-transpose as involution). If w is a partial isometry in A, say $w^*w=e$, $ww^*=f$, then the matrix

$$u=\begin{pmatrix}w & 1-f\\ 1-e & w^*\end{pmatrix}$$

is a unitary element of A_2 (that is, $u^*u=uu^*=1$, the identity matrix).

21A. Does the Schröder-Bernstein theorem (i.e., the conclusion of Theorem 1) hold in every $*$-ring?

§ 2. $*$-Rings with Proper Involution

If A is a $*$-ring, the 'inner product' $(x, y)=xy^*$ $(x, y\in A)$ has properties reminiscent of a Hermitian bilinear form: it is additive in x and y, and it is Hermitian in the sense that $(y, x)=(x, y)^*$. Nondegeneracy is a special event:

Definition 1. The involution of a $*$-ring is said to be *proper* if $x^*x=0$ implies $x=0$.

Proposition 1. *In a $*$-ring with proper involution, $xy=0$ if and only if $x^*xy=0$.*

Proof. If $x^*xy=0$, then $y^*x^*xy=0$, $(xy)^*(xy)=0$, $xy=0$. ∎

The theory of equivalence of projections is slightly simplified in a ring with proper involution:

Proposition 2. *In a $*$-ring with proper involution, w is a partial isometry if and only if w^*w is a projection.*

Proof. If $w^*w=e$, e a projection, straightforward computation yields $(we-w)^*(we-w)=0$, hence $we=w$; thus w is a partial isometry [§ 1, Def. 6]. ∎

This is a good moment to introduce a famous example:

Definition 2. A *C^*-algebra* is a (complex) Banach ∗-algebra whose norm satisfies the identity $\|x^*x\| = \|x\|^2$.

Remarks and Examples. 1. The involution of a C^*-algebra is obviously proper.

2. If $\mathscr{H}$ is a Hilbert space then the algebra $\mathscr{L}(\mathscr{H})$ of all bounded linear operators in $\mathscr{H}$, with the usual operations and norm (and with the adjoint operation as involution), is a C^*-algebra; so is any closed ∗-subalgebra of $\mathscr{L}(\mathscr{H})$, and this example is universal:

3. If A is any C^*-algebra, then there exists a Hilbert space $\mathscr{H}$ such that A is isometrically ∗-isomorphic to a closed ∗-subalgebra of $\mathscr{L}(\mathscr{H})$ (Gel'fand-Naĭmark theorem; cf. [**75**, Th. 4.8.11], [**24**, Th. 2.6.1]).

4. If A is a C^*-algebra without unity, and A_1 is the usual algebra unitification of A [§ 5, Def. 3], then A_1 can be normed to be a C^*-algebra [cf. **75**, Lemma 4.1.13].

5. If T is a locally compact (Hausdorff) space and $C_\infty(T)$ is the ∗-algebra of continuous, complex-valued functions on T that 'vanish at infinity', then $C_\infty(T)$ is a commutative C^*-algebra; in order that A have a unity element, it is necessary and sufficient that T be compact (in which case we write simply $C(T)$). Conversely, if A is a commutative C^*-algebra and $\mathscr{M}$ is the character space of A (i.e., the suitably topologized space of modular maximal ideals of A), then the Gel'fand transform maps A isometrically and ∗-isomorphically onto $C_\infty(\mathscr{M})$ (commutative Gel'fand-Naĭmark theorem [cf. **75**, Th. 4.2.2]).

Exercises

1A. In a ∗-ring with proper involution, if e is a normal idempotent (that is, $e^*e = ee^*$ and $e^2 = e$) then e is a projection.

2A. A partial converse to Proposition 1: If A is a ∗-ring in which $xy = 0$ iff $x^*xy = 0$, and if $x \in xA$ for every x (e.g., if A has a unity element, or if A is regular in the sense of von Neumann [§ 51, Def. 1]), then the involution of A is proper.

3A. In a ∗-ring A with proper involution, $x^*xAy = 0$ implies $xx^*Ay = 0$.

4A. The complete direct product of a family of ∗-rings [§ 1, Exer. 13] has a proper involution if and only if every factor does.

5A. If R is a commutative ring $\neq \{0\}$ and if A is the ring of all 2×2 matrices over R, then the correspondence

$$\begin{pmatrix} a & b \\ c & d \end{pmatrix} \mapsto \begin{pmatrix} d & -b \\ -c & a \end{pmatrix}$$

defines an improper involution on A.

6B. The involution is proper in a ∗-ring satisfying the (VWEP)-axiom [§ 7, Def. 3].

§ 3. Rickart $*$-Rings

To motivate the next definitions, suppose A is a $*$-ring with unity, and let w be a partial isometry in A. If $e=w^*w$, it results from $w=ww^*w$ that $wy=0$ iff $ey=0$ iff $(1-e)y=y$ iff $y\in(1-e)A$, thus the elements that right-annihilate w form a principal right ideal generated by a projection. The idea of a Rickart $*$-ring (defined below) is that such a projection exists for every element w (not just the partial isometries). It is useful first to discuss some generalities on annihilators in a ring (always associative):

Definition 1. If A is a ring and S is a nonempty subset of A, we write

$$R(S)=\{x\in A: sx=0 \text{ for all } s\in S\}$$

and call $R(S)$ the *right-annihilator* of S. Similarly,

$$L(S)=\{x\in A: xs=0 \text{ for all } s\in S\}$$

denotes the *left-annihilator* of S.

Proposition 1. *Let S, T and S_ι $(\iota\in I)$ be nonempty subsets of a ring A. Then:*

(1) $S\subset L(R(S))$, $S\subset R(L(S))$;

(2) $S\subset T$ *implies* $R(S)\supset R(T)$ *and* $L(S)\supset L(T)$;

(3) $R(S)=R(L(R(S)))$, $L(S)=L(R(L(S)))$;

(4) $R\left(\bigcup_{\iota\in I} S_\iota\right)=\bigcap_{\iota\in I} R(S_\iota)$, $L\left(\bigcup_{\iota\in I} S_\iota\right)=\bigcap_{\iota\in I} L(S_\iota)$;

(5) *$R(S)$ is a right ideal of A, $L(S)$ a left ideal.*

(6) *If J is a left ideal of A, then $L(J)$ is an ideal of A, in other words, the left-annihilator of a left ideal is a two-sided ideal. Similarly, the right-annihilator of a right ideal is an ideal.*

(7) *If A is an algebra, then $R(S)$ and $L(S)$ are linear subspaces (hence are subalgebras).*

(8) *If A is a $*$-ring then $L(S)=(R(S^*))^*$, where $S^*=\{s^*: s\in S\}$. Similarly, $R(S)=(L(S^*))^*$.*

Proof. There is nothing deeper here than the associative law for multiplication. ∎

Definition 2. A *Rickart $*$-ring* is a $*$-ring A such that, for each $x\in A$, $R(\{x\})=gA$ with g a projection (note that such a projection is unique [§ 1, Prop. 1]). It follows that $L(\{x\})=(R(\{x^*\}))^*=(hA)^*=Ah$ for a suitable projection h.

The example that motivates the terminology:

Definition 3. A C^*-algebra that is a Rickart $*$-ring will be called a *Rickart C^*-algebra.* {These are the 'B_p^*-algebras', first studied by C. E. Rickart [**74**].}

Proposition 2. *If A is a Rickart $*$-ring, then A has a unity element and the involution of A is proper.*

Proof. Write $R(\{0\})=gA$, g a projection. Since $R(\{0\})=A$, we have $A=gA$, thus g is a left unity for A; since A is a $*$-ring, g is a (two-sided) unity element.

Suppose $xx^*=0$. Write $R(\{x\})=hA$, h a projection. By assumption, $x^*\in R(\{x\})$, thus $hx^*=x^*$, $xh=x$; then $h\in R(\{x\})$ yields $0=xh=x$. ∎

Proposition 3. *Let A be a Rickart $*$-ring, $x\in A$. There exists a unique projection e such that* (1) $xe=x$, *and* (2) $xy=0$ *iff* $ey=0$. *Similarly, there exists a unique projection f such that* (3) $fx=x$, *and* (4) $yx=0$ *iff* $yf=0$. *Explicitly,* $R(\{x\})=(1-e)A$ *and* $L(\{x\})=A(1-f)$. *The projections e and f are minimal in the properties* (1) *and* (3), *respectively.*

Proof. Let g be the projection with $R(\{x\})=gA$, and set $e=1-g$; clearly e has the properties (1) and (2). If h is any projection such that $xh=x$, then $x(1-h)=0$, $e(1-h)=0$, $e\le h$. ∎

Definition 4. With notation as in Proposition 3, we write

$$e=\mathrm{RP}(x), \qquad f=\mathrm{LP}(x),$$

called the *right projection* and the *left projection* of x.

Proposition 4. *In a Rickart $*$-ring,*

(i) $\mathrm{LP}(x)=\mathrm{RP}(x^*)$,

(ii) $xy=0$ *iff* $\mathrm{RP}(x)\mathrm{LP}(y)=0$,

(iii) *if w is a partial isometry, then* $w^*w=\mathrm{RP}(w)$ *and* $ww^*=\mathrm{LP}(w)$.

Proof. (i) is obvious, (ii) is immediate from Proposition 3, and (iii) follows from the discussion at the beginning of the section. ∎

The following example is too important to be omitted from the mainstream of propositions (see also [§ 4, Prop. 3]):

Proposition 5. *If $\mathscr{H}$ is a Hilbert space, then $\mathscr{L}(\mathscr{H})$ is a Rickart C^*-algebra. Explicitly, if $T\in\mathscr{L}(\mathscr{H})$ then $\mathrm{LP}(T)$ is the projection on the closure of the range of T, and $I-\mathrm{RP}(T)$ is the projection on the null space of T.*

Proof. Let F be the projection on $\overline{T(\mathscr{H})}$. For an operator S, the following conditions are equivalent: $ST=0$, $S=0$ on $T(\mathscr{H})$, $S=0$ on $\overline{T(\mathscr{H})}$, $SF=0$, $S(I-F)=S$, $S\in\mathscr{L}(\mathscr{H})(I-F)$. Thus

$$L(\{T\})=\mathscr{L}(\mathscr{H})(I-F).$$

This shows that $\mathscr{L}(\mathscr{H})$ is a Rickart $*$-ring (dualize Definition 2). It follows that $\mathrm{RP}(T)=\mathrm{LP}(T^*)$ is the projection on $\overline{T^*(\mathscr{H})}$, therefore $I-\mathrm{RP}(T)$ is the projection on $(\overline{T^*(\mathscr{H})})^{\perp}=(T^*(\mathscr{H}))^{\perp}$, which is the null space of T. ∎

In the preceding example, the projections happen to form a complete lattice (because the closed subspaces do); this is not true in every Rickart $*$-ring (or even in every Rickart C^*-algebra—see Example 2 below), but when a particular supremum does exist, it has annihilation properties:

Proposition 6. *Let A be a Rickart $*$-ring and suppose (e_ι) is a family of projections that has a supremum e. If $x\in A$, then $xe=0$ iff $xe_\iota=0$ for all ι.*

Proof. In view of Proposition 3, the following conditions are equivalent: $xe=0, \mathrm{RP}(x)e=0, e\le 1-\mathrm{RP}(x), e_\iota\le 1-\mathrm{RP}(x)$ for all ι, $\mathrm{RP}(x)e_\iota=0$ for all ι, $xe_\iota=0$ for all ι. ∎

Although the projections of a Rickart $*$-ring need not form a complete lattice, they always form a lattice:

Proposition 7. *The projections of a Rickart $*$-ring form a lattice. Explicitly,*

$$e\cup f=f+\mathrm{RP}[e(1-f)], \qquad e\cap f=e-\mathrm{LP}[e(1-f)]$$

for every pair of projections e, f.

Proof. Write $x=e(1-f)=e-ef$ and set $g=\mathrm{RP}(x)$. Since $xf=0$, one has $gf=0$, thus $f+g$ is a projection. Write $h=f+g$; it is to be shown that h serves as $\sup\{e,f\}$.

Obviously $f\le h$. Also, $e\le h$; for, $0=xg-x=(e-ef)g-(e-ef)$ $=eg-e(fg)-e+ef=eg-e+ef$, thus $e=ef+eg=eh$.

On the other hand, suppose $e\le k$ and $f\le k$; it is to be shown that $h\le k$. Clearly $xk=x$, thus $g\le k$; combined with $f\le k$, this yields $f+g\le k$.

Thus $h=e\cup f$. It follows at once that $e\cap f$ exists and is equal to $1-(1-f)\cup(1-e)$, and that

$$\begin{aligned} e\cap f&=1-\{(1-e)+\mathrm{RP}[(1-f)(1-(1-e))]\}\\ &=e-\mathrm{RP}[(1-f)e]=e-\mathrm{LP}[e(1-f)]. \end{aligned}$$ ∎

If A is a Rickart $*$-ring and B is a $*$-subring of A, then B need not be a Rickart $*$-ring: an obvious example is when B has no unity element, but adjoining a unity element may not be a remedy (see, e.g., Example 1 below). There is, nevertheless, a useful positive result:

Proposition 8. *Let A be a Rickart $*$-ring and let B be a $*$-subring such that* (1) *B has a unity element, and* (2) *$x \in B$ implies* $\mathrm{RP}(x) \in B$. *Then B is also a Rickart $*$-ring; moreover, if $x \in B$, then the right projection of x is the same whether calculated in A or in B.*

Proof. Write e for the unity element of B (we do not require that $e=1$). Let $x \in B$ and let $g=\mathrm{RP}(x)$. By assumption, $g \in B$; since $xe=x$, we have $g \leq e$. If $y \in B$ then $xy=0$ iff $gy=0$ iff $(e-g)y=y$, thus the right-annihilator of x in B is $(e-g)B$. ∎

A useful application:

Corollary. *If A is a Rickart $*$-ring and e is a projection in A, then eAe is also a Rickart $*$-ring (with unambiguous* RP*'s and* LP*'s).*

We interrupt the general theory for two instructive examples.

Proposition 7 suggests the question: If the projections of a $*$-ring form a lattice, is it a Rickart $*$-ring? An obvious counterexample is given in Exercise 2; a subtler example is the following:

Example 1. *There exists a C^*-algebra with unity, whose projections form a lattice, but in which, for certain elements x, there exists no smallest projection e such that $xe=x$.*

Proof. The set of all compact operators a on an infinite-dimensional Hilbert space $\mathscr{H}$ forms a C^*-algebra A; let A_1 be its unitification:

$$A_1=\{a+\lambda 1 \colon a \in A, \lambda \text{complex}\},$$

where 1 is the identity operator on $\mathscr{H}$.

The set of projections in A is the set of all projections of finite rank; this is a lattice. It is easy to see that the projections of A_1 are the projections e, $1-e$, where e is a projection in A (cf. [§ 5, Lemma 2]); it follows that the projections of A_1 form a lattice too. (Sample calculation: if e, f are projections of finite rank, then $e \cup (1-f)=1-(1-e) \cap f$ is the complement of a projection of finite rank.)

Fix a compact operator a with infinite-dimensional range and infinite-dimensional null space, and let g be the projection on the null space of a. Both g and $1-g$ have infinite rank (in other words, $g \notin A_1$).

Suppose h is any projection in A_1 with $ah=a$; we show that there exists a smaller projection k in A_1 with $ak=a$ (thereby thwarting the bid of h to be the 'right projection of a in A_1'). From $a(1-h)=0$ we have $1-h \leq g$, $1-g \leq h$, therefore h has infinite rank; since $h \in A_1$ it follows that $1-h$ has finite rank. Choose a projection e of finite rank such that $1-h \leq e \leq g$ and $1-h \neq e$, and set $k=1-e$. ∎

Example 2. *There exists a Rickart C^*-algebra whose projection lattice is not complete.*

Proof. The motivating idea is that the closed linear span of a family of separable subspaces may be inseparable.

Let $\mathscr{H}$ be an inseparable Hilbert space, $B=\mathscr{L}(\mathscr{H})$ the algebra of all bounded operators on $\mathscr{H}$, and let A be the set of all operators with separable range:

$$A=\{a\in B\colon a(\mathscr{H}) \text{ has a countable dense subset}\}.$$

Obviously A_1 is a C^*-algebra; we show that it is a Rickart $*$-ring with incomplete projection lattice.

Regard B as a Rickart $*$-ring (Proposition 5); thus, if $x\in B$ then $1-\mathrm{RP}(x)$ is the projection on the null space of x. To prove that A_1 is a Rickart $*$-ring, it will suffice, by Proposition 8, to show that $x\in A_1$ implies $\mathrm{RP}(x)\in A_1$. Suppose $x\in A_1$, say $x=a+\lambda 1$. If $\lambda=0$ then $\mathrm{RP}(x)=\mathrm{RP}(a)=\mathrm{LP}(a^*)\in A$ (because the closure of a separable linear subspace is separable). If $\lambda\neq 0$, write $\mathrm{RP}(x)=1-g$, where g is the projection on the null space of x; then $(a+\lambda 1)g=0$, $g=a(-\lambda^{-1}g)\in A$ (because A is an ideal in B), thus $1-g\in A_1$.

Finally, the projection lattice of A_1 is incomplete; for example, if $e\in B$ is any projection such that both e and $1-e$ have inseparable range, then the separable subprojections of e can have no supremum in A_1.

Incidentally, the projections of A, and therefore of A_1, do form a σ-lattice (that is, countable sups and infs exist).

{We remark that if $\dim\mathscr{H}>\aleph_1$ then the game can be played with ideals of B other than A.} ∎

Returning to the general theory, we assemble some elementary facts about commutants in a ring (always assumed associative), with a view to applying them to obtain new Rickart $*$-rings from old.

Definition 5. If A is a ring and S is a nonempty subset of A, the *commutant* of S in A, denoted S', is the set of elements of A that commute with every element of S:

$$S'=\{x\in A\colon xs=sx \text{ for all } s\in S\}.$$

We write $S''=(S')'$, called the *bicommutant* of S; and $S'''=(S'')'$ (but see the following proposition).

Proposition 9. *Let A be a ring, S and T nonempty subsets of A. Then:*

(1) *S' is a subring of A;*

(2) *if $S\subset T$ then $S'\supset T'$;*

(3) *$S\subset S''$;*

(4) *$S'=S'''$;*

(5) *S is commutative iff S'' is commutative.*

(6) *If A has a unity element, then $1\in S'$ and S' contains inverses.*

(7) *If A is an algebra then S' is a subalgebra of A.*

(8) *If A is a $*$-ring and S is a $*$-subset of A, then S' is a $*$-subring of A.*

Proof. (1)–(4) and (7) are routine.

(5) If S is commutative, that is, if $S \subset S'$, then $S'' \subset S' = (S'')'$ by (2) and (4), thus S'' is commutative. The converse follows trivially from (3).

(6) Assuming A has a unity element and $x \in S'$ is invertible in A, say $xy = yx = 1$, it is to be shown that $y \in S'$. For all $s \in S$, we have $xs = sx$; left- and right-multiplication by y yields $sy = ys$.

(8) The assumption is that $s \in S$ implies $s^* \in S$; the easily inferred conclusion is that $x \in S'$ implies $x^* \in S'$. ∎

A very useful application:

Proposition 10. *Let A be a Rickart $*$-ring, and let B be a $*$-subring of A such that $B = B''$ (equivalently, $B = S'$ for some $*$-subset S of A). Then* (i) *$x \in B$ implies* RP$(x) \in B$, *and* (ii) *B is also a Rickart $*$-ring (with unambiguous* RP*'s and* LP*'s).*

Proof. For a $*$-subring B, the equivalence of the condition $B = B''$ with the parenthetical condition follows from (4) and (8) of Proposition 9.

By Proposition 8 we need only prove (i). Let $x \in B$, $e = \mathrm{RP}(x)$, $y \in B'$; since $B = (B')'$ it will suffice to show that $ey = ye$. Since $xy = yx$ and $xe = x$, we have

$$x(y - ye) = xy - xye = xy - yxe = xy - yx = 0,$$

therefore $e(y - ye) = 0$, $ey = eye$. Replacing y by y^* (which is also in B') we have $ey^* = ey^*e$, that is, $ye = eye$; thus $ye = eye = ey$. ∎

Corollary 1. *If A is a Rickart $*$-ring, then the center of A is also a Rickart $*$-ring (with unambiguous* RP*'s). Explicitly, if $xy = yx$ for all $y \in A$, then* RP(x) *is a central projection.*

Proof. Take $B = A'$ in Proposition 10. ∎

Corollary 2. *If A is a Rickart $*$-ring and $x^* = x \in A$, then* RP$(x) \in \{x\}''$.

Proof. $S = \{x\}'$ is a $*$-ring; take $B = S'$ in Proposition 10. ∎

Propositions 1 and 9 show an analogy between annihilators and commutants; the commutant analogue of Proposition 6 is as follows:

Proposition 11. *Let A be a Rickart $*$-ring, B a $*$-subring of A such that $B = B''$. If (e_ι) is a family of projections in B that possesses a supremum e in A, then $e \in B$.*

Proof. Assuming $y \in B'$, it is to be shown that $ey = ye$. Since $e_\iota \in B$ for all ι, we have

$$e_\iota(y - ye) = e_\iota y - e_\iota ye = e_\iota y - ye_\iota e = e_\iota y - ye_\iota = 0,$$

therefore $e(y-ye)=0$ by Proposition 6. The proof continues as in Proposition 10. ∎

Exercises

1A. In a Rickart $*$-ring, every idempotent is similar to a projection; in particular, every central idempotent is a projection.

2A. If A is a $*$-ring such that, for each element $x \in A$, there exists a smallest projection e with $xe=x$, it does not follow that A is a Rickart $*$-ring. (Consider, for example, $A=C(T)$, where T is a connected compact space with more than one point.)

3C. Let A_1 be the algebra constructed in Example 1. For any $x \in A_1$, write $\mathrm{RP}(x)$ for the right projection of x as calculated in the algebra of all bounded operators on $\mathscr{H}$. (i) If $x=a+\lambda 1$ with $a \in A$ and $\lambda \neq 0$, then $\mathrm{RP}(x) \in A_1$ (Riesz-Schauder theory). (ii) If $x=a \in A$ and a has finite rank, then $\mathrm{RP}(x) \in A_1$. (iii) If $x=a \in A$ and a has finite-dimensional null space, then $\mathrm{RP}(x) \in A_1$. (iv) Thus, the pathology exhibited in the proof of Example 1 is the only kind possible.

4A. Let A be a Rickart $*$-ring, I an ideal in A. In order that I be a direct summand of A, it is necessary and sufficient that there exist a central projection h such that $I=hA$. (An ideal I in a ring A is said to be a *direct summand* of A if there exists an ideal J such that $A=I \oplus J$, in the sense that $A=I+J$ and $I \cap J=\{0\}$.)

5A. Let A be a Rickart $*$-ring, and suppose (h_ι) is a family of central projections such that (i) each $h_\iota A$ is a commutative ring, and (ii) $h=\sup h_\iota$ exists. Then hA is also commutative.

6A. A ring A is called *regular* if, for each $x \in A$, there exists an element $y \in A$ such that $xyx=x$; a $*$-ring A is called $*$-*regular* if it is a regular ring with proper involution. For a $*$-ring A with unity, the following conditions are equivalent: (a) A is $*$-regular; (b) for each $x \in A$ there exists a projection e such that $Ax=Ae$; (c) A is regular and is a Rickart $*$-ring.

7A. An associative ring A is called a *Rickart ring* if, for each $x \in A$, there exist idempotents e, f such that $R(\{x\})=eA$, $L(\{x\})=Af$.

(i) Every Rickart ring has a unity element.

(ii) If A is a symmetric $*$-ring [§1, Exer. 7] and if A is a Rickart ring, then A is a Rickart $*$-ring.

8A. If A is a Rickart $*$-ring, B is a $*$-subring satisfying $B=B''$, S is a $*$-subset of B, and C is the commutant of S in B, then C is also a Rickart $*$-ring, with unambiguous RP's and LP's; indeed, $C=C''$.

9A. A set U of nonzero projections in a $*$-ring is said to be *ubiquitous* if, for each nonzero projection e, there exists $f \in U$ with $f \leq e$. If U is a ubiquitous set of projections in a Rickart $*$-ring, then, given any nonzero projection e, there exists an orthogonal family (f_ι) with $f_\iota \in U$ and $e=\sup f_\iota$.

10A. Suppose A is a $*$-ring such that, for each nonzero x in A, there exists $y \in A$ with $xy=g$, g a nonzero projection. Let U be a ubiquitous set of projections in A (see Exercise 9). If e is any nonzero projection in A, then there exists an orthogonal family (f_ι) with $f_\iota \in U$ and $e=\sup f_\iota$.

11A. Let A be a Rickart $*$-ring and suppose e, f are projections in A that are algebraically equivalent (see [§1, Exer. 6]). Then the projection lattices of eAe and

fAf are isomorphic. Explicitly, if $x \in fAe$ and $y \in eAf$ satisfy $yx=e$, $xy=f$, then the formula $\varphi(g)=\mathrm{RP}(xgy)$ defines an order-preserving bijection φ of the projection lattice of eAe onto the projection lattice of fAf. Moreover, if $g_1 \leq g_2 \leq e$, then the projections g_2-g_1 and $\varphi(g_2)-\varphi(g_1)$ are algebraically equivalent.

12A. If $(A_\iota)_{\iota\in I}$ is a family of $*$-rings and A is their complete direct product [§1, Exer. 13], then A is a Rickart $*$-ring iff every A_ι is a Rickart $*$-ring.

13A. If A is a Rickart $*$-ring and B is a $*$-subring that contains every projection of A, then B is also a Rickart $*$-ring (with unambiguous RP's).

14C. A *Boolean ring* is a ring in which every element x is idempotent ($x^2=x$). A Boolean ring with unity is called a *Boolean algebra*.

(i) In a Boolean ring, $x+x=0$ for all elements x.

(ii) A Boolean ring is commutative.

(iii) Every Boolean ring with unity is a Rickart $*$-ring, with the identity involution $x^*=x$; explicitly, $\mathrm{RP}(x)=x$ for all elements x.

(iv) If T is a separated (i.e., Hausdorff) topological space, then the set of all compact open subsets $P, Q, \ldots$ is a Boolean ring with respect to the operations $P\cap Q$ (intersection) and $P \triangle Q=(P-Q)\cup(Q-P)$ (symmetric difference) as product and sum.

(v) If A is any Boolean ring, there exists a topological space T such that A is isomorphic to the Boolean ring of all compact open subsets of T. One can suppose, in addition, that T is locally compact and that the compact open sets are basic for the topology; then T is unique up to homeomorphism (it is called the *Stone representation space* of A; it is compact iff A has a unity element).

(vi) If A is a Boolean ring, then A is a lattice with respect to the operations $x\cap y=xy$ and $x\cup y=x+y+xy$. (In this connection, the term *complete Boolean algebra* is self-explanatory.) A Boolean algebra is complete iff its Stone representation space is Stonian (i.e., the closure of every open set is open).

(vii) Let B be any $*$-ring all of whose projections commute [cf. §15, Exer. 2]. Let A be the set of all projections in B and define products and sums in A by the formulas ef and $e+f-ef$. Then A is a Boolean ring.

15A. Let $S\subset B\subset A$, where A is a ring and B is a subring of A. Write $S^\dagger$ for the commutant of S in B; thus $S^\dagger=B\cap S'$, where S' is the commutant of S in A. Then $S^{\dagger\dagger}\supset B\cap S''$.

16A. If u is an idempotent in a Rickart $*$-ring, such that $\mathrm{LP}(u)$ commutes with $\mathrm{RP}(u)$, then u is a projection.

17A. If $(A_\iota)_{\iota\in I}$ is a family of Rickart $*$-rings, then the P^*-sum of the A_ι [§1, Exer. 16] is also a Rickart $*$-ring.

18A. If A is a $*$-ring, we denote by A° the ring generated by the projections of A, and call it the *reduced ring* of A; explicitly, A° is the set of all finite sums of elements of the form $\pm e_1 e_2 \cdots e_n$, where the e_i are projections in A. We write $e \overset{\circ}{\sim} f$ if there exists a partial isometry w in A° such that $w^*w=e$, $ww^*=f$ (that is, $e\sim f$ relative to A°).

(i) If A is a Rickart $*$-ring, then A° is also a Rickart $*$-ring (with unambiguous RP's).

(ii) If A is a Rickart $*$-ring and e, f are orthogonal projections in A such that $e\sim f$, then $e \overset{\circ}{\sim} f$.

(iii) If A is a Rickart $*$-ring in which 2 is invertible, and if e, f are projections such that $ueu=f$ with u a symmetry (= self-adjoint unitary), then $e \overset{\circ}{\sim} f$.

(iv) If A is the $*$-ring of 2×2 matrices over the field of three elements, then $A^\circ = A$. Explicitly, if e, f are the projections described in [§1, Exer. 17], and if

$$w=\begin{pmatrix}0 & 0\\ 1 & 0\end{pmatrix},$$

then $w=2e-fe$.

§ 4. Baer $*$-Rings

The definition of a Rickart $*$-ring [§ 3, Def. 2] involves the annihilators of singletons; for a Baer $*$-ring, one considers arbitrary subsets:

Definition 1. A *Baer $*$-ring* is a $*$-ring A such that, for every nonempty subset S of A, $R(S)=gA$ for a suitable projection g. (It follows that $L(S)=(R(S^*))^*=(hA)^*=Ah$ for a suitable projection h.)

The relation between Rickart $*$-rings and Baer $*$-rings is the relation between lattices and complete lattices:

Proposition 1. *The following conditions on a $*$-ring A are equivalent:*
(a) *A is a Baer $*$-ring;*
(b) *A is a Rickart $*$-ring whose projections form a complete lattice;*
(c) *A is a Rickart $*$-ring in which every orthogonal family of projections has a supremum.*

Proof. (a) implies (b): Suppose A is a Baer $*$-ring and S is any nonempty set of projections in A. Write $R(S)=(1-e)A$, e a projection; we show that e is a supremum for S.

Since $1-e\in R(S)$, we have $f(1-e)=0$ for all $f\in S$, that is, $f\le e$ for all $f\in S$. On the other hand, if g is a projection such that $f\le g$ for all $f\in S$, then $1-g\in R(S)=(1-e)A$, thus $1-g\le 1-e$, $e\le g$. This shows that $\sup S$ exists and is equal to e. It follows trivially that $\inf S$ exists and is equal to $1-\sup\{1-f: f\in S\}$.

(b) implies (c) trivially.

(c) implies (a): Let S be a nonempty subset of A. Note that $x\in R(S)$ iff $\mathrm{LP}(x)\in R(S)$ [§3, Prop. 4, (ii)]. It follows that if the only projection in $R(S)$ is 0 then $R(S)=\{0\}=0A$. Otherwise, let (e_ι) be a maximal orthogonal family of nonzero projections in $R(S)$. By hypothesis, $e=\sup e_\iota$ exists; we conclude the proof by showing that $R(S)=eA$. At any rate, $e\in R(S)$ [§3, Prop. 6], therefore $eA\subset R(S)$. Conversely, assuming $x\in R(S)$, we assert that $x\in eA$. Setting $y=x-ex$, it is to be shown that $y=0$. Note that $y\in R(S)$, therefore $\mathrm{LP}(y)\in R(S)$. Also, for all ι, we have $e_\iota y=e_\iota x-e_\iota ex=e_\iota x-e_\iota x=0$, therefore $e_\iota \mathrm{LP}(y)=0$. It follows from maximality that $\mathrm{LP}(y)=0$, thus $y=0$. ∎

Completeness of the projection lattice permits a useful characterization of annihilators in a Baer $*$-ring:

Proposition 2. *If A is a Baer $*$-ring and S is a nonempty subset of A, then $R(S)=(1-g)A$, where $g=\sup\{\mathrm{RP}(s)\colon s\in S\}$.*

Proof. The point is that $x\in R(S)$ iff $gx=0$, and this is immediate from [§3, Prop. 6]. ∎

The concept of Baer $*$-ring is due to I. Kaplansky (see [**52**], [**54**]). It arose from his study of two special cases: the 'complete $*$-regular rings' (this is the class of rings described in Exercise 4, (ii)) and the class of C^*-algebras described in the following definition:

Definition 2. An *AW^*-algebra* is a C^*-algebra that is a Baer $*$-ring. {A term consistent with [§3, Def. 3] would be 'Baer C^*-algebra'; but, with most of the literature of such algebras probably already written, it is too late to change.}

AW^*-algebras were proposed by Kaplansky ([**47**], [**48**], [**49**]) as an appropriate setting for certain parts of the algebraic theory of von Neumann algebras (defined below). There are two elemental examples, of which most other examples are refinements or derivatives: (1) the algebras $C(T)$, where T is a Stonian space (these are discussed in Section 7), and (2) the following:

Proposition 3. *If $\mathscr{H}$ is a Hilbert space, then $\mathscr{L}(\mathscr{H})$ is an AW^*-algebra.*

Proof. We know already that $\mathscr{L}(\mathscr{H})$ is a Rickart C^*-algebra [§3, Prop. 5]; in view of Proposition 1, it suffices to check that its projection lattice is complete. Indeed, the correspondence $E\mapsto E(\mathscr{H})$ is an isomorphism of the projection lattice of $\mathscr{L}(\mathscr{H})$ onto the lattice of closed linear subspaces of $\mathscr{H}$, and the latter is obviously complete.

{More explicitly, if $\mathscr{S}$ is a nonempty subset of $\mathscr{L}(\mathscr{H})$ and if F is the projection on the closed linear span of the ranges of the $S\in\mathscr{S}$, it is easy to see directly that $L(\mathscr{S})=\mathscr{L}(\mathscr{H})(I-F)$.} ∎

A capsule résumé of the rest of the section is 'subrings of Baer $*$-rings'. Other 'catogorical' matters requiring attention are direct products (see, e.g., Exercises 7 and 8, and Section 10), quotient rings (see Chapter 7) and matrix rings (see Chapter 9); this is not the place to go into detail, but it ought to be reported here that the general results obtained are shallow, and deeper results are obtained only at the cost of restrictive hypotheses.

In the category of Baer $*$-rings [cf. §23, Exer. 6], what is the appropriate notion of 'subobject'? To be precise, suppose A is a Baer $*$-ring; what is the appropriate notion of 'sub Baer $*$-ring'? An obvious choice would be to consider $*$-subrings B that are themselves Baer $*$-rings. The disadvantage of this choice is that there is ambiguity in the lattice operations: a set S of projections in B has two competing suprema—$\sup S$ as calculated in the projection lattice of A, and $\sup_B S$ as calculated in the projection lattice of B—and they may be different [§7, Exer. 4]. The difficulty is avoided by choosing the right definition:

Definition 3. Let A be a Baer $*$-ring, B a $*$-subring of A. We say that B is a *Baer $*$-subring* of A provided that (i) $x \in B$ implies $\mathrm{RP}(x) \in B$, and (ii) if S is any nonempty set of projections in B, then $\sup S \in B$.

It follows that B is itself a Baer $*$-ring, with unambiguous lattice constructs:

Proposition 4. *If A is a Baer $*$-ring and B is a Baer $*$-subring of A, then B is also a Baer $*$-ring, and* RP*'s,* LP*'s, sups and infs in B are unambiguous.*

Proof. Let $e = \sup\{\mathrm{RP}(x): x \in B\}$; by hypothesis $e \in B$, and it is clear that e is a unity element for B. It follows from [§3, Prop. 8] that B is a Rickart $*$-ring, with unambiguous RP's and LP's. In addition, since the projection lattice of A is complete (Proposition 1), it follows from Definition 3 that the projection lattice of B is also complete, therefore B is a Baer $*$-ring (Proposition 1) with unambiguous sups and infs. ∎

Right-annihilators in a Baer $*$-subring are 'unambiguous' too:

Corollary. *Let A be a Baer $*$-ring, B a Baer $*$-subring of A, e the unity element of B, and let T be a nonempty subset of B. Write $R(T) = (1-g)A$, g a projection. Then the right-annihilator of T in B is $(e-g)B$.*

Proof. From Proposition 2, we know that $g = \sup\{\mathrm{RP}(t): t \in T\}$; since RP's and sups in B are unambiguous, it follows from Proposition 2 (applied to B) that the right-annihilator of T in B is $(e-g)B$. ∎

A convenience of Definition 3 is that it is obviously stable under intersection:

Proposition 5. *If A is a Baer $*$-ring and (B_ι) is any family of Baer $*$-subrings of A, then $\bigcap B_\iota$ is also a Baer $*$-subring.*

{It follows in the obvious way that if S is any subset of a Baer $*$-ring A, there exists a unique smallest Baer $*$-subring of A that contains S, called the Baer $*$-subring *generated* by S. However, the latter concept will play no direct role in the present text; the Baer $*$-subrings that do occur are more conveniently described in other ways.}

The next two propositions yield the major examples of Baer $*$-subring:

Proposition 6. *If A is a Baer $*$-ring and e is a projection in A, then eAe is a Baer $*$-subring of A.*

Proof. Since the projections of eAe are precisely the projections of A that are $\leq e$, the criteria of Definition 3 are trivially verified. ∎

Proposition 7. *If A is a Baer $*$-ring and B is a $*$-subring such that $B = B''$ (equivalently, $B = S'$ for some $*$-subset S of A), then B is a Baer $*$-subring of A.*

Proof. The criteria (i) and (ii) of Definition 3 are fulfilled, by [§ 3, Prop. 10] and [§ 3, Prop. 11], respectively. ∎

Corollary 1. *Let A be a Baer $*$-ring, B a $*$-subring such that $B = B''$. If T is a $*$-subset of B and if C is the commutant of T in B, then C is a Baer $*$-subring of A; indeed, $C = C''$.*

Proof. By definition, $C = B \cap T' = (B' \cup T)'$, where $B' \cup T$ is a $*$-subset of A; quote Proposition 7. ∎

Corollary 2. *If A is a Baer $*$-ring, then the center Z of A is a Baer $*$-subring of A. More generally, if B is a $*$-subring of A such that $B = B''$ and if C is the center of B, then C is a Baer $*$-subring of A; indeed, $C = C''$.*

Proof. Put $T = B$ in Corollary 1; explicitly, $C = B \cap B' = (B' \cup B)'$. ∎

The foregoing results apply, in particular, to AW^*-algebras, modulo the following definition:

Definition 4. Let A be an AW^*-algebra, B a $*$-subalgebra of A. We say that B is an *AW^*-subalgebra* of A provided that (1) B is a norm-closed subset of A (hence B is a C^*-algebra), and (2) B is a Baer $*$-subring of A in the sense of Definition 3.

Proposition 8. *Let A be an AW^*-algebra.*

(i) *If B is an AW^*-subalgebra of A, then B is also an AW^*-algebra, and* RP's, LP's, *sups and infs in B are unambiguous; moreover, if T is a non-empty subset of B and if $R(T) = (1-g)A$, g a projection, then the right-annihilator of T in B is $(e-g)B$, where e is the unity element of B.*

(ii) *If (B_ι) is any family of AW^*-subalgebras of A, then $\bigcap B_\iota$ is also an AW^*-subalgebra.*

(iii) *If e is a projection in A, then eAe is an AW^*-subalgebra of A.*

(iv) *If B is a $*$-subalgebra of A such that $B = B''$ (equivalently, $B = S'$ for some $*$-subset S of A), then B is an AW^*-subalgebra of A.*

(v) *The center of A is an AW^*-subalgebra of A; more generally,*

(vi) *if B is a $*$-subalgebra of A such that $B = B''$, if T is a $*$-subset of B, and if C is the commutant of T in B, then C is an AW^*-subalgebra of A; indeed, $C = C''$.*

Proof. (i) Since B is a C^*-algebra, this is immediate from Proposition 4 and its corollary.

(ii) All that needs to be added to Proposition 5 is the remark that the intersection of a family of closed sets in closed.

(iii) This is immediate from Proposition 6, and the observation that eAe is a closed subset of A.

(iv)—(vi) are immediate from Proposition 7 and its corollaries, plus the observation that commutants in A are closed subsets. ∎

The single most important application of all of this:

Definition 5. If $\mathscr{H}$ is a Hilbert space, a *von Neumann algebra* on $\mathscr{H}$ is a $*$-subalgebra $\mathscr{A}$ of $\mathscr{L}(\mathscr{H})$ such that $\mathscr{A}=\mathscr{A}''$.

Proposition 9. *Every von Neumann algebra is an AW^*-algebra.*

Proof. Immediate from Proposition 3, and Part (iv) of Proposition 8. ∎

Remarks. The converse of Proposition 9 was shown to be false by J. Dixmier [**20**], who showed that there exist commutative AW^*-algebras that cannot be represented ($*$-isomorphically) as von Neumann algebras on any Hilbert space (see [§7, Exer. 2, 3]). At the other end of the commutativity spectrum, it has recently been announced by J. Dyer [**25**] that there exist AW^*-algebras with one-dimensional center that are not $*$-isomorphic to any von Neumann algebra. In between, it is known that certain types of AW^*-algebras are von Neumann-representable provided that their centers are (cf. [§18, Exer. 7, 8]).

Historical note. If $\mathscr{A}$ is a von Neumann algebra on a Hilbert space $\mathscr{H}$, it is elementary that $\mathscr{A}$ is closed in $\mathscr{L}(\mathscr{H})$ for the weak operator topology. Von Neumann showed, conversely, that if $\mathscr{A}$ is a weakly closed $*$-subalgebra of $\mathscr{L}(\mathscr{H})$ containing the identity operator, then $\mathscr{A}=\mathscr{A}''$ [cf. **23**, Ch. I, §3, No. 4]. These algebras have also been called 'W^*-algebras' (von Neumann called them 'rings of operators'), the W referring to closure in the weak operator topology; thus 'AW^*-algebras' are proposed as an abstract (or algebraic) generalization of W^*-algebras (see I. Kaplansky [**47**]).

Exercises

1A. If S is a nonempty set of projections in a Baer $*$-ring A, then

$$(\inf S)A=\bigcap_{e\in S} eA\,,$$

$$(1-\sup S)A=\bigcap_{e\in S}(1-e)A=R(S)\,.$$

2A. (i) A $*$-ring with unity and without divisors of 0 (that is, $xy=0$ implies $x=0$ or $y=0$) is trivially a Baer $*$-ring. Examples: any division ring with involution; any integral domain (in particular, any field) with the identity involution.

(ii) A Baer $*$-ring need not be semisimple.

3A. (i) The following conditions on a Rickart $*$-ring A are equivalent: (a) 0, 1 are the only idempotents in A; (b) 0, 1 are the only projections in A; (c) A has no divisors of 0. Such a ring is trivially a Baer $*$-ring, in which $yx=1$ implies $xy=1$; if, in addition, A has finitely many elements, then A is a (commutative) field.

(ii) A nonzero projection e in a Rickart $*$-ring A is minimal if and only if eAe has no divisors of 0. (A nonzero projection e in a $*$-ring is said to be *minimal* if 0 and e are its only subprojections.)

4A. (i) If A is an associative ring, any two of the following conditions imply the third: (a) if S is any subset of A, then $R(S)=eA$ for a suitable idempotent e; (b) if S is any subset of A, then $L(S)=Af$ for a suitable idempotent f; (c) A has a unity element. A ring with these properties is called a *Baer ring.*

(ii) The following conditions on a $*$-ring A are equivalent: (a) A is a regular Baer $*$-ring; (b) A is a $*$-regular Baer ring. (See [§ 3, Exer. 6] for the definitions of regular and $*$-regular.) Such rings are also called *complete $*$-regular rings.*

5A. If A is a symmetric $*$-ring [§ 1, Exer. 7] and if A is a Baer ring (see Exercise 4), then A is a Baer $*$-ring. In particular, the AW^*-algebras may be described as the C^*-algebras that are Baer rings.

6A. If A is a Baer $*$-ring and B is any $*$-subring of A that contains all projections of A, then B is a Baer $*$-subring of A. In particular, the reduced ring A° of A is a Baer $*$-subring [§ 3, Exer. 18].

7A. If $(A_\iota)_{\iota\in I}$ is a family of Baer $*$-rings and $A=\prod_{\iota\in I} A_\iota$ is their complete direct product, then A is also a Baer $*$-ring. The supremum of a nonempty set of projections in A may be calculated coordinatewise.

8A. If $(A_\iota)_{\iota\in I}$ is a family of Baer $*$-rings, then the P^*-sum of the A_ι [§ 1, Exer. 16] is also a Baer $*$-ring.

9A. Let A be a Baer $*$-ring, B a $*$-subring such that $B=B''$. Let T be a nonempty subset of B, and write $R(T)=eA$, e a projection. Prove that $e\in B$ from first principles.

10A. If A is a Baer$*$-ring and J is a right ideal of A, then $R(J)$ is a direct summand of A [cf. § 3, Exer. 4].

11C. If $\mathscr{H}$ is a Hilbert space and $\mathscr{A}$ is an AW^*-subalgebra of $\mathscr{L}(\mathscr{H})$ containing the identity operator, then $\mathscr{A}$ is a von Neumann algebra on $\mathscr{H}$. (See also [§ 18, Exer. 8].)

12A. Let A be a $*$-ring, $x\in A$, e a projection in A. In analogy with operators on Hilbert space, the condition $exe=xe$ may be interpreted as "e is invariant under x", and $ex=xe$ as "e reduces x".

(i) If $exe=xe$ then $(1-e)x^*(1-e)=x^*(1-e)$. So to speak, if e is invariant under x then its orthogonal complement is invariant under x^*. (If A has no unity element, then the use of 1 is formal: $x^*(1-e)$ stands for x^*-x^*e.)

(ii) Let A be a Baer $*$-ring, $x\in A$, and let $(e_\iota)_{\iota\in I}$ be an orthogonal family of projections such that $e_\iota x e_\iota = x e_\iota$ for all $\iota\in I$. For any subset J of I, define $e_J=\sup\{e_\iota\colon \iota\in J\}$. Then $e_J x e_J = x e_J$.

(iii) Suppose, in addition to the hypothesis of (ii), that $\sup e_\iota=1$. Then $e_\iota x^* e_\iota = x^* e_\iota$ for all ι, hence $xe_\iota=e_\iota x$ for all ι.

13A. Assume (i) C is a $*$-ring with unity and proper involution, (ii) A is a $*$-subring of C, (iii) all projections of C are in A, and (iv) for every $x\in C$, $1+x^*x$

is invertible and $(1+x^*x)^{-1} \in A$. Then A is a Rickart $*$-ring [Baer $*$-ring] iff C is a Rickart $*$-ring [Baer $*$-ring].

14A. A *real AW^*-algebra* is a Banach $*$-algebra A over the field of real numbers, such that $\|x^*x\| = \|x\|^2$ for all $x \in A$, and such that A is a Baer $*$-ring. Every AW^*-algebra is trivially a real AW^*-algebra (by restriction of scalars). A nontrivial example is the algebra of bounded sequences that are 'real at infinity', defined as follows.

Let A be the set of all sequences $x = (\lambda_n)$ of complex numbers, such that $|\lambda_n|$ is bounded and $\operatorname{Im} \lambda_n \to 0$.

(i) With the coordinatewise operations, A is a commutative algebra over the field of real numbers.

(ii) Setting $x^* = (\lambda_n^*)$, A is a $*$-ring with proper involution.

(iii) The projections in A are the sequences of 0's and 1's.

(iv) Every orthogonal family of nonzero projections in A is countable.

(v) A is a Baer $*$-ring.

(vi) Setting $\|x\| = \sup |\lambda_n|$, A is a real Banach algebra, such that $\|x^*x\| = \|x\|^2$. Thus A is a real AW^*-algebra.

(vii) A is symmetric [§ 1, Exer. 7].

(viii) Write $x \geq 0$ in case $\lambda_n \geq 0$ for all n (equivalently, $x = y^*y$ for some $y \in A$). If $x \geq 0$ then there exists a unique $y \geq 0$ in A such that $x = y^2$. It follows that A satisfies that (UPSR)-axiom [§ 13, Def. 10] (note that $S' = A$ for every nonempty subset S of A).

(ix) If $x \in A$, $x \neq 0$, there exists $y \in A$ such that $xy = e$, e a nonzero projection. It follows that A satisfies the (EP)-axiom [§ 7, Def. 1].

15A. The algebra of all bounded operators on a real Hilbert space is a real AW^*-algebra (see Exercise 14).

16A. (i) Every complete Boolean algebra is a commutative Baer $*$-ring [cf. § 3, Exer. 14].

(ii) Let B be a Rickart $*$-ring all of whose projections commute, and let A be the set of all projections in B. Endow A with the Boolean algebra structure described in [§ 3, Exer. 14, (vii)] and the identity involution. Then B is a Baer $*$-ring iff A is a complete Boolean algebra iff A is a Baer $*$-ring.

17A. If A is a Baer$*$-ring and S is a nonempty subset of the center of A, then $R(S)$ is a direct summand of A.

18A. A Rickart $*$-ring with finitely many elements is a Baer $*$-ring.

19A. The ring A of all 2×2 matrices over the field of three elements, with transpose as involution, is a Baer $*$-ring [cf. § 1, Exer. 17].

20C. Let D be a division ring with involution $\lambda \mapsto \lambda^*$, let n be a positive integer, and let D_n be the $*$-ring of all $n \times n$ matrices over D (with $*$-transpose as involution). The following conditions are equivalent: (a) D_n is a Baer $*$-ring; (b) D_n is a Rickart $*$-ring; (c) if $\lambda_i \in D$ $(i = 1, \ldots, n)$ and $\lambda_1^* \lambda_1 + \cdots + \lambda_n^* \lambda_n = 0$, then $\lambda_1 = \cdots = \lambda_n = 0$. (See also [§ 56, Exer. 1].)

21B. An AW^*-algebra B is said to be *AW^*-embedded* in an AW^*-algebra A if (1) B is a closed $*$-subalgebra of A, and (2) if (e_ι) is any orthogonal family of projections in B, then $\sup e_\iota$ (the supremum as calculated in A) is also in B. {It follows that $\sup e_\iota$ is the same whether calculated in A or in B.}

(i) If B is an AW^*-subalgebra of A, then B is AW^*-embedded in A.

(ii) If B is AW^*-embedded in A, then $x \in B$ implies $\mathrm{RP}(x) \in B$ (RP as calculated in A).

In fact, the concepts "AW^*-embedded" and "AW^*-subalgebra" are equivalent (Exercise 27).

22C. Let $\mathscr{H}$ be a Hilbert space and let $\mathscr{A}$ be a norm-closed $*$-subalgebra of $\mathscr{L}(\mathscr{H})$ such that (i) $\mathscr{A}$ is an AW^*-algebra, (ii) $I\in\mathscr{A}$, and (iii) $\mathscr{A}$ is AW^*-embedded in $\mathscr{L}(\mathscr{H})$. Assume, in addition, that (iv) $\mathscr{A}$ is $*$-isomorphic to a von Neumann algebra. Then $\mathscr{A}=\mathscr{A}''$ in $\mathscr{L}(\mathscr{H})$, that is, $\mathscr{A}$ is already a von Neumann algebra on $\mathscr{H}$. {Better yet, the assumption (iv) can be omitted—see Exercise 24.}

23C. Let A be an AW^*-algebra. A linear form φ on A is said to be *positive* if $\varphi(x^*x)\geq 0$ for all $x\in A$. {It follows that if $(e_\iota)_{\iota\in I}$ is any orthogonal family of projections in A, then $\varphi(e_\iota)=0$ for all but countably many $\iota\in I$, and the sum $\sum_{\iota\in I}\varphi(e_\iota)$—that is, the supremum of all finite subsums—is finite and $\leq\varphi(\sup e_\iota)$.} A positive linear form φ on A is said to be *completely additive on projections* (CAP) if $\varphi(\sup e_\iota)=\sum\varphi(e_\iota)$ for every orthogonal family (e_ι) of projections in A. Let $\mathscr{P}$ be the set of all such φ (conceivably, $\mathscr{P}=\{0\}$).

The following conditions on A are equivalent: (a) A is $*$-isomorphic to a von Neumann algebra; (b) there exists a Hilbert space $\mathscr{H}$ and a $*$-monomorphism $\theta\colon A\to\mathscr{L}(\mathscr{H})$ such that $\theta(A)$ is an AW^*-subalgebra of $\mathscr{L}(\mathscr{H})$; (c) there exists a Hilbert space $\mathscr{H}$ and a $*$-monomorphism $\theta\colon A\to\mathscr{L}(\mathscr{H})$ such that the AW^*-algebra $\theta(A)$ is AW^*-embedded in $\mathscr{L}(\mathscr{H})$; (d) for each nonzero $x\in A$, there exists $\varphi\in\mathscr{P}$ with $\varphi(x)\neq 0$; (e) for each nonzero $x\in A$, there exists $\varphi\in\mathscr{P}$ with $\varphi(x^*x)>0$.

24C. Let $\mathscr{H}$ be a Hilbert space and let $\mathscr{A}$ be a norm-closed $*$-subalgebra of $\mathscr{L}(\mathscr{H})$ such that (i) $\mathscr{A}$ is an AW^*-algebra, and (ii) $I\in\mathscr{A}$.

The following conditions on $\mathscr{A}$ are equivalent: (a) $\mathscr{A}$ is an AW^*-subalgebra of $\mathscr{L}(\mathscr{H})$; (b) $\mathscr{A}$ is AW^*-embedded in $\mathscr{L}(\mathscr{H})$; (c) $\mathscr{A}$ is a von Neumann algebra on $\mathscr{H}$ (i.e., $\mathscr{A}=\mathscr{A}''$ in $\mathscr{L}(\mathscr{H})$).

25C. Let $\mathscr{H}$ be a Hilbert space and let $\mathscr{A}$ be a norm-closed $*$-subalgebra of $\mathscr{L}(\mathscr{H})$ such that (i) $I\in\mathscr{A}$, and (ii) if (E_ι) is any orthogonal family of projections in $\mathscr{A}$, then $\sup E_\iota$ (as calculated in $\mathscr{L}(\mathscr{H})$) is also in $\mathscr{A}$.

The following conditions are equivalent: (a) $\mathscr{A}=\mathscr{A}''$ in $\mathscr{L}(\mathscr{H})$, that is, $\mathscr{A}$ is a von Neumann algebra on $\mathscr{H}$; (b) $\mathscr{A}$ is $*$-isomorphic to a von Neumann algebra; (c) $\mathscr{A}$ is an AW^*-algebra; (d) $\mathscr{A}$ is an Rickart C^*-algebra; (e) $T\in\mathscr{A}$ implies $\mathrm{RP}(T)\in\mathscr{A}$ (RP as calculated in $\mathscr{L}(\mathscr{H})$).

26C. Let $\mathscr{H}$ be a Hilbert space and let $\mathscr{A}$ be a norm-closed $*$-subalgebra of $\mathscr{L}(\mathscr{H})$ with $I\in\mathscr{A}$.

The following conditions on $\mathscr{A}$ are equivalent: (a) $\mathscr{A}=\mathscr{A}''$ in $\mathscr{L}(\mathscr{H})$; (b) if $\mathscr{S}$ is any set of positive elements of $\mathscr{A}$ that is increasingly directed (i.e., $S,T\in\mathscr{S}$ implies there exists $U\in\mathscr{S}$ with $S\leq U$ and $T\leq U$) and bounded in norm, then $\sup\mathscr{S}$ (as calculated in the partially ordered set of self-adjoint elements of $\mathscr{L}(\mathscr{H})$) is also in $\mathscr{A}$.

27A. Let A be a Baer $*$-ring and let B be a $*$-subring of A. The following conditions on B are equivalent: (a) B is a Baer $*$-subring of A; (b) $x\in B$ implies $\mathrm{RP}(x)\in B$ (RP as calculated in A), and if (e_ι) is any orthogonal family of projections in B, then $\sup e_\iota$ (as calculated in A) is also in B.

§ 5. Weakly Rickart $*$-Rings

What about the unitless case? The answer, in principle, is to modify the arguments or attempt to adjoin a unity element, and in practice it is congenial and instructive to do a little of both. First, let us review the unit case:

Proposition 1. *A $*$-ring A is a Rickart $*$-ring if and only if* (i) *A has a unity element, and* (ii) *for each $x \in A$, there exists a projection e such that $R(\{x\}) = R(\{e\})$.*

Proof. If A is a Rickart $*$-ring, then A has a unity element [§3, Prop. 2], and $e = \mathrm{RP}(x)$ meets the requirement of (ii) [§3, Prop. 3].

Conversely, if A satisfies (i) and (ii), then the relations $R(\{x\}) = R(\{e\}) = (1-e)A$ show that A is a Rickart $*$-ring. ∎

Condition (ii) alone is of no help (Exercise 1). The appropriate extension to the unitless case is as follows:

Definition 1. Let A be a $*$-ring, $x \in A$, e a projection in A. We say that e is an *annihilating right projection* (briefly, ARP) of x in case (i) $xe = x$, and (ii) $xy = 0$ implies $ey = 0$. The notion of *annihilating left projection* (ALP) is defined dually. (Thus e is an ARP of x iff it is an ALP of x^*.) We say that A is a *weakly Rickart $*$-ring* if every x in A has an ARP.

If x and e are related as in Definition 1 then e is uniquely determined by x; that is, if e and f are ARP's for x then $e = f$. Indeed, $x(e-f) = xe - xf = x - x = 0$, therefore $e(e-f) = 0$ and $f(e-f) = 0$, thus $e = ef = fe = f$.

Definition 2. If x and e are related as in Definition 1, we write $e = \mathrm{RP}(x) = \mathrm{LP}(x^*)$. This is consistent with the usage for Rickart $*$-rings, in view of the following:

Proposition 2. *The following conditions on a $*$-ring A are equivalent:*
(a) *A is a Rickart $*$-ring;*
(b) *A is a weakly Rickart $*$-ring with unity.*

Proof. Immediate from Proposition 1 and Definition 1. ∎

A characterization of weakly Rickart $*$-rings:

Proposition 3. *The following conditions on a $*$-ring A are equivalent:*
(a) *A is a weakly Rickart $*$-ring;*
(b) *A has proper involution, and, for each $x \in A$, there exists a projection e such that $R(\{x\}) = R(\{e\})$.*

Proof. (a) implies (b): Suppose $xx^* = 0$ and let $e = \mathrm{RP}(x)$. On the one hand, $xx^* = 0$ implies $ex^* = 0$, thus $xe = 0$; on the other hand, $xe = x$, thus $x = 0$.

(b) implies (a): Assuming $R(\{x\}) = R(\{e\})$, e a projection, it will suffice to show that $xe = x$. For all $y \in A$, $e(y - ey) = 0$, hence $x(y - ey) = 0$

by hypothesis. Putting $y=x^*$, we have $xex^*=xx^*$; a straightforward computation yields $(x-xe)(x-xe)^*=0$, therefore $x-xe=0$. ∎

The obvious question posed by Proposition 2: Can every weakly Rickart $*$-ring be embedded in a Rickart $*$-ring (e.g., by some process of 'adjoining a unity')? In the last part of the section we show that the answer is often yes, but in general the problem is open; thus there is some point in developing the basic properties of weakly Rickart $*$-rings directly from the definition, as we do in the next several propositions.

Proposition 4. *In a weakly Rickart $*$-ring,*
(i) $xy=0$ *if and only if* $\mathrm{RP}(x)\mathrm{LP}(y)=0$;
(ii) $e=\mathrm{RP}(x)$ *is the smallest projection such that* $xe=x$.

Proof. (i) is immediate from the definitions.
(ii) If f is any projection with $xf=x$, then $x(f-e)=xf-xe=x-x=0$, therefore $e(f-e)=0$, $e=ef$, thus $e\leq f$. ∎

Proposition 5. *Let A be a weakly Rickart $*$-ring, and let B be a $*$-subring of A such that $B=B''$ (equivalently, $B=S'$ for some $*$-subset S of A). Then* (i) $x\in B$ *implies* $\mathrm{RP}(x)\in B$, (ii) *B is also a weakly Rickart $*$-ring (with unambiguous* RP*'s and* LP*'s).*

Proof. Same as [§ 3, Prop. 10]. ∎

Corollary 1. *If A is a weakly Rickart $*$-ring, then the center of A is also a weakly Rickart $*$-ring. Explicitly, if x is in the center of A, then* $\mathrm{RP}(x)$ *is a central projection.*

Corollary 2. *If A is a weakly Rickart $*$-ring and $x^*=x\in A$, then* $\mathrm{RP}(x)\in\{x\}''$.

A useful technique for reducing to the Rickart $*$-ring case:

Proposition 6. *If A is a weakly Rickart $*$-ring and e is a projection in A, then eAe is a Rickart $*$-ring, with unambiguous* RP*'s and* LP*'s.*

Proof. If $x\in eAe$ then $\mathrm{RP}(x)\in eAe$ by Proposition 4, (ii), thus A is a weakly Rickart $*$-ring with unambiguous RP's and LP's; quote Proposition 2. ∎

Proposition 7. *The projections of a weakly Rickart $*$-ring form a lattice. Explicitly,*

$$e\cup f=f+\mathrm{RP}[e-ef], \quad e\cap f=e-\mathrm{LP}[e-ef]$$

for every pair of projections e, f.

Proof. The first formula is proved exactly as in [§ 3, Prop. 7]; whereupon, dropping down to $(e\cup f)A(e\cup f)$, the whole matter may be referred back to the case of a Rickart $*$-ring (Proposition 6). ∎

Definition 3. Let A be a $*$-ring. We define a *unitification* A_1 of A, provided there exists an *auxiliary ring* K, called the ring of *scalars* (denoted $\lambda, \mu, \ldots$), such that

(1) K is an integral domain with involution (necessarily proper), that is, K is a commutative $*$-ring with unity and without divisors of zero (the identity involution is permitted);

(2) A is a $*$-algebra over K (that is, A is a left K-module such that, identically, $1a=a$, $\lambda(ab)=(\lambda a)b=a(\lambda b)$, and $(\lambda a)^*=\lambda^* a^*$); and

(3) A is a torsion-free K-module (that is, $\lambda a=0$ implies $\lambda=0$ or $a=0$).

In two special situations, we agree on the following choices of auxiliary ring:

(I) If A is a $*$-algebra (in the usual sense) over a field F with involution, then we take $K=F$ (note that (3) is satisfied automatically, since A is a vector space over F).

(II) If A is not a $*$-algebra over a field with involution, but the additive group of A is torsion-free, then we take $K=\mathbb{Z}$ (the integers, equipped with the identity involution).

Returning to the general case, we define

$$A_1 = A \oplus K \qquad \text{(the additive group direct sum)};$$

thus $(a, \lambda)=(b, \mu)$ means, by definition, that $a=b$ and $\lambda=\mu$, and addition in A_1 is defined by the formula

$$(a, \lambda)+(b, \mu)=(a+b, \lambda+\mu).$$

Defining

$$(a, \lambda)(b, \mu)=(ab+\mu a+\lambda b, \lambda\mu),$$
$$\mu(a, \lambda)=(\mu a, \mu\lambda),$$
$$(a, \lambda)^*=(a^*, \lambda^*),$$

evidently A_1 is also a $*$-algebra over K, has unity element $(0, 1)$, and is torsion-free; moreover, A is a $*$-ideal in A_1. We write $a+\lambda 1$ in place of (a, λ).

The following lemmas are aimed at the proof that if A is a weakly Rickart $*$-ring possessing a unitification A_1 in the sense of Definition 3, then A_1 is a Rickart $*$-ring. The proofs of the first two are straightforward.

Lemma 1. *Notation as in Definition 3. If the involution of A is proper, then so is the involution of A_1.*

Lemma 2. *Notation as in Definition 3. The projections of A_1 are the projections e, $1-e$, where e is a projection in A.*

Lemma 3. *Notation as in Definition 3. Let $a \in A$, e a projection in A. Then e is an* ARP *of a in A if and only if it is an* ARP *of a in A_1.*

Proof. The "if" part is trivial.

Conversely, assume e is an ARP of a in A, and suppose $ay=0$, where $y=b+\mu 1\in A_1$. Then

$$0=ay=(ae)y=a(ey)=a(eb+\mu e),$$

where $eb+\mu e\in A$, therefore $e(eb+\mu e)=0$, thus $ey=0$. ∎

It is in the following key lemma that the torsion-free hypothesis (condition (3) of Definition 3) is used:

Lemma 4. *Notation as in Definition 3. Assume, in addition, that A is a weakly Rickart $*$-ring. If $a^*=a\in A$ and $\lambda\in K$, $\lambda\neq 0$, then there exists a largest projection g in A such that $ag=\lambda g$.*

Proof. Let $e=\mathrm{RP}(a)=\mathrm{LP}(a)$; then $a-\lambda e\in eAe$. Define $h=\mathrm{RP}(a-\lambda e)$; thus $h\leq e$ and h is the smallest projection in A such that $(a-\lambda e)h=a-\lambda e$ (Proposition 4). Setting $g=e-h$, we have

$$0=(a-\lambda e)g=ag-\lambda eg=ag-\lambda g,$$

thus $ag=\lambda g$.

On the other hand, suppose k is a projection in A such that $ak=\lambda k$. Since $ea=a$, we have $e(\lambda k)=\lambda k$, $\lambda(k-ek)=0$; since A is a torsion-free K-module, we have $k-ek=0$, thus $k\leq e$. Then

$$(a-\lambda e)k=ak-\lambda ek=\lambda k-\lambda k=0,$$

therefore $hk=0$, thus $k\leq e-h=g$. {The condition that $a^*=a$ can be dropped (see Theorem 1 below).} ∎

Lemma 5. *Let B be a $*$-ring with proper involution, let $x\in B$, and let e be a projection in B. Then e is an* ARP *of x if and only if it is an* ARP *of x^*x.*

Proof. Since $R(\{x\})=R(\{x^*x\})$ [§ 2, Prop. 1], the essential point is that $x^*xe=x^*x$ implies $(xe-x)^*(xe-x)=0$. ∎

Theorem 1. *If A is a weakly Rickart $*$-ring, and if A has a unitification A_1 in the sense of Definition 3, then A_1 is a Rickart $*$-ring.*

Proof. Since the involution of A is proper (Proposition 3) so is the involution of A_1 (Lemma 1), hence it will suffice to show that every self-adjoint element of A_1 has an ARP (see Lemma 5 and Proposition 2). Let $x^*=x\in A_1$. Say $x=a+\lambda 1$; thus $a^*=a$ and $\lambda^*=\lambda$.

If $\lambda=0$, that is, if $x=a\in A$, and if e is the ARP of a in A, then e is also an ARP of a in A_1 (Lemma 3).

Assume $\lambda\neq 0$. By Lemma 4, there exists a largest projection g in A such that $ag=(-\lambda)g$, that is, $xg=0$. Set $e=1-g$; we show that e

is an ARP of x in A_1. Of course $xe=x$. Assuming $xy=0$, where $y=b+\mu 1 \in A_1$, it is to be shown that $ey=0$. We have

$$0=(a+\lambda 1)(b+\mu 1)=(ab+\mu a+\lambda b)+(\lambda\mu)1,$$

therefore $0=\lambda\mu$ (hence $\mu=0$ since K has no divisors of zero) and $0=ab+\mu a+\lambda b=ab+\lambda b$, thus

$$(*) \qquad ab+\lambda b=0.$$

Let $f=\mathrm{LP}(b)$ in A. Writing $(*)$ as $(a+\lambda f)b=0$, we infer that $(a+\lambda f)f=0$, thus $af=(-\lambda)f$; it follows that $f\leq g$, hence $f(1-g)=0$, that is, $fe=0$. Then

$$ey=e(b+\mu 1)=e(b+01)=eb=e(fb)=(ef)b=0b=0. \quad \blacksquare$$

Exercises

1A. Let A be a $*$-ring with $A^2=\{0\}$. Then 0 is the only projection in A, and $R(\{x\})=R(\{0\})=A$ for every element x, but $x0\neq x$ when $x\neq 0$.

2A. Let A be a weakly Rickart $*$-ring, and let $(B_\iota)_{\iota\in I}$ be a family of $*$-subrings of A such that, for each ι, $x\in B_\iota$ implies $\mathrm{RP}(x)\in B_\iota$. Then $B=\bigcap_{\iota\in I} B_\iota$ is also a $*$-subring such that $x\in B$ implies $\mathrm{RP}(x)\in B$, hence B is a weakly Rickart $*$-ring with unambiguous RP's and LP's.

3A. Let A be a weakly Rickart $*$-ring, B a $*$-subring of A such that $B=B''$. If (e_ι) is a family of projections in B that has a supremum e in A, then $e\in B$.

4A. Let A be a weakly Rickart $*$-ring, and suppose (e_ι) is a family of projections in A that has a supremum e. If $x\in A$ and $xe_\iota=0$ for all ι, then $xe=0$.

5A. If A is a weakly Rickart $*$-ring such that every orthogonal family of projections in A has a supremum, then A is a Baer $*$-ring.

6A. A weakly Rickart $*$-ring with finitely many elements is a Baer $*$-ring.

7A. Let A be a weakly Rickart $*$-ring, let $x,y,z\in A$, and let $e=\mathrm{RP}(x)$, $f=\mathrm{RP}(y)$, $g=\mathrm{RP}(z)$. If $x^*x+y^*y=0$, then $e=f$. If $x^*x+y^*y+z^*z=0$, then $e\cup f=e\cup g=f\cup g$.

8A. Let A be a weakly Rickart $*$-ring such that $x^*x+y^*y=0$ implies $x=y=0$. (i) If $e=\mathrm{RP}(a)$ and $f=\mathrm{RP}(b)$, then $e\cup f=\mathrm{RP}(a^*a+b^*b)$. (ii) In particular, $e\cup f=\mathrm{RP}(e+f)$ for every pair of projections e, f in A. (iii) If e, f are projections in A such that $f-e=a^*a$ for some $a\in A$, then $e\leq f$.

9C. If A is a weakly Rickart C^*-algebra, then A_1 (with suitably defined norm) is a Rickart C^*-algebra.

10A. If A is a weakly Rickart $*$-ring and $a\in A$, there exists a largest projection g in A such that $xg=g$.

11A. Let A be the set of all complex sequences $x=(\lambda_n)$ such that $\lambda_n=0$ ultimately, with the usual $*$-algebra structure. Thus A_1 is the set of all 'ultimately constant' sequences. Then A is a weakly Rickart $*$-ring whose projection lattice is conditionally complete, but A_1 is not a Baer $*$-ring. (See also the ring A described in [§ 3, Example 2].)

12A. Let A be a weakly Rickart $*$-ring without unity, and suppose $(e_\iota)_{\iota\in I}$ is a maximal orthogonal family of nonzero projections in A. Then (i) I is infinite; (ii) if A has a unitification A_1 in the sense of Definition 3, then the family $(e_\iota)_{\iota\in I}$ has supremum 1 in A_1.

13A. Let A be a weakly Rickart $*$-ring without unity, and suppose that the projection lattice of A is countable (see, e. g., Exercise 11). If A admits a unitification A_1 in the sense of Definition 3, then A_1 cannot be a Baer $*$-ring.

14A. Let B be a Boolean algebra [§ 3, Exer. 14] containing more than two elements. In order that there exist a Boolean ring A without unity, such that $A_1=B$, it is necessary and sufficient that B be infinite.

15A. Does there exist a weakly Rickart $*$-ring $A\neq\{0\}$ without unity, possessing a unitification A_1 in the sense of Definition 3, such that A_1 is a Baer $*$-ring?

16A. If A is a $*$-ring in which $\sum_1^n x_i^* x_i=0$ implies $x_1=\cdots=x_n=0$ (for any n), then the additive group of A is torsion-free.

17A. Let A be a $*$-ring, let K be a commutative $*$-ring with unity, and suppose that A is an algebra over K, that is, A is a left K-module satisfying $1a=a$ and $\lambda(ab)=(\lambda a)b=a(\lambda b)$ for all $a,b\in A$, $\lambda\in K$. Write $\mathscr{E}=\mathscr{E}(A,+)$ for the endomorphism ring of the additive group of A; each $a\in A$ determines an element L_a of $\mathscr{E}$ via $L_a x=ax$; and each $\lambda\in K$ determines an element λI of $\mathscr{E}$ via $(\lambda I)x=\lambda x$.

Let $A_1=A\oplus K$ with the $*$-algebra operations described in Definition 3. Each (a,λ) in A_1 determines an element $L_a+\lambda I$ of $\mathscr{E}$, and the mapping $(a,\lambda)\mapsto L_a+\lambda I$ is a ring homomorphism of A_1 onto a subring $\mathscr{A}_1$ of $\mathscr{E}$, namely, the subring of $\mathscr{E}$ generated by the left-multiplications L_a and the homotheties λI. Defining $\mu(L_a+\lambda I)$ to be the ring product $(\mu I)(L_a+\lambda I)$, $\mathscr{A}_1$ becomes an algebra over K, and the mapping $(a,\lambda)\mapsto L_a+\lambda I$ an algebra homomorphism of A_1 onto $\mathscr{A}_1$. Let N be the kernel of this mapping, and write $\hat{A}_1=A_1/N$ for the quotient algebra. Denote the coset $(a,\lambda)+N$ by $[a,\lambda]$; thus $[a,\lambda]$ is the equivalence class of (a,λ) under the equivalence relation $(a,\lambda)\equiv(b,\mu)$ defined by $ax+\lambda x=bx+\mu x$ for all $x\in A$.

(i) Write $\bar{a}=[a,0]$ for $a\in A$; the mapping $a\mapsto\bar{a}$ is an algebra homomorphism of A into $\hat{A}_1$. Writing $u=[0,1]$ for the unity element of $\hat{A}_1$, we have $[a,\lambda]=\bar{a}+\lambda u$.

(ii) If $L(A)=\{0\}$ (e. g., if the involution of A is proper), then the mapping $a\mapsto\bar{a}$ is injective, and we may regard A as 'embedded' in $\hat{A}_1$.

(iii) If the involution of A is proper, then $[a,\lambda]=0$ implies $[a^*,\lambda^*]=0$, and the formula $[a,\lambda]^*=[a^*,\lambda^*]$ defines unambiguously a proper involution in $\hat{A}_1$.

(iv) If A is a weakly Rickart $*$-ring, $a\in A$, and $e=\mathrm{RP}(a)$, then $\bar{e}$ is an ARP of $\bar{a}$ in $\hat{A}_1$.

18D. *Problem:* Can every weakly Rickart $*$-ring be embedded in a Rickart $*$-ring? with preservation of RP's?

§ 6. Central Cover

The concept of central cover is most useful in Baer $*$-rings, but it can be formulated in any $*$-ring:

Definition 1. Let A be a $*$-ring, $x \in A$. We say that x possesses a *central cover* if there exists a smallest central projection h such that $hx = x$; that is, (i) h is a central projection, (ii) $hx = x$, and (iii) if k is a central projection with $kx = x$ then $h \le k$. If such a projection h exists, it is clearly unique; we call it the central cover of x, denoted $h = C(x)$; when it is necessary to emphasize the ring A, we write $h = C_A(x)$.

Central cover works best when the lattice of central projections is complete, but a few remarks can be made in a more general setting:

Proposition 1. *Let A be a $*$-ring in which every projection e has a central cover $C(e)$.*

(i) *If e is a projection in A, then $C(e)$ is the smallest central projection h such that $e \le h$.*

(ii) *If e and f are projections such that $e \le f$, then $C(e) \le C(f)$.*

(iii) $C(he) = hC(e)$ *if h is a central projection.*

(iv) $e = \sup e_\iota$ *implies* $C(e) = \sup C(e_\iota)$.

(v) *If the involution of A is proper, then $e \sim f$ implies $C(e) = C(f)$, and $e \lesssim f$ implies $C(e) \le C(f)$.*

Suppose, in addition, that A is a Rickart $*$-ring, with center Z. Then:

(vi) *Every element x of A has a central cover $C(x)$; explicitly,* $C(x) = C(\mathrm{LP}(x)) = C(\mathrm{RP}(x))$.

(vii) $C(xx^*) = C(x^*x)$ *for all* $x \in A$.

(viii) $C(hx) = hC(x)$ *if h is a central projection.*

(ix) *If $x \in A$ and $z \in Z$, then $zx = 0$ iff $zC(x) = 0$.*

Proof. (i) is obvious, and (ii) follows immediately from it.

(iii) Let $k = C(he)$; it is to be shown that $k = hC(e)$. By (ii) we have $k \le h$ and $k \le C(e)$, thus $k \le hC(e)$. On the other hand, $k(he) = he$, $(h - kh)e = 0$, $(h - kh)C(e)e = 0$; then $C(e) - (h - kh)C(e)$ is a central projection such that

$$[C(e) - (h - kh)C(e)]e = e,$$

therefore $C(e) \le C(e) - (h - kh)C(e)$. It follows that $C(e)[(h - kh)C(e)] = 0$, thus $hC(e) = khC(e) = k$.

(iv) The hypothesis is that (e_ι) is a family of projections in A that possesses a supremum e; it is asserted that the family $(C(e_\iota))$ admits $C(e)$ as supremum. By (ii), $C(e_\iota) \le C(e)$ for all ι. Assuming h is a central projection such that $C(e_\iota) \le h$ for all ι, it remains to show that $C(e) \le h$; indeed, for all ι we have $e_\iota \le C(e_\iota) \le h$, therefore $e \le h$, thus $C(e) \le h$ by (i).

(v) In view of (ii), it suffices to consider the case that $e \sim f$. Say $x^*x = e$, $xx^* = f$. It is to be shown that if h is a central projection, then $he = e$ iff $hf = f$. Since the involution is proper, it is routine to check that $hx^*x=x^*x$ iff $(x-hx)^*(x-hx)=0$ iff $x-hx=0$ iff $(x-hx)(x-hx)^*=0$ iff $hxx^*=xx^*$.

(vi) Assume, in addition, that A is a Rickart $*$-ring, let $x\in A$, and let $f=\mathrm{LP}(x)$. If h is a central projection, then, citing [§ 3, Prop. 3], $hx=x$ iff $h\geq f$ iff $hf=f$; by hypothesis, there exists a smallest such h (namely, $C(f)$), thus x has a central cover, and $C(x)=C(f)$. Similarly $C(x)=C(\mathrm{RP}(x))$.

(vii) Since the involution in a Rickart $*$-ring is proper [§ 3, Prop. 2], the calculation in (v) shows that $C(x^*x)=C(xx^*)$ for every $x\in A$.

(viii) Let $x\in A$, let h be a central projection, and write $f=\mathrm{LP}(x)$. Clearly hx and hf have the same left-annihilator, therefore $\mathrm{LP}(hx)=\mathrm{LP}(hf)$; citing (vi) and (iii), we have $C(hx)=C(hf)=hC(f)=hC(x)$.

(ix) Let $x\in A$, $z\in Z$. If $zC(x)=0$ then $zx=zC(x)x=0$. Conversely, suppose $zx=0$. Writing $h=\mathrm{RP}(z)$, we have $hx=0$. Since h is a central projection [§ 3, Cor. 1 of Prop. 10], it follows from (viii) that $0=C(hx)=hC(x)$, hence $zC(x)=0$. ∎

The key to the theory of central cover in Baer $*$-rings is the following result:

Proposition 2. *If A is a Baer $*$-ring and J is a right ideal in A, then $R(J)=hA$, where h is a central projection.*

Proof. Write $R(J)=hA$, h a projection. Since J is a right ideal, $R(J)$ is an ideal (two-sided) [§ 3, Prop. 1]; in particular, $A(hA)\subset hA$, therefore $Ah\subset hA$. Thus $h(xh)=xh$ for all $x\in A$; then also $hx^*h=x^*h$, thus $hx=hxh=xh$ for all $x\in A$. ∎

Proposition 3. *In a Baer $*$-ring A, every element x has a central cover. Explicitly, if $e=\mathrm{RP}(x)$ and $f=\mathrm{LP}(x)$, then $C(x)=C(e)=C(f)$ and*

$$(1-C(x))A=L(Ax)=L(Ae)=L(Af)\,,$$
$$(1-C(x))A=R(xA)=R(eA)=R(fA)\,.$$

Proof. The existence of central covers is immediate from [§ 3, Prop. 6] and the fact that the central projections form a complete lattice [§ 4, Cor. 2 of Prop. 7]. We give here an alternate proof that exhibits the desired formulas.

Write $R(xA)=(1-h)A$, h a central projection (Proposition 2). In particular, $1-h$ right-annihilates x, thus $xh=x$. On the other hand, if k is a central projection such that $kx=x$, then $x(1-k)=0$, $xA(1-k)=x(1-k)A=0$, $1-k\in(1-h)A$, $1-k\leq 1-h$, $h\leq k$. This proves that $C(x)$ exists and is equal to h (see Definition 1).

A similar proof shows that if $L(Ax)=(1-k)A$, then $k=C(x)$. Thus

$$R(xA)=L(Ax)=(1-C(x))A\,.$$

Since $C(x)=C(e)=C(f)$ by Proposition 1, the remaining formulas follow at once. ∎

Corollary 1. *If A is a Baer $*$-ring and $x, y \in A$, then $xAy=0$ if and only if $C(x)C(y)=0$.*

Proof. Let $h=C(x)$, $k=C(y)$. If $hk=0$ then $xAy=(xh)A(ky)=xA(hk)y=0$. Conversely, if $xAy=0$ then $x\in L(Ay)=(1-k)A$, therefore $(1-k)x=x$, $h=C(x)\le 1-k$, $hk=0$. ∎

The phenomenon in Corollary 1 occurs often enough to merit a name:

Definition 2. Projections e, f in a Baer $*$-ring are called *very orthogonal* if $C(e)C(f)=0$ (equivalently, in view of Corollary 1, $eAf=0$).

Corollary 2. *Let A be a Baer $*$-ring whose only central projections are* 0 *and* 1. *If $x, y \in A$ then $xAy=0$ if and only if $x=0$ or $y=0$.*

Proof. If $x\neq 0$ and $y\neq 0$ then $C(x)=C(y)=1$ (0 and 1 are the only available values), thus $C(x)C(y)\neq 0$; citing Corollary 1, we have $xAy\neq 0$. ∎

The condition on A in Corollary 2 is an import recurring theme:

Definition 3. If A is a $*$-ring with unity, whose only central projections are 0 and 1, we say that A is *factorial*, or that A is a *factor*. A factorial Baer $*$-ring is called a *Baer $*$-factor*; a factorial AW^*-algebra is called an *AW^*-factor.*

Remarks and Examples. 1. In a factorial $*$-ring, every element has a central cover in the sense of Definition 1.

2. If A is a $*$-algebra with unity, over an involutive field K, and if the center of A is one-dimensional over K, then A is factorial. {Proof: In any $*$-algebra, orthogonal nonzero projections are linearly independent.}

3. A factorial Baer $*$-algebra need not have one-dimensional center. {Consider, for example, the complex polynomial ring $A=\mathbb{C}[t]$, t indeterminate, with the identity involution.}

4. If $\mathscr{H}$ is a Hilbert space, then $\mathscr{L}(\mathscr{H})$ is a factor. {Suppose E is a nonzero projection such that $TE=ET$ for every $T\in\mathscr{L}(\mathscr{H})$, and let $\mathscr{M}$ be the range of E; it is to be shown that $\mathscr{M}=\mathscr{H}$. Fix a unit vector x in $\mathscr{M}$. Given any $y\in\mathscr{H}$, consider the operator $Tz=(z|x)y$ $(z\in\mathscr{H})$. In particular, $Tx=(x|x)y=y$, thus $y\in T(\mathscr{M})$. By hypothesis, $\mathscr{M}$ is invariant under T (indeed, $\mathscr{M}$ reduces every operator in $\mathscr{L}(\mathscr{H})$), thus $y\in T(\mathscr{M})\subset\mathscr{M}$.}

5. If A is a Baer $*$-ring and B is a $*$-subring such that $B=B''$, and if $x\in B$, then $C_B(x)\le C_A(x)$ (because B contains every central projection of A), but in general equality does not hold. {For example, let $\mathscr{A}$ be

a commutative von Neumann algebra on a Hilbert space $\mathscr{H}$, thus $\mathscr{A}' \supset \mathscr{A} = \mathscr{A}''$ in $\mathscr{L}(\mathscr{H})$; if E is a projection in $\mathscr{A}$ different from 0 and I, then its central cover in $\mathscr{A}$ is E, but its central cover in $\mathscr{L}(\mathscr{H})$ is I (by Example 4 above).} Such $*$-subrings B are Baer $*$-subrings of A in the sense of [§4, Def. 3]; for Baer $*$-subrings of the form eAe, the situation is clearer:

Proposition 4. *Let A be a Baer $*$-ring, e a projection in A, and f a projection in eAe (that is, $f \le e$). The following conditions on f are equivalent*:

(a) *f is a central projection in eAe*;

(b) $f = eC(f)$;

(c) *$f = eh$ for some central projection h of A.*

If A is a factor, then so is eAe.

Proof. Obviously (b) implies (c), and (c) implies (a).

(a) implies (b): For all $a \in A$, we have $f(eae) = (eae)f$, that is, $fae = eaf$; it follows that $fae(1-f) = 0$ for all $a \in A$, thus $fAe(1-f) = 0$. Citing Proposition 3, we have

$$e(1-f) \in R(fA) = (1 - C(f))A,$$

therefore $e(1-f)C(f) = 0$, $eC(f) = efC(f) = fC(f) = f$.

It follows that if 0 and 1 are the only central projections in A, then 0 and e are the only central projections in eAe. ∎

Corollary 1. *Let A be a Baer $*$-ring, with center Z, and let e be a projection in A.*

(i) *The center of eAe contains eZe, that is, $eZ \subset (eAe) \cap (eAe)'$.*

(ii) *eZ and $(eAe) \cap (eAe)'$ contain the same projections.*

(iii) *If $x \in eAe$ then $eC(x)$ is the central cover of x in eAe.*

Proof. (i) is obvious.

(ii) If f is a projection in $(eAe) \cap (eAe)'$, then, by Proposition 4, there exists a projection $h \in Z$ with $f = eh$.

(iii) Let $f = C_{eAe}(x)$, $h = C(x)$; it is to be shown that $f = eh$. Since eh is central in eAe, and $(eh)x = e(hx) = ex = x$, we have $f \le eh$. On the other hand, since f is central in eAe, we have $f = eC(f)$ by Proposition 4; then $xC(f) = xeC(f) = xf = x$, therefore $h \le C(f)$, $he \le eC(f) = f$. ∎

When A is an AW^*-algebra, the inclusion in Corollary 1, (i) is an equality; the proof borrows a result from Section 8:

Corollary 2. *If A is an AW^*-algebra, with center Z, and if e is a projection in A, then eZ is the center of eAe, that is, $eZ = (eAe) \cap (eAe)'$.*

Proof. It is immediate from [§ 4, Prop. 8] that eZ and $(eAe) \cap (eAe)'$ are AW^*-algebras, and by Corollary 1 they contain the same projections; the desired conclusion follows from the fact that an AW^*-algebra is the closed linear span of its projections [§ 8, Prop. 1]. ∎

Definition 4. A projection f in a Baer $*$-ring A is said to be *faithful* (in A) if $C(f)=1$.

Corollary 3. *Let f be a faithful projection in a Baer $*$-ring A.*
(i) *If e is any projection in A such that $f \leq e$, then f is faithful in eAe.*
(ii) *If h is any central projection in A, then hf is faithful in hA.*

Proof. (i) Citing (iii) of Corollary 1, we have $C_{eAe}(f)=eC(f)=e$.
(ii) Similarly, $C_{hA}(hf)=hC(hf)=h(hC(f))=h$. ∎

The final result of the section is for application later on (cf. [§ 20, Prop. 5], [§ 29, Lemma 1], [§ 30, Lemma 1]); the gist of it is that, in a certain sense, the consideration of finitely many projections can be reduced to the case of faithful projections:

Proposition 5. *Let A be a $*$-ring with unity, in which every projection has a central cover, and let e_i $(i=1,\ldots,n)$ be projections in A. Then there exist orthogonal central projections k_v $(v=1,\ldots,r)$, with $k_1+\cdots+k_r=1$, such that, for each pair of indices i, v, either $k_v e_i=0$ or $C(k_v e_i)=k_v$.*

Proof. Let $k=\sup\{C(e_i): i=1,\ldots,n\}$ and note that $(1-k)e_i=0$ for all i; dropping down to kA (cf. Exercise 4), we can suppose that $\sup C(e_i)=1$.

Disjointify the $C(e_i)$, that is, let $k_1,\ldots,k_r$ be orthogonal central projections with $k_1+\cdots+k_r=1$, such that each $C(e_i)$ is the sum of certain of the k_v. {For example, consider the projections

$$C(e_1)^{\varepsilon_1} C(e_2)^{\varepsilon_2} \ldots C(e_n)^{\varepsilon_n},$$

where $\varepsilon_i=+1$ or -1, with the understanding that $C(e_i)^{+1}=C(e_i)$ and $C(e_i)^{-1}=1-C(e_i)$.} For any pair of indices i, v we have either $k_v C(e_i)=0$ or $k_v \leq C(e_i)$ (because $C(e_i)$ is the sum of those k_v that it contains, and is therefore orthogonal to the rest). It follows that if $k_v C(e_i) \neq 0$ then $k_v \leq C(e_i)$, hence, citing Proposition 1, $C(k_v e_i)=k_v C(e_i)=k_v$. On the other hand, if $k_v C(e_i)=0$ then $k_v e_i=0$. ∎

Exercises

1A. A projection e in a $*$-ring A is central iff $eA(1-e)=0$.

2A. Let A be a Baer $*$-ring, and suppose that S and J are nonempty subsets of A such that $JS \subset J$. Write $R(J)=hA$, h a projection. Prove: (i) $SR(J) \subset R(J)$; (ii) $hsh=sh$ for all $s \in S$; (iii) if S is a $*$-subset of A, then $h \in S'$. Proposition 2 is a special case.

3A. Proposition 1 holds with "weakly Rickart $*$-ring" in place of "Rickart $*$-ring".

4A. Let h be a central projection in the $*$-ring A. The central projections of hA are the central projections k in A such that $k \le h$; in other words, a projection in hA is central in hA if and only if it is central in A.

5A. Let A be a $*$-ring in which every projection has a central cover, and let h be a central projection in A. If e is a projection in hA, then $C_{hA}(e) = C(e)$.

6C. Let A be an AW^*-algebra, e a projection in A. Then $C(e) = \sup\{ueu^*: u \text{ unitary}\}$.

7A. Let A be a Baer $*$-ring, let e be a projection in A, and define

$$h = \sup\{f: f \sim e\}, \qquad k = \sup\{f: f \precsim e\}.$$

It is easy to see that $e \le h \le k \le C(e)$, and that $uhu^* = h$ for every unitary $u \in A$. Write $U = \{u \in A: u \text{ unitary}\}$, $P = \{g \in A: g \text{ a projection}\}$. Prove that $h = k = C(e)$ under either of the following three hypotheses:

(i) Every projection in U' is central.

(ii) Every projection in P' is central, and $2x = 0$ implies $x = 0$.

(iii) A is an AW^*-algebra.

8A. If A is a Rickart $*$-ring with PC [§ 14, Def. 3], then a projection in A is central iff it commutes with every projection of A (thus a projection is central in A iff it is central in the reduced ring A° [§ 3, Exer. 18]).

9A. If A is a $*$-ring with the property that A and A° have the same central projections (cf. Exercise 8), then every direct summand has this property.

10A. In a Baer $*$-ring with PC [§ 14, Def. 3], the following conditions on a pair of projections e, f are equivalent: (a) $C(e) \le C(f)$; (b) $e = \sup e_\iota$ with (e_ι) an orthogonal family of projections such that $e_\iota \precsim f$ for all ι; (c) $e = \sup e_\iota$ with (e_ι) a family of projections such that $e_\iota \precsim f$ for all ι.

11C. In a Rickart $*$-ring with orthogonal GC [§ 14, Def. 4], a projection is central if and only if it has a unique complement. {Projections e, f are said to be *complementary* if $e \cup f = 1$ and $e \cap f = 0$; either is called a *complement* of the other. Every projection e has at least $1 - e$ as a complement.}

12A. If e is a minimal projection [§ 4, Exer. 3] in a Baer $*$-ring A, then the direct summand $C(e)A$ is a factor.

13A. Let B be a Baer $*$-ring, A a $*$-subring such that $A = A''$. (Model: B the algebra of all bounded operators on a Hilbert space $\mathscr{H}$, A any von Neumann algebra on $\mathscr{H}$ [§ 4, Def. 5].)

Let $D = (A \cap A')' = (A' \cup A)''$; thus D is the smallest $*$-subring of B such that $A \subset D$, $A' \subset D$ and $D = D''$ (in this sense, it is 'generated' by A and A').

Then A, A' and D have common center $D' = (A' \cup A)''' = (A' \cup A)' = A \cap A'$; hence if $a \in A$ then $C_A(a) = C_D(a)$, and if $a' \in A'$ then $C_{A'}(a') = C_D(a')$.

14A. Let B be a Baer $*$-ring and adopt the notation of Exercise 13. If $a \in A$ and $a' \in A'$, the following conditions are equivalent: (a) there exists a projection c in the center of A such that $ac = 0$ and $ca' = a'$; (b) there exists an element c in the center of A such that $ac = 0$ and $ca' = a'$; (c) $aDa' = 0$; (d) $C_D(a)C_D(a') = 0$; (e) $C_A(a)C_{A'}(a') = 0$; (f) $aa' = 0$.

15A. Let B be a Baer $*$-ring, with notation as in Exercise 13. Let e be a projection in A. Define a mapping φ: $A' \to eA'e$ $(= eA' = A'e)$ by the formula $\varphi(a') = a'e$. Thus φ is a $*$-epimorphism, with kernel

$$\ker \varphi = \{a' \in A': a'e = 0\}.$$

Let $h=C_A(e)$ $(=C_D(e))$. Prove: $\ker\varphi=(1-h)A'$. In particular, φ is injective if and only if $h=1$, that is, e is faithful in A.

16A. Let B be a Baer *-ring, with notation as in Exercise 13. Write Z to indicate center; thus $Z_A=Z_{A'}=Z_D=A\cap A'$. Let e be a projection in A. Then (i) $Z_{eAe}\supset eZ_A$; (ii) Z_{eAe} and eZ_A contain the same projections. (iii) If, moreover, B is an AW^*-algebra, then $Z_{eAe}=eZ_A$, in other words (calculating commutants in B),

$$(eAe)\cap(eAe)'=e(A\cap A').$$

17A. Let B be a Baer *-ring, with notation as in Exercise 13. Let e be a projection in A. Then

$$(eBe)\cap(eAe)'\supset eA'e(=eA').$$

If f is a projection in B, then $f\in(eBe)\cap(eAe)'$ iff $f\le e$ and $fA(e-f)=0$.

18A. Let A be a ring with unity, $B=A_n$ the ring of $n\times n$ matrices over A.

(i) The center of B is the set of all 'scalar' matrices

$$(\delta_{ij}z)=\operatorname{diag}(z,\dots,z)$$

with z in the center of A. In particular, the central idempotents of A and B may be identified.

(ii) Suppose, in addition, that A is a *-ring, and B is given the natural involution (*-transposition). If every $a\in A$ has a central cover $C(a)$ (in the sense of Definition 1) then every $x=(a_{ij})\in B$ has a central cover, namely, $C(x)=\sup\{C(a_{ij}): i,j=1,\dots,n\}$. (This is a reason for considering central cover for elements other than projections.)

(iii) If A is a Baer *-ring and $x,y\in B$, then $xBy=0$ iff $C(x)C(y)=0$.

19A. If A is a factorial Rickart *-ring, then the center of A is an integral domain.

20A. If h is a central projection in a weakly Rickart *-ring, then $\mathrm{RP}(hx)=h\mathrm{RP}(x)$ for every element x.

§ 7. Commutative AW^*-Algebras

The commutative C^*-algebras A with unity are the algebras $A=C(T)$, T compact [§ 2, Example 5]. What properties of T are needed to make A an AW^*-algebra? The following answer is the main result of the section:

Theorem 1. *Let A be a commutative C^*-algebra with unity and write $A=C(T)$, T compact. In order that A be an AW^*-algebra, it is necessary and sufficient that T be a Stonian space.*

A compact space is said to be *Stonian* (or *extremally disconnected*) if the closure of every open set is open (hence closed and open—'clopen').

For clarity, we separate the proof of sufficiency (Proposition 1) and necessity (Proposition 2). The key to the analysis is the observation that the projections of $A=C(T)$ correspond to the clopen sets in T: if $e\in A$, then $e^*=e=e^2$ iff $e(t)=0$ or 1 for all $t\in T$ iff $e^2=e$ iff e is the characteristic function of a clopen set P in T (namely, $P=\{t: e(t)=1\}$). Thus, the projection lattice of $A=C(T)$ [§ 1, Prop. 3] is isomorphic to the lattice of clopen subsets P of T (ordered by inclusion), via the correspondence $P \leftrightarrow \chi_P$ (the characteristic function of P).

The first lemma is valid in any topological space:

Lemma. *Let (P_ι) be a family of clopen sets, let $U = \bigcup P_\iota$, and suppose that $\bar{U}$ is clopen. Then the family (P_ι) has $\bar{U}$ as supremum in the class of all clopen sets (ordered by inclusion).*

Proof. Let $P = \bar{U}$. On the one hand, $P_\iota \subset P$ for all ι. On the other hand, if Q is clopen and $P_\iota \subset Q$ for all ι, then $U \subset Q$, therefore $P = \bar{U} \subset \bar{Q} = Q$. ∎

Proposition 1. *If T is Stonian, then $A = C(T)$ is an AW^*-algebra.*

Proof. Since T is Stonian, it is clear from the lemma that the lattice of clopen sets is complete, in other words, the projection lattice of A is complete. To show that A is a Baer $*$-ring, it will therefore suffice to show that it is a Rickart $*$-ring [§ 4, Prop. 1]. Let $x \in A$; we seek a projection e such that $R(\{x\}) = (1-e)A$. Set $U = \{t: x(t) \neq 0\}$, $P = \bar{U}$. Since U is open, P is clopen; let $e = \chi_P$. If $y \in A$, then $xy = 0$ iff $y = 0$ on U iff $y = 0$ on $\bar{U} = P$ iff $ey = 0$. Thus $R(\{x\}) = R(\{e\}) = (1-e)A$. ∎

We approach the converse through a pair of lemmas. If $A = C(T)$ is an AW^*-algebra, we know that the clopen sets of T form a complete lattice [§ 4, Prop. 1]; what remains to be verified is the explicit formula for supremum indicated in the above lemma.

Lemma 1. *If $A = C(T)$ is a Rickart $*$-ring, then the clopen sets of T separate the points of T.*

Proof. Suppose $s, t \in T$, $s \neq t$; we seek a clopen set P such that $s \in P$ and $t \in T - P$. Choose neighborhoods U, V of s, t with $U \cap V = \varnothing$, and choose $x, y \in A$ such that $x(s) \neq 0$, $x = 0$ on $T - U$, and $y(t) \neq 0$, $y = 0$ on $T - V$. Obviously $xy = 0$; thus, setting $e = \mathrm{RP}(x)$, we have $ey = 0$. Write $e = \chi_P$, P clopen. Since $xe = x$ and $x(s) \neq 0$, we have $e(s) = 1$, $s \in P$; on the other hand, $ey = 0$ and $y(t) \neq 0$, therefore $e(t) = 0$, $t \in T - P$. ∎

In essence, the next lemma is a result about the Stone representation space of a Boolean algebra:

Lemma 2. *If $A = C(T)$ is a Rickart $*$-ring, then the clopen sets in T are basic for the topology.*

Proof. Let U be an open set, $t \in U$; we seek a clopen set P such that $t \in P \subset U$. For each $s \in T - U$, choose a clopen set P_s such that $s \in P_s$ and $t \in T - P_s$ (Lemma 1). Since $T - U$ is compact,

$$T - U \subset P_{s_1} \cup \cdots \cup P_{s_n}$$

for suitable $s_1,\ldots,s_n$; the clopen set $P=(T-P_{s_1})\cap\cdots\cap(T-P_{s_n})$ has the required properties. ∎

Proposition 2. *If $A=C(T)$ is an AW^*-algebra, then T is Stonian.*

Proof. Let U be an open set in T; it is to be shown that $\bar{U}$ is open. By Lemma 2, there exists a family (P_ι) of clopen sets such that $U=\bigcup P_\iota$. As remarked following Proposition 1, the clopen sets form a complete lattice; let P be the supremum of the family (P_ι). Since $P_\iota\subset P$ for all ι, we have $U\subset P$, therefore $\bar{U}\subset P$; it will suffice to show that $\bar{U}=P$.

Assume to the contrary that the open set $P-\bar{U}$ is nonempty. Let Q be a nonempty clopen set such that $Q\subset P-\bar{U}$ (Lemma 2). Since $Q\cap P_\iota=\varnothing$ for all ι, that is, $P_\iota\subset T-Q$ for all ι, it follows that $P\subset T-Q$ (because $T-Q$ is clopen and P is the supremum of the P_ι). Thus $P\cap Q=\varnothing$; but this is contrary to $Q\subset P$, $Q\neq\varnothing$. ∎

In many situations, Proposition 2 passes for 'spectral theory' in an AW^*-algebra. Form example:

Proposition 3. *Let A be any AW^*-algebra, $x\in A$, $x\geq 0$, $x\neq 0$. Given any $\varepsilon>0$, there exists $y\in\{x\}''$, $y\geq 0$, such that* (i) *$xy=e$, e a nonzero projection, and* (ii) $\|x-xe\|<\varepsilon$.

Proof. The condition $x\geq 0$ means that $x^*=x$ and the spectrum of x (as an element of A) is nonnegative.

Since $\{x\}''$ is a commutative AW^*-algebra [§ 4, Prop. 8], we have

$$\{x\}''=C(T),$$

T a Stonian space (Proposition 2). The spectrum of x as an element of $\{x\}''$ is the same as its A-spectrum [§ 3, Prop. 9, (6)], therefore $x\geq 0$ as a function on T.

We can suppose that $0<\varepsilon<\|x\|$. Define

$$U=\left\{t\colon x(t)>\frac{\varepsilon}{2}\right\};$$

U is a nonempty open set, $P=\bar{U}$ is clopen, and the characteristic function of P is a nonzero projection e. Since $x(t)\leq\varepsilon/2$ on $T-U$, and therefore on $T-\bar{U}$, we have

$$0\leq x(1-e)\leq\left(\frac{\varepsilon}{2}\right)(1-e)\leq\left(\frac{\varepsilon}{2}\right)1,$$

thus $\|x(1-e)\|\leq\varepsilon/2$. On the other hand, since $x(t)>\varepsilon/2$ on U, we have $x(t)\geq\varepsilon/2$ on $\bar{U}=P$, therefore the function $t\mapsto 1/x(t)$ $(t\in P)$ is continuous; it follows that the function y on T defined by

$$y(t)=\frac{1}{x(t)}\quad\text{for } t\in P,\qquad y(t)=0\quad\text{for } t\in T-P$$

is continuous, that is, $y \in \{x\}''$. Since $x(t)y(t)=1$ on P, and $=0$ on $T-P$, we have $xy=e$. ∎

Corollary. *Let A be any AW^*-algebra, $x \in A$, $x \neq 0$. Given any $\varepsilon > 0$, there exists $y \in \{x^* x\}''$, $y \geq 0$, such that* (i) $(x^* x)y^2 = e$, *e a nonzero projection, and* (ii) $\|x - xe\| < \varepsilon$.

Proof. One knows from C^*-algebra theory that $x^* x \geq 0$. By Proposition 3, there exists $z \in \{x^* x\}''$, $z \geq 0$, such that $(x^* x)z = e$, e a nonzero projection, and $\|x^* x - (x^* x)e\| < \varepsilon^2$; set $y = z^{\frac{1}{2}}$. Since e commutes with $x^* x$, we have $\|x - xe\|^2 = \|(x - xe)^*(x - xe)\| = \|x^* x - x^* xe\| < \varepsilon^2$. ∎

The conclusion of the corollary is a mixture of algebra and analysis: the ideal generated by x contains nonzero projections e, and x can be approximated from the ideal generated by such projections. The purely algebraic part can be formulated in any $*$-ring:

Definition 1. A $*$-ring A is said to satisfy the *existence of projections axiom* (briefly, the (EP)-*axiom*) if for every $x \in A$, $x \neq 0$, there exists $y \in \{x^* x\}''$ such that $y^* = y$ and $(x^* x)y^2 = e$, e a nonzero projection.

As long as we're on the subject, this is a convenient place to record two weaker axioms that sometimes suffice:

Definition 2. A $*$-ring A is said to satisfy the *weak* (EP)-*axiom* (briefly, the (WEP)-*axiom*) if for every $x \in A$, $x \neq 0$, there exists $y \in \{x^* x\}''$ (y is necessarily normal, but not necessarily self-adjoint) such that $(x^* x)(y^* y) = e$, e a nonzero projection.

Definition 3. A $*$-ring A is said to satisfy the *very weak* (EP)-*axiom* (briefly, the (VWEP)-*axiom*) if for every $x \in A$, $x \neq 0$, there exists $y \in \{x^* x\}'$ such that $(x^* x)(y^* y) = e$, e a nonzero projection.

Obviously (EP) $\Rightarrow$ (WEP) $\Rightarrow$ (VWEP) $\Rightarrow$ the involution is proper.

Exercises

1C. A C^*-algebra A is an AW^*-algebra if and only if (A) in the partially ordered set of projections of A, every nonempty set of orthogonal projections has a supremum, and (B) every masa [§1, Exer. 14] in A is the closed linear span of its projections.

2C. Let A be a commutative AW^*-algebra. In order that A be $*$-isomorphic to some (commutative) von Neumann algebra, it is necessary and sufficient that there exist a family $\mathscr{P}$ of linear forms on A having the following three properties: (i) each $f \in \mathscr{P}$ is positive, that is, $f(x^* x) \geq 0$ for all $x \in A$; (ii) each $f \in \mathscr{P}$ is completely additive on projections, that is, $f(\sup e_\iota) = \sum f(e_\iota)$ for every orthogonal family of projections (e_ι); (iii) $\mathscr{P}$ is total, that is, if $x \in A$ is nonzero then $f(x^* x) > 0$ for some $f \in \mathscr{P}$.

3C. Let T be a compact space. In order that $C(T)$ be $*$-isomorphic to some (commutative) von Neumann algebra, it is necessary and sufficient that T be hyperstonian. {A Stonian space is said to be *hyperstonian* if the supports of its normal measures have dense union (a measure is *normal* iff it vanishes on every closed set with empty interior).}

4C. There exists a Hilbert space $\mathscr{H}$ and a commutative $*$-algebra $\mathscr{A}$ of operators on $\mathscr{H}$ such that (i) the identity operator belongs to $\mathscr{A}$, (ii) $\mathscr{A}$ is an AW^*algebra, and (iii) $\mathscr{A}$ is not a von Neumann algebra. It follows that there exists a family (E_ι) of projections in $\mathscr{A}$ such that $\sup E_\iota$, as calculated in the projection lattice of $\mathscr{A}$, is distinct from the projection on the closed linear span of the ranges of the E_ι.

5A. Let A be a $*$-ring, B a $*$-subring of A such that $B=B''$. If A satisfies the (EP)-axiom [(WEP)-axiom] then so does B.

6A. Let A be a $*$-ring, e a projection in A. If A satisfies the (EP)-axiom [(WEP)-axiom] then so does eAe.

7A. Let A be a Baer $*$-ring satisfying the (WEP)-axiom, and let $x\in A$, $x\neq 0$. Let (e_ι) be a maximal orthogonal family of nonzero projections such that, for each ι, there exists $y_\iota\in\{x^*x\}''$ with $(x^*x)(y_\iota^*y_\iota)=e_\iota$. Then $\sup e_\iota=\mathrm{RP}(x)$.

8A. Let A be a Baer $*$-ring, and let B be a $*$-subring of A such that (1) if S is any nonempty set of orthogonal projections in B, then $\sup S\in B$, and (2) B satisfies the (WEP)-axiom. Then the following conditions are equivalent: (a) $x\in B$ implies $\mathrm{RP}(x)\in B$; (b) B is a Baer $*$-ring; (c) B is a Baer $*$-subring of A.

9A. Let A be an AW^*-algebra, and let B be a closed $*$-subalgebra of A such that $\sup S\in B$ whenever S is a nonempty set of orthogonal projections in B. Then the following conditions are equivalent: (a) $x\in B$ implies $\mathrm{RP}(x)\in B$; (b) B is an AW^*-algebra; (c) B is an AW^*-subalgebra of A.

10A. A compact space is Stonian if and only if (i) the clopen sets are basic for the topology, and (ii) the set of all clopen sets, partially ordered by inclusion, is a complete lattice.

11C. A commutative AW^*-algebra A is 'algebraically closed' in the following sense: If $p(t)=t^n+a_1t^{n-1}+\cdots+a_{n-1}t+a_n$ is a monic polynomial with coefficients $a_1,\dots,a_n$ in A, then $p(a)=0$ for some $a\in A$.

§ 8. Commutative Rickart C^*-Algebras

If T is a compact space, when is $C(T)$ a Rickart C^*-algebra?; precisely when the clopen sets are basic and form a σ-lattice:

Theorem 1. *Let A be a commutative C^*-algebra with unity, and write $A=C(T)$, T compact. The following conditions are necessary and sufficient for A to be a Rickart C^*-algebra:* (1) *the clopen sets in T are basic for the topology, and* (2) *if P_n is any sequence of clopen sets, and if $U=\bigcup_{n=1}^{\infty} P_n$, then $\bar{U}$ is clopen.*

The proofs of necessity (Proposition 1) and sufficiency (Proposition 2) are separated for greater clarity. We begin with three general lemmas:

Lemma 1. *If, in a weakly Rickart $*$-ring, every orthogonal sequence of projections has a supremum, then every sequence of projections has a supremum.*

Proof. If e_n is any sequence of projections, consider the orthogonal sequence f_n defined by $f_1=e_1$ and

$$f_n=(e_1\cup\cdots\cup e_n)-(e_1\cup\cdots\cup e_{n-1})$$

for $n>1$. ∎

Lemma 2. *If B is a weakly Rickart Banach $*$-algebra (real or complex scalars), then every sequence of projections in B has a supremum.*

Proof. By Lemma 1, it suffices to show that any orthogonal sequence of nonzero projections e_n has a supremum. Define

$$x_n=\sum_{k=1}^{n} 2^{-k}\|e_k\|^{-1}e_k \qquad (n=1,2,3,\ldots);$$

since $2^{-k}\|e_k\|^{-1}e_k$ has norm 2^{-k}, it follows that $\|x_m-x_n\|\to 0$ as $m,n\to\infty$, thus we may define $x=\lim x_n$. (Formally, $x=\sum_1^\infty 2^{-n}\|e_n\|^{-1}e_n$.) Let $e=\mathrm{RP}(x)$; we show that $e=\sup e_n$.

If f is any projection such that $e_n\le f$ for all n, then $x_nf=x_n$ for all n; passing to the limit, we have $xf=x$, therefore $e\le f$. It remains to show that $e_n\le e$ for all n. Fix an index m. By orthogonality,

$$e_mx_n=2^{-m}\|e_m\|^{-1}e_m \quad \text{for all } n\ge m,$$

therefore $e_mx=2^{-m}\|e_m\|^{-1}e_m$, that is, $e_m=2^m\|e_m\|e_mx$. Since $xe=x$, it follows that $e_me=e_m$, thus $e_m\le e$. ∎

In particular:

Lemma 3. *In a weakly Rickart C^*-algebra, every sequence of projections has a supremum.*

Proposition 1. *Let T be a compact space such that $A=C(T)$ is a Rickart C^*-algebra. Then:*

(1) *The clopen sets in T are basic for the topology.*

(2) *If P_n is a sequence of clopen sets and if $U=\bigcup_1^\infty P_n$, then $\overline{U}$ is clopen.*

(3) *A is the closed linear span of its projections.*

(4) *If $x\in A$ and $U=\{t\colon x(t)\ne 0\}$, then U is the union of a sequence of clopen sets, $\overline{U}$ is clopen, and the characteristic function of $\overline{U}$ is* $\mathrm{RP}(x)$.

Proof. (1) See [§ 7, Lemma 2].

(2) By Lemma 3, there exists a clopen set P which is a supremum for the P_n. Since $P_n \subset P$ for all n, we have $U \subset P$, therefore $\bar{U} \subset P$; the proof that $\bar{U} = P$ proceeds as in [§7, Prop. 2].

(3) Let B be the linear span of the projections of A. Evidently B is a $*$-subalgebra of A, with $1 \in B$; moreover, it is clear from (1) that B separates the points of T, therefore B is dense in A by the Weierstrass-Stone theorem.

(4) Note first that if $C \subset U$, with C compact (i.e., closed) and U open, then there exists a clopen set P such that $C \subset P \subset U$ (this follows from (1) and an obvious open covering argument).

Let $x \in A$ and let $U = \{t: x(t) \neq 0\}$. For $n = 1, 2, 3, \ldots$ set

$$C_n = \left\{t: |x(t)| \geq \frac{1}{n}\right\};$$

then C_n is compact and $U = \bigcup_1^\infty C_n$. For each n choose a clopen set P_n such that $C_n \subset P_n \subset U$; then

$$U = \bigcup_1^\infty C_n \subset \bigcup_1^\infty P_n \subset U,$$

thus $U = \bigcup_1^\infty P_n$, therefore $\bar{U}$ is clopen by (2).

Let $e = \mathrm{RP}(x)$ and let f be the characteristic function of $\bar{U}$; it is to be shown that $e = f$. Since $xe = x$, clearly $e(t) = 1$ for $t \in U$, hence for $t \in \bar{U}$, thus $e \geq f$. On the other hand, since $f = 1$ on $\bar{U}$, hence on U, we have $xf = x$, therefore $e \leq f$. ∎

In the reverse direction:

Proposition 2. *If T is a compact space satisfying conditions* (1) *and* (2) *of Proposition 1, then $A = C(T)$ is a Rickart C^*-algebra.*

Proof. If $x \in A$ and $U = \{t: x(t) \neq 0\}$, then $\bar{U}$ is clopen by the argument in the proof of (4) above; writing e for the characteristic function of $\bar{U}$, we have $R(\{x\}) = R(\{e\}) = (1 - e)A$ as in the proof of [§7, Prop. 1]. ∎

This completes the proof of Theorem 1. Another characterization of these algebras is as follows:

Proposition 3. *Let A be a commutative C^*-algebra with unity. Then A is a Rickart C^*-algebra if and only if* (i) *A is the closed linear span of its projections, and* (ii) *every orthogonal sequence of projections in A has a supremum.*

Proof. Write $A = C(T)$, T compact.

If A is a Rickart C^*-algebra, then (i) and (ii) hold by Proposition 1 and Lemma 3, respectively.

Conversely, suppose A satisfies (i) and (ii). Let B be the linear span of the projections of A; by hypothesis (i), B is dense in A. Since A separates the points of T, so does B; it follows that if s and t are distinct points of T, then there exists a projection e such that $e(s) \neq e(t)$. In other words, the clopen sets in T are separating, and the argument in [§ 7, Lemma 2] shows that they are basic for the topology. To complete the proof that A is a Rickart C^*-algebra, it will suffice, by Proposition 2, to show that if $U = \bigcup_1^\infty P_n$, where P_n is a sequence of clopen sets, then $\bar{U}$ is clopen. Replacing P_n, for $n > 1$, by the clopen set

$$\left(\bigcup_1^n P_k\right) - \left(\bigcup_1^{n-1} P_k\right),$$

we can suppose without loss of generality that the P_n are mutually disjoint. By hypothesis (ii), there exists a clopen set P which is a supremum for the P_n; then $\bar{U} = P$ by the argument in [§ 7, Prop. 2], thus $\bar{U}$ is clopen. ∎

The foregoing results are the basis for 'spectral theory' in Rickart C^*-algebras. For example:

Proposition 4. *Let A be any Rickart C^*-algebra, $x \in A$, $x \geq 0$, $x \neq 0$. Given any $\varepsilon > 0$, there exists $y \in \{x\}''$, $y \geq 0$, such that* (i) $xy = e$, *e a nonzero projection, and* (ii) $\|x - xe\| < \varepsilon$.

Proof. Since $\{x\}''$ is a commutative Rickart C^*-algebra [§ 3, Prop. 10], we have

$$\{x\}'' = C(T),$$

where T is a compact space with the properties (1), (2) of Proposition 1. As argued in [§ 7, Prop. 3], $x \geq 0$ as a function on T. We can suppose $0 < \varepsilon < \|x\|$. Define

$$U = \left\{t : x(t) > \frac{\varepsilon}{2}\right\};$$

since $\|x\| > \varepsilon/2$, the open set U is nonempty. Writing $z = x - (\varepsilon/2)1 + |x - (\varepsilon/2)1|$, we have $z \in A$ and

$$U = \{t : z(t) \neq 0\},$$

therefore $\bar{U}$ is clopen by (4) of Proposition 1; let e be the characteristic function of $\bar{U}$. The proof continues as in [§ 7, Prop. 3]. ∎

Corollary. *Let A be any Rickart C^*-algebra, $x \in A$, $x \neq 0$. Given any $\varepsilon > 0$, there exists $y \in \{x^*x\}''$, $y \geq 0$, such that* (i) $(x^*x)y^2 = e$, *e a nonzero projection, and* (ii) $\|x - xe\| < \varepsilon$.

Proof. Same as [§7, Cor. of Prop. 3]. ∎

In particular: *Every Rickart C^*-algebra satisfies the* (EP)*-axiom* [§7, Def. 1].

Exercise

1A. If A is a Rickart C^*-algebra, in which every orthogonal family of nonzero projections is countable (e.g., if A can be represented faithfully as operators on a separable Hilbert space), then A is an AW^*-algebra.

§ 9. Commutative Weakly Rickart C^*-Algebras

In this section the results of the preceding section are generalized to the unitless case. The unitless commutative C^*-algebras A are the algebras $C_\infty(T)$, where T is a noncompact, locally compact space, and $C_\infty(T)$ denotes the algebra of all continuous, complex-valued functions x on T that 'vanish at infinity' in the sense that

$$\{t: |x(t)| \geq \varepsilon\}$$

is compact for every $\varepsilon > 0$; clearly the projections $e \in A$ are the characteristic functions $e = \chi_P$, where P is *compact* and open. We seek conditions on T necessary and sufficient for $A = C_\infty(T)$ to be a weakly Rickart C^*-algebra. A natural strategy is to adjoin a unity element [§5, Exer. 9]; the effect of this on the character space is to adjoin a 'point at infinity' to T (the 'one-point compactification'), and the relation between A and T can be studied by applying the results of the preceding section to their enlargements A_1 and $T \cup \{\infty\}$. It is equally easy—and more instructive—to work out the unitless case directly, as we now do. The central result is as follows:

Theorem 1. *Let A be a commutative C^*-algebra without unity, and write $A = C_\infty(T)$, where T is locally compact and noncompact. The following conditions are necessary and sufficient for A to be a weakly Rickart C^*-algebra:* (1) *the compact-open sets in T are basic for the topology, and* (2) if *P_n is any sequence of compact-open sets, and if $U = \bigcup_1^\infty P_n$, then $\bar{U}$ is compact-open.*

The proofs of necessity (Proposition 1) and sufficiency (Proposition 2) are separated for greater clarity; throughout these results, we assume that $A=C_\infty(T)$, where T is a noncompact, locally compact space, with extra assumptions on A or T as needed.

Lemma 1. *If $A=C_\infty(T)$ is a weakly Rickart C^*-algebra, then any two points of T may be separated by disjoint compact-open sets.*

Proof. Assuming $s, t\in T$, $s\neq t$, we seek compact-open sets P and Q such that $s\in P$, $t\in Q$ and $P\cap Q=\emptyset$. Let U, V be neighborhoods of s, t with $U\cap V=\emptyset$. Choose $x, y\in A$ so that $x(s)\neq 0$, $x=0$ on $T-U$, and $y(t)\neq 0$, $y=0$ on $T-V$. Evidently $xy=0$; writing $e=\mathrm{RP}(x)$, $f=\mathrm{RP}(y)$, we have $ef=0$. Let P and Q be the compact-open sets such that $e=\chi_P$, $f=\chi_Q$. The orthogonality of e and f means that P and Q are disjoint. Since $xe=x$ and $x(s)\neq 0$, we have $e(s)=1$, $s\in P$; similarly $t\in Q$. ∎

Lemma 2. *If any two points of T can be separated by disjoint compact-open sets, then $A=C_\infty(T)$ is the closed linear span of its projections.*

Proof. Let B be the linear span of the projections of A; B is a $*$-subalgebra of A. If $s, t\in T$, $s\neq t$, by hypothesis there exist projections e, f such that $ef=0$ and $e(s)=f(t)=1$; it follows that B separates the points of T, and no point of T is annihilated by every function in B, therefore B is dense in A by the Weierstrass-Stone theorem. ∎

Lemma 3. *If $A=C_\infty(T)$ is the closed linear span of its projections, then the compact-open sets of T are basic for the topology.*

Proof. Let U be an open set, $s\in U$; we seek a compact-open set P such that $s\in P\subset U$.

Choose $x\in A$ with $x(s)=1$ and $x=0$ on $T-U$. By hypothesis, there exists an element $y\in A$, y a linear combination of projections, such that $\|x-y\|<1/2$. Say

$$y=\lambda_1 e_1+\cdots+\lambda_n e_n,$$

where the e_n are projections; we can suppose that the e_i are orthogonal, and that $e_i\neq 0$, $\lambda_i\neq 0$. Say $e_i=\chi_{P_i}$, P_i compact-open. Since

$$\tfrac{1}{2}>|x(s)-y(s)|=|1-y(s)|,$$

necessarily $y(s)\neq 0$, thus there exists an index j such that $e_j(s)\neq 0$, that is, $s\in P_j$. The proof will be concluded by showing that $P_j\subset U$.

If $t\in P_j$ then $y(t)=\lambda_j$ by the assumed orthogonality, thus

$$(*)\qquad \tfrac{1}{2}>|x(t)-y(t)|=|x(t)-\lambda_j|;$$

in particular,

$$(**)\qquad \tfrac{1}{2}>|x(s)-\lambda_j|=|1-\lambda_j|.$$

It follows from (∗) and (∗∗) that $t \in P_j$ implies $|x(t)-1| < \frac{1}{2}+\frac{1}{2}$, therefore $x(t) \neq 0$, hence $t \in U$. ∎

Proposition 1. *Let T be a noncompact, locally compact space such that $A = C_\infty(T)$ is a weakly Rickart C^*-algebra. Then:*

(1) *The compact-open sets in T are basic for the topology.*

(2) *If P_n is a sequence of compact-open sets and if $U = \bigcup_1^\infty P_n$, then $\bar{U}$ is compact-open.*

(3) *A is the closed linear span of its projections.*

(4) *If $x \in A$ and $U = \{t: x(t) \neq 0\}$, then U is the union of a sequence of compact-open sets, $\bar{U}$ is compact-open, and the characteristic function of $\bar{U}$ is* RP(x).

Proof. (1) and (3) are covered by Lemmas 1–3.

(2) Since every sequence of projections in A has a supremum [§ 8, Lemma 3], there exists a compact-open set P which is a supremum for the P_n. Since $P_n \subset P$ for all n, we have $\bar{U} \subset P$; it will suffice to show that $\bar{U} = P$. Assume to the contrary that the open set $P - \bar{U}$ is nonempty. By (1), choose a nonempty compact-open set Q with $Q \subset P - \bar{U}$; thus $Q \cap P = Q \neq \varnothing$ and $Q \cap P_n = \varnothing$ for all n. It follows that if e_n, e and f are the characteristic functions of P_n, P and Q, then $fe = f \neq 0$ and $fe_n = 0$ for all n. Thus $e - f$ is a projection, and $(e-f)e_n = ee_n - fe_n = e_n - 0$ shows that $e_n \leq e - f$ for all n, therefore $e \leq e - f$; it follows that $f = 0$, a contradiction.

(4) The argument in [§ 8, Prop. 1, (4)] may be used verbatim, provided 'clopen' is replaced by 'compact-open'. ∎

Conversely:

Proposition 2. *If T is a noncompact, locally compact space satisfying conditions* (1) *and* (2) *of Proposition 1, then $A = C_\infty(T)$ is a weakly Rickart C^*-algebra.*

Proof. If $x \in A$ and $U = \{t: x(t) \neq 0\}$, then $\bar{U}$ is compact-open by the proof of (4) above. If e is the characteristic function of $\bar{U}$, then $xe = x$ and $R(\{x\}) = R(\{e\})$ as in the proof of [§ 7, Prop. 1], thus e is an ARP of x [§ 5, Def. 1]. ∎

Another characterization:

Proposition 3. *Let A be a commutative C^*-algebra. Then A is a weakly Rickart C^*-algebra if and only if* (i) *A is the closed linear span of its projections, and* (ii) *every orthogonal sequence of projections in A has a supremum.*

Proof. We can assume A is unitless (the unity case is covered by [§ 8, Prop. 3]).

If A is a weakly Rickart C^*-algebra, then (i) holds by Proposition 1, and (ii) holds by [§ 8, Lemma 3].

Conversely, suppose (i) and (ii) hold. Write $A=C_\infty(T)$, T locally compact. By Lemma 3, the compact-open sets in T are basic, thus condition (1) of Proposition 2 holds; to complete the proof that A is a weakly Rickart C^*-algebra, it will suffice to verify condition (2). Let $U=\bigcup_1^\infty P_n$, where P_n is a sequence of compact-open sets; as argued in [§ 8, Prop. 3], we can suppose the P_n to be mutually disjoint, and hypothesis (ii) yields a compact-open set P which is a supremum for the P_n. The proof that $\bar{U}=P$ proceeds as in the proof of (2) in Proposition 1. ∎

In a compact space, 'clopen' means the same as 'compact-open', and every continuous function 'vanishes at infinity'. Since the term 'weakly Rickart' does not rule out the presence of a unity element, it follows that the results of this and the preceding section can be stated in unified form; the details are left to the reader.

An application to 'spectral theory':

Proposition 4. *Let A be any weakly Rickart C^*-algebra, $x\in A$, $x\geq 0$, $x\neq 0$. Given any $\varepsilon>0$, there exists $y\in\{x\}''$, $y\geq 0$, such that* (i) *$xy=e$, e a nonzero projection, and* (ii) $\|x-xe\|<\varepsilon$.

Proof. Let $g=\mathrm{RP}(x)$, drop down to the Rickart C^*-algebra gAg, and apply [§ 8, Prop. 4]; a minor technical point—that $y\in\{x\}''$—is settled by the elementary observation that the bicommutant of x relative to gAg is contained in $\{x\}''$. ∎

Corollary. *Let A be any weakly Rickart C^*-algebra, $x\in A$, $x\neq 0$. Given any $\varepsilon>0$, there exists $y\in\{x^*x\}''$, $y\geq 0$, such that* (i) $(x^*x)y^2=e$, *e a nonzero projection, and* (ii) $\|x-xe\|<\varepsilon$.

Proof. Same as [§ 7, Cor. of Prop. 3]. ∎

In particular: *Every weakly Rickart C^*-algebra satisfies the* (EP)-*axiom* [§ 7, Def. 1].

Exercises

1A. If A is a weakly Rickart C^*-algebra in which every orthogonal family of nonzero projections is countable, then A is an AW^*-algebra (in particular, A has a unity element).

2A. Let A be a commutative C^*-algebra that is the closed linear span of its projections. If $x\in A$, $x\neq 0$, and if $\varepsilon>0$, then there exists a nonzero projection e such that (i) $e=xy$ for some $y\in A$, and (ii) $\|x-xe\|<\varepsilon$.

3A. Let A be a C^*-algebra in which every masa [§ 1, Exer. 14] is the closed linear span of its projections. Suppose that (e_ι) is a family of projections in A that possesses a supremum e. Let $x \in A$. (i) If $xe_\iota = 0$ for all ι, then $xe = 0$. (ii) If $xe_\iota = e_\iota x$ for all ι, then $xe = ex$.

4A. Let A be a weakly Rickart C^*-algebra and let B be a closed $*$-subalgebra of A such that if (e_n) is any orthogonal sequence of projections in B, then $\sup e_n$ (as calculated in A) is also in B.

The following conditions are equivalent: (a) $x \in B$ implies $\mathrm{RP}(x) \in B$ (RP as calculated in A); (b) B is a weakly Rickart C^*-algebra. In this case, RP's and countable sups in B are unambiguous—i.e., they are the same whether calculated in B or in A.

§ 10. C^*-Sums

If $(A_\iota)_{\iota \in I}$ is a family of Baer $*$-rings and $A = \prod_{\iota \in I} A_\iota$ is their complete direct product [§ 1, Exer. 13], it is easy to see that A is also a Baer $*$-ring. However, if $(A_\iota)_{\iota \in I}$ is a family of AW^*-algebras, and A is their complete direct product (as $*$-algebras), it may not be possible to norm A so as to make it an AW^*-algebra (Exercise 1); in other words, for AW^*-algebras, the complete direct product is the wrong notion of 'direct product'. The right notion is the C^*-sum:

Definition 1. If $(A_\iota)_{\iota \in I}$ is a family of C^*-algebras, the C^*-*sum* of the family is the C^*-algebra B defined as follows. Let B be the set of all families $x = (a_\iota)_{\iota \in I}$ with $a_\iota \in A_\iota$ and $\|a_\iota\|$ bounded; equip B with the coordinatewise $*$-algebra operations, and the norm $\|x\| = \sup \|a_\iota\|$. (It is routine to check that B is a C^*-algebra.) Notation: $B = \bigoplus_{\iota \in I} A_\iota$.

Proposition 1. *If $(A_\iota)_{\iota \in I}$ is a family of weakly Rickart C^*-algebras [Rickart C^*-algebras, AW^*-algebras], then their C^*-sum $B = \bigoplus_{\iota \in I} A_\iota$, is also a weakly Rickart C^*-algebra [Rickart C^*-algebra, AW^*-algebra].*

Proof. Let $A = \prod_{\iota \in I} A_\iota$ be the complete direct product of the A_ι, equipped with the coordinatewise $*$-algebra operations [cf. § 1, Exer. 13]. Since the projections of A are the families $e = (e_\iota)$, with e_ι a projection in A_ι for each $\iota \in I$, and since the projections in a C^*-algebra have norm 0 or 1, it is clear that B contains every projection of A.

Suppose every A_ι is a weakly Rickart C^*-algebra. It is routine to check that A is a weakly Rickart $*$-ring; explicitly, if $x = (a_\iota) \in A$ and if, for each ι, $e_\iota = \mathrm{RP}(a_\iota)$, then the projection $e = (e_\iota)$ is an ARP of x in A. Since B contains all projections of A, it follows that B is a weakly Rickart C^*-algebra. If, in addition, every A_ι has a unity element, then so does B; this proves the assertion concerning Rickart C^*-algebras [cf. § 3, Exer. 12, 13].

Finally, suppose every A_ι is an AW^*-algebra; it is to be shown that B is a Baer $*$-ring. Since B contains every projection in A, it is sufficient to show that A is a Baer $*$-ring [cf. § 4, Exer. 6, 7]. Let S be a nonempty subset of A; we seek a projection $e \in A$ such that $R(S)=eA$. Write $\pi_\iota: A \to A_\iota$ for the canonical projection, and let $S_\iota=\pi_\iota(S)$. Clearly $R(S)=\{x \in A: \pi_\iota(x) \in R(S_\iota) \text{ for all } \iota \in I\}$. Write $R(S_\iota)=e_\iota A_\iota$, e_ι a projection in A_ι, and set $e=(e_\iota)$; evidently $x \in R(S)$ iff $e_\iota \pi_\iota(x)=\pi_\iota(x)$ for all $\iota \in I$ iff $ex=x$, thus $R(S)=eA$. ∎

Proposition 1 is a result about 'external' direct sums; let us now look at 'internal' ones. If $(A_\iota)_{\iota \in I}$ is a family of C^*-algebras with unity, and if, for each $\varkappa \in I$, $h_\varkappa=(\delta_{\iota\varkappa} 1)$ is the element of $B=\bigoplus_{\iota \in I} A_\iota$ with 1 in the $\varkappa$th place and 0's elsewhere, it is clear that the $h_\varkappa$ are orthogonal central projections in B, and that $\sup h_\varkappa$ exists and is equal to 1. Conversely, under favorable conditions, a central partition of unity in an algebra induces a representation as a C^*-sum; the next two propositions are important examples.

Proposition 2. *Let A be an AW^*-algebra and suppose $(h_\iota)_{\iota \in I}$ is an orthogonal family of central projections in A with $\sup h_\iota=1$. Then A is $*$-isomorphic to the C^*-sum of the $h_\iota A$; explicitly, the mapping $\varphi: A \to \bigoplus_{\iota \in I} h_\iota A$ defined by $\varphi(a)=(h_\iota a)$ is a norm-preserving $*$-isomorphism of A onto $\bigoplus_{\iota \in I} h_\iota A$.*

Proof. Since the $h_\iota A$ are AW^*-algebras [§ 4, Prop. 8, (iii)], their C^*-sum $B=\bigoplus_{\iota \in I} h_\iota A$ is also an AW^*-algebra (Proposition 1).

If $a \in A$ then $\|h_\iota a\| \le \|a\|$ shows that the family $(h_\iota a)_{\iota \in I}$ is in B, and, defining $\varphi(a)=(h_\iota a)$, we have $\|\varphi(a)\| \le \|a\|$. Clearly $\varphi: A \to B$ is a $*$-homomorphism, and $\|\varphi(a)\| \le \|a\|$ shows that φ is continuous. Moreover, φ is injective (if $h_\iota a=0$ for all ι then $a=0$ [§ 3, Prop. 6]). It follows that $\|\varphi(a)\|=\|a\|$ for all $a \in A$ (see Exercise 3). Since A is complete in norm, so is its isometric image $\varphi(A)$, therefore $\varphi(A)$ is a closed $*$-subalgebra of B. It remains to show that $\varphi(A)=B$; since B is the closed linear span of its projections [cf. § 8, Prop. 1], it will suffice to show that $\varphi(A)$ contains every projection of B. Suppose $e \in B$ is a projection, say $e=(e_\iota)$, where e_ι is a projection in $h_\iota A$, that is, $e_\iota \le h_\iota$. Define $f=\sup e_\iota$; it is elementary that $h_\iota f=e_\iota$ for all ι, thus $\varphi(f)=e$. ∎

The above argument requires that the projection lattice of A be complete (via the definition of f); this hypothesis can be weakened provided that we retreat from families to sequences:

Proposition 3. *Let A be a Rickart C^*-algebra and suppose (h_n) is an orthogonal sequence of central projections in A with $\sup h_n = 1$. Then A is isometrically $*$-isomorphic to the C^*-sum of the $h_n A$, via the mapping $\varphi: A \to \bigoplus_{n=1}^{\infty} h_n A$ defined by $\varphi(a) = (h_n a)$.*

Proof. In the proof of Proposition 2, replace the index set I by the set of positive integers, and read "Rickart C^*-algebra" in place of "AW^*-algebra"; no other change is necessary, except to note that every sequence of projections in A has a supremum [§ 8, Lemma 3]. ∎

Exercises

1C. (i) If a Banach algebra is a regular ring [§3, Exer. 6], it must be finite-dimensional.

(ii) If, for $n = 1, 2, 3, \ldots$, A_n is the algebra of all 2×2 complex matrices, and if $A = \prod_{n=1}^{\infty} A_n$ is the complete direct product, then A cannot be normed to be a Banach algebra.

2A. If $(A_\iota)_{\iota \in I}$ is a family of C^*-algebras, the $C^*(\infty)$-*sum* of the family is the closed $*$-subalgebra of $\bigoplus_{\iota \in I} A_\iota$ consisting of those $x = (a_\iota)_{\iota \in I}$ such that, for every $\varepsilon > 0$, $\|a_\iota\| < \varepsilon$ for all but finitely many indices. (This amounts to putting the discrete topology on I and requiring that $\|a_\iota\| \to 0$ at ∞, in the sense of the one-point compactification of I.)

3A. (i) In a C^*-algebra, if $x = \sum_1^n \lambda_i e_i$, where the e_i are orthogonal, nonzero projections, then $\|x\| = \max |\lambda_i|$.

(ii) If $\varphi: A \to B$ is a $*$-homomorphism, where A is a Banach $*$-algebra with continuous involution and B is a C^*-algebra, then φ is continuous.

(iii) If A and B are C^*-algebras, and if $\varphi: A \to B$ is a $*$-monomorphism, then $\|\varphi(x)\| = \|x\|$ for all $x \in A$. When A and B are weakly Rickart C^*-algebras, a simple proof can be based on (i) and (ii).

4A. If $(A_\iota)_{\iota \in I}$ is a family of C^*-algebras, then their P^*-sum [§1, Exer. 16] is a subalgebra of their C^*-sum.

5A. Let $(T_\iota)_{\iota \in I}$ be a family of connected, compact spaces, let $A_\iota = C(T_\iota)$, and let A be the P^*-sum of the A_ι. If $x = (a_\iota)_{\iota \in I}$ is in A, then, for all but finitely many ι, a_ι is a scalar multiple of the identity of A_ι; moreover, only finitely many scalars can occur as coordinates of x.

Chapter 2

Comparability of Projections

§ 11. Orthogonal Additivity of Equivalence

Let A be a Baer $*$-ring, let $(e_\iota)_{\iota\in I}$ and $(f_\iota)_{\iota\in I}$ be orthogonal families of projections indexed by the same set I, let $e=\sup e_\iota$, $f=\sup f_\iota$, and suppose that $e_\iota \sim f_\iota$ for all $\iota\in I$. Does it follow that $e\sim f$?

I don't know (see Exercise 3). If the index set I is finite, the question is answered affirmatively by trivial algebra [§ 1, Prop. 8]. The present section settles the question affirmatively under the added restriction that $ef=0$; this restriction is removed in Section 20, but only under an extra hypothesis on A. Some terminology helps to simplify the statements of these results:

Definition 1. Let A be a Baer $*$-ring (or, more generally, a $*$-ring in which the suprema in question are assumed to exist). If the answer to the question in the first paragraph is always affirmative, we say that equivalence in A is *additive* (or 'completely additive'); if it is affirmative whenever $\operatorname{card} I \le \aleph$, we say that equivalence in A is *$\aleph$-additive;* if it is affirmative whenever $ef=0$, we say that equivalence in A is *orthogonally additive* (see Theorem 1). The term *orthogonally $\aleph$-additive* is self-explanatory.

Suppose, more precisely, that the equivalences $e_\iota \sim f_\iota$ in question are implemented by partial isometries $w_\iota\,(\iota\in I)$. We say that partial isometries in A are *addable* if $e\sim f$ via a partial isometry w such that $we_\iota = w_\iota = f_\iota w$ for all $\iota\in I$. The terms *$\aleph$-addable*, *orthogonally addable*, and *orthogonally $\aleph$-addable* are self-explanatory. The main result of the section:

Theorem 1. *In any Baer $*$-ring, partial isometries are orthogonally addable; in particular, equivalence is orthogonally additive.*

Four lemmas prepare the way for the proof of Theorem 1.

Lemma 1. *In a weakly Rickart $*$-ring, suppose $(h_\iota)_{\iota\in I}$ is an orthogonal family of projections, and $(e_\iota)_{\iota\in I}$ is a (necessarily orthogonal) family* *of projections such that* $e_\iota \le h_\iota$ *for all* ι *and such that* $e = \sup e_\iota$ *exists. Then* $e_\iota = h_\iota e$ *for all* ι.

Proof. Fix ι and set $x = h_\iota e - e_\iota$; obviously $xe = x$. If $\varkappa \ne \iota$ then $xe_\varkappa = h_\iota e_\varkappa - e_\iota e_\varkappa = 0$ by the assumed orthogonality; also $xe_\iota = h_\iota e_\iota - e_\iota = 0$; therefore $xe = 0$ [§ 5, Exer. 4], that is, $x = 0$. ∎

Lemma 2. *If* A *is a Rickart* $*$*-ring containing a projection* e *such that* $e \sim 1-e$, *then* $2 = 1+1$ *is invertible in* A.

Proof. Let w be a partial isometry such that $w^*w = e$, $ww^* = 1-e$, and write $R(\{e-w^*\}) = fA$, f a projection; we show that $x = fe + wf - w$ satisfies $2x = 1$.

From $(e-w^*)f = 0$, we have $fw = fe$. Since $w \in (1-e)Ae$, it follows that $(e-w^*)(e+w) = 0$, therefore $f(e+w) = e+w$; noting that $fw = fe$, this yields

$$(*) \qquad e = 2fe - w.$$

Right-multiplying $(*)$ by w^*, we have $w^* = 2fw^* - (1-e)$; taking adjoints, we obtain

$$(**) \qquad 1-e = 2wf - w.$$

Addition of $(*)$ and $(**)$ yields $1 = 2x$. ∎

Lemma 3. *Let* A *be a weakly Rickart* $*$*-ring in which every orthogonal family of projections of cardinality* $\le \aleph$ *has a supremum.*

Let $(h_\iota)_{\iota \in I}$ *be an orthogonal family of central projections, with* $\operatorname{card} I \le \aleph$, *and suppose that, for each* ι, e_ι *and* f_ι *are orthogonal projections with* $e_\iota + f_\iota = h_\iota$, $e_\iota \sim f_\iota$. *Let* $e = \sup e_\iota$, $f = \sup f_\iota$.

Then $e \sim f$. *More precisely, if the equivalences* $e_\iota \sim f_\iota$ *are implemented by partial isometries* w_ι, *then there exists a partial isometry* w *implementing* $e \sim f$, *such that* $we_\iota = w_\iota = f_\iota w$ *for all* ι.

Proof. Since $h_\iota A$ is a Rickart $*$-ring [§ 5, Prop. 6] and $w_\iota \in h_\iota A$, we have $e_\iota \sim f_\iota$ in $h_\iota A$. By Lemma 2, $2h_\iota$ is invertible in $h_\iota A$; say $a_\iota \in h_\iota A$ with $h_\iota = (2h_\iota)a_\iota = 2a_\iota$. Since $2h_\iota$ is self-adjoint and central in $h_\iota A$, so is a_ι.

Write $u_\iota = w_\iota + w_\iota^*$; clearly u_ι is a symmetry (= self-adjoint unitary) in $h_\iota A$, that is, $u_\iota^* = u_\iota$, $u_\iota^2 = h_\iota$. Defining

$$g_\iota = a_\iota(h_\iota + u_\iota)$$

(informally, $g_\iota = (1/2)(h_\iota + w_\iota + w_\iota^*)$), it is easy to check that g_ι is a projection in $h_\iota A$.

Define $g = \sup g_\iota$. Citing Lemma 1, we have $h_\iota g = g_\iota$, $h_\iota e = e_\iota$, $h_\iota f = f_\iota$. Finally, define

$$w = 2fge;$$

the proof will be concluded by showing that w is a partial isometry having the desired properties.

Note that $f_\iota u_\iota = w_\iota = u_\iota e_\iota$; it follows easily that

$$f_\iota g_\iota e_\iota = a_\iota w_\iota,$$

therefore $2 f_\iota g_\iota e_\iota = 2 a_\iota w_\iota = h_\iota w_\iota = w_\iota$. Then

$$h_\iota w = w_\iota; \tag{1}$$

for, $h_\iota w = h_\iota(2fge) = 2(h_\iota f)(h_\iota g)(h_\iota e) = 2 f_\iota g_\iota e_\iota = w_\iota$. It follows that

$$w e_\iota = w_\iota = f_\iota w; \tag{2}$$

for example, $w e_\iota = w(h_\iota e_\iota) = (h_\iota w) e_\iota = w_\iota e_\iota = w_\iota$.

It remains to show that $w^* w = e$ and $w w^* = f$. Let $h = \sup h_\iota$. Since $e_\iota \le h_\iota \le h$, we have $e \le h$, and similarly $f \le h$, $g \le h$. It follows that $wh = w$, $w^* h = w^*$. For all ι, we have

$$(w^* w - e) h_\iota = (h_\iota w)^* (h_\iota w) - h_\iota e = w_\iota^* w_\iota - e_\iota = 0,$$

therefore $(w^* w - e) h = 0$ [§ 5, Exer. 4], thus $w^* w - e = 0$. Similarly $w w^* - f = 0$. ∎

Lemma 4. *If A is a weakly Rickart $*$-ring in which every orthogonal family of projections of cardinality $\le \aleph$ has a supremum, then partial isometries in A are orthogonally $\aleph$-addable.*

Proof. Let $(w_\iota)_{\iota \in I}$ be a family of partial isometries, $\operatorname{card} I \le \aleph$, with orthogonal initial projections e_ι, orthogonal final projections f_ι, and such that, setting $e = \sup e_\iota$, $f = \sup f_\iota$, we have $ef = 0$. We seek a partial isometry w such that $w^* w = e$, $w w^* = f$ and $w e_\iota = w_\iota = f_\iota w$ for all ι.

Write $h_\iota = e_\iota + f_\iota$, $S = \{h_\iota : \iota \in I\}$, and let $B = S'$. According to [§ 5, Prop. 5], B is also a weakly Rickart $*$-ring, with unambiguous RP's. Since the h_ι are orthogonal, S is a commutative set; it follows that $B \supset B'$, and that B has center $B \cap B' = B' = S''$. In particular, the h_ι are central projections in B.

Each w_ι belongs to B; for, $h_\iota w_\iota = w_\iota h_\iota$ $(= w_\iota)$ and $h_\varkappa w_\iota = w_\iota h_\varkappa$ $(= 0)$ whenever $\varkappa \ne \iota$, thus $w_\iota \in S' = B$. In particular, $e_\iota \sim f_\iota$ in B. Finally, it is clear that $e, f \in B$ (cf. the proof of [§ 3, Prop. 11]), thus the desired w exists by Lemma 3. ∎

Proof of Theorem 1. When A is a Baer $*$-ring, the hypothesis of Lemma 4 is verified for every $\aleph$. ∎

We now take up several other useful consequences of Lemma 4.

Proposition 1. *Let A be a weakly Rickart $*$-ring in which every orthogonal family of projections of cardinality $\le \aleph$ has a supremum.*

Let $(e_\iota)_{\iota\in I}$ *be an infinite family of mutually equivalent, orthogonal projections, with* $\operatorname{card} I \le \aleph$, *and let* $J \subset I$ *with* $\operatorname{card} J = \operatorname{card} I$. *Define*

$$e = \sup\{e_\iota : \iota \in I\}, \qquad f = \sup\{e_\iota : \iota \in J\}.$$

Then $e \sim f$.

Proof. Dropping down to eAe, we can suppose that A is a Rickart $*$-ring [§ 5, Prop. 6]. Write $J = J' \cup J''$, where $J' \cap J'' = \varnothing$ and $\operatorname{card} J' = \operatorname{card} J'' = \operatorname{card} J$ $(= \operatorname{card} I)$. Define

$$f' = \sup\{e_\iota : \iota \in J'\},$$
$$f'' = \sup\{e_\iota : \iota \in J''\},$$
$$g = \sup\{e_\iota : \iota \in J'' \cup (I - J)\}.$$

Since the e_ι are orthogonal, clearly $f'g = 0$, thus $\sup\{f', g\} = f' + g$; but $J' \cup [J'' \cup (I - J)] = I$, therefore

$$e = f' + g \tag{1}$$

by the associativity of suprema. Similarly,

$$f = f' + f''. \tag{2}$$

Since J' and $J'' \cup (I - J)$ have the same cardinality, and since $f'g = 0$, we have

$$f' \sim g \tag{3}$$

by Lemma 4. Similarly,

$$f'' \sim f'. \tag{4}$$

Adding (3) and (4), we have $f' + f'' \sim g + f'$, that is, citing (1) and (2), $f \sim e$. ∎

Proposition 2. *Let A be a Baer $*$-ring, e and f projections in A with $ef = 0$, (h_α) an orthogonal family of central projections, and $h = \sup h_\alpha$.*

If $h_\alpha e \precsim h_\alpha f$ for all α, then $he \precsim hf$. More precisely, if, for each α, w_α is a partial isometry such that $w_\alpha^* w_\alpha = h_\alpha e$ and $w_\alpha w_\alpha^* = f_\alpha \le h_\alpha f$, then there exists a partial isometry w such that $w^* w = he$ and $w(h_\alpha e) = w_\alpha = f_\alpha w$ for all α.

Proof. Since $(\sup h_\alpha e)(\sup f_\alpha) = he \sup f_\alpha \le hef = 0$, and since the f_α are also mutually orthogonal (because the h_α are), Lemma 4 is applicable to the partial isometries w_α. ∎

In a weakly Rickart C^*-algebra, countable suprema are available [§ 8, Lemma 3], thus Lemma 4 holds with $\aleph = \aleph_0$, implying the obvious sequential forms of Propositions 1 and 2.

Exercises

1A. Let e, f be orthogonal projections in a Rickart $*$-ring. In order that $e \sim f$, it is necessary and sufficient that there exist a projection g such that $2ege=e$, $2fgf=f$, $2geg=2gfg=g$.

2A. Let C be a Baer $*$-ring possessing a projection e such that $e \sim 1-e$ (for example, the $*$-ring of all 2×2 complex matrices). Let B be the complete direct product of $\aleph_0$ copies of C, that is, $B=\prod_1^\infty A_n$ with $A_n=C$ for all n [cf. § 1, Exer. 13]. Let B_0 be the weak direct product of the A_n, that is, the ideal of all $x=(a_n)$ in B such that $a_n=0$ for all but finitely many n. Write $f=1-e$, $\bar{e}=(e,e,e,\dots)$, $\bar{f}=(f,f,f,\dots)=1-\bar{e}$, and let S be the $*$-subring of B generated by $\bar{e}$ and 1; thus, $S=\{m\bar{e}+n\bar{f}: m,n \text{ integers}\}$. Define $A=B_0+S$; thus, A is the $*$-subring of B generated by B_0, $\bar{e}$ and 1. If the additive group of C is torsion-free, then A is a Rickart $*$-ring.

For $m=1,2,3,\dots$ write $e_m=(\delta_{mn}e)$, $f_m=(\delta_{mn}f)$. Then, relative to the $*$-ring A, we have $e_m \sim f_m$ for all m, $\sup e_m=\bar{e}$, $\sup f_m=\bar{f}$, $\bar{e}\bar{f}=0$, but $\bar{e}$ is not equivalent to $\bar{f}$. Thus Theorem 1 does not generalize to Rickart $*$-rings.

3D. Let C be a Baer $*$-ring, let C° be the reduced ring of C [§ 3, Exer. 18], and suppose there exists an equivalence $e \sim f$ in C which cannot be implemented by any partial isometry in C° (that is, $e \sim f$ but not $e \overset{\circ}{\sim} f$).

Let A be the P^*-sum of $\aleph_0$ copies of C [cf. § 4, Exer. 8]. For $m=1,2,3,\dots$ let e_m and f_m be the projections in A defined by the sequences $e_m=(\delta_{mn}e)$, $f_m=(\delta_{mn}f)$. Then $e_m \sim f_m$ for all m, but $\sup e_m$ is not equivalent to $\sup f_m$.

Problem: Does there exist such a Baer $*$-ring C? (Cf. [§ 17, Exer. 20].).

4D. *Problem:* Is equivalence $\aleph_0$-additive (i.e., 'countably additive') in a Rickart C^*-algebra?

5A. If A is a Baer $*$-ring such that the ring A_2 of all 2×2 matrices over A is also a Baer $*$-ring (with $*$-transpose as involution), then partial isometries in A are addable.

6A. In the notation of Definition 1, if there exists an element x such that $xe_\iota=w_\iota=f_\iota x$ for all ι, then the partial isometries w_ι are addable (the desired partial isometry is $w=xe$).

§ 12. A General Schröder-Bernstein Theorem

We say that *the Schröder-Bernstein theorem holds* in a $*$-ring if the relations $e \precsim f$ and $f \precsim e$ imply $e \sim f$. In Section 1, it was shown that the Schröder-Bernstein theorem holds in any $*$-ring whose set of projections is conditionally complete [§ 1, Th. 1]—in particular, it holds in any Baer $*$-ring [cf. § 4, Prop. 1]. The fixed-point theorem employed there requires lattice completeness; by reverting to the format of the classical set-theoretic proof, one can get along with countable lattice operations:

Proposition 1. *If A is a weakly Rickart $*$-ring in which every countable family of orthogonal projections has a supremum, then the Schröder-Bernstein theorem holds in A.*

It is convenient to separate out an elementary lemma:

Lemma. *If e_n is a decreasing sequence of projections in such a $*$-ring, that is, if $e_1 \geq e_2 \geq e_3 \geq \cdots$, then $\inf e_n$ exists. Explicitly,*

$$\inf e_n = e_1 - g,$$

where $g = \sup\{e_n - e_{n+1} : n = 1, 2, 3, \ldots\}$.

Proof of Proposition 1. Assuming $e \sim f' \leq f$ and $f \sim e' \leq e$, it is to be shown that $e \sim f$.

Let w be a partial isometry such that $w^* w = f$, $w w^* = e'$. Setting $v = wf'$, we have $v^* v = f'$; thus v is a partial isometry, and, writing $e'' = v v^*$, we have

$$f' \sim e'' \leq e'.$$

Combining this with $e \sim f'$, we have the following situation:

$$e'' \leq e' \leq e \quad \text{and } e'' \sim e. \tag{1}$$

On the basis of (1), it will be shown that $e' \sim e$ (the observation $f \sim e' \sim e$ then ends the proof); no further reference to f is necessary.

By (1), there exists a partial isometry u such that

$$u^* u = e'', \qquad u u^* = e.$$

Since $g \mapsto \varphi(g) = u^* g u$ is an order-preserving bijection of $[0, e]$ onto $[0, e'']$ (see [§ 1, Prop. 9]), we may define a sequence $e_0, e_2, e_4, \ldots, e_{2n}, \ldots$ of subprojections of e as follows:

$$\begin{aligned}
&e_0 = e \\
&e_2 = u^* e_0 u = u^* e u = u^* u = e'' \leq e \\
&e_4 = u^* e_2 u \\
&\ldots \\
&e_{2n} = u^* e_{2(n-1)} u \qquad (n = 1, 2, 3, \ldots).
\end{aligned}$$

Define another sequence $e_1, e_3, e_5, \ldots, e_{2n-1}, \ldots$ of subprojections of e by the same technique, starting with e':

$$\begin{aligned}
&e_1 = e' \leq e \\
&e_3 = u^* e_1 u = u^* e' u \\
&e_5 = u^* e_3 u \\
&\ldots \\
&e_{2n+1} = u^* e_{2n-1} u \qquad (n = 1, 2, 3, \ldots).
\end{aligned}$$

Observe that

$$e_0 \geq e_1 \geq e_2 \geq e_3 \geq \cdots . \tag{2}$$

(Indeed, (1) may be written $e_0 \geq e_1 \geq e_2$; application of φ yields $e_2 \geq e_3 \geq e_4$; etc.)

We now look at the 'gaps' in the decreasing sequence (2). Since, by definition, $u^* e_n u = e_{n+2}$ $(n=0,1,2,3,\ldots)$, we have $u^*(e_n - e_{n+1})u = e_{n+2} - e_{n+3}$, thus

$$e_n - e_{n+1} \sim e_{n+2} - e_{n+3} \qquad (n=0,1,2,3,\ldots) \tag{3}$$

(the equivalence (3) is implemented by the partial isometry $u^*(e_n - e_{n+1})$).

By the lemma, we may define

$$e_\infty = \inf\{e_n : n=0,1,2,3,\ldots\}. \tag{4}$$

Obviously any truncation of the sequence e_n has the same infimum, in particular,

$$e_\infty = \inf\{e_n : n=1,2,3,\ldots\}. \tag{5}$$

Consider the following two sequences of orthogonal projections:

$$e_\infty,\ e_0 - e_1,\ e_1 - e_2,\ e_2 - e_3, \ldots \tag{$*$}$$

$$e_\infty,\ e_1 - e_2,\ e_2 - e_3,\ e_3 - e_4, \ldots \tag{$**$}$$

(the second sequence merely omits the second term of the first sequence). In view of (4), it follows from the lemma that

$$e_0 = e_\infty + \sup\{e_0 - e_1, e_1 - e_2, e_2 - e_3, \ldots\};$$

thus, by the associativy of suprema, the sequence $(*)$ has supremum $e_\infty + (e_0 - e_\infty) = e_0 = e$. It follows similarly from (5) that the sequence $(**)$ has supremum $e_\infty + (e_1 - e_\infty) = e_1 = e'$.

The desired equivalence $e \sim e'$ is obtained by putting together the pieces in $(*)$ and $(**)$ in another way. We define

$$g = \sup\{e_0 - e_1, e_2 - e_3, e_4 - e_5, \ldots\},$$
$$g' = \sup\{e_2 - e_3, e_4 - e_5, e_6 - e_7, \ldots\},$$
$$h = e_\infty + \sup\{e_1 - e_2, e_3 - e_4, e_5 - e_6, \ldots\}.$$

By the associativity of suprema, $g + h$ coincides with the supremum of the sequence $(*)$, thus

$$e = g + h; \tag{6}$$

similarly, $g' + h$ is the supremum of the sequence $(**)$, thus

$$e' = g' + h. \tag{7}$$

It follows from (3), and the definitions of g and g', that $g \sim g'$ [§11, Prop. 1]; in view of (6) and (7), this implies $e \sim e'$. ▮

The principal applications:

Corollary. *The Schröder-Bernstein theorem holds* (i) *in any Baer $*$-ring, and* (ii) *in any weakly Rickart C^*-algebra.*

Proof. Of course (i) is also covered by [§1, Th. 1]; (ii) follows from the fact that every sequence of projections in a weakly Rickart C^*-algebra has a supremum [§8, Lemma 3]. ▮

Exercises

1A. Let A be a weakly Rickart $*$-ring in which every countable family of orthogonal projections has a supremum. If e is any projection, write $[e]$ for the equivalence class of e with respect to $\sim$, that is, $[e]=\{f: f\sim e\}$. Define $[e]\leq[f]$ iff $e\lesssim f$. This is a partial ordering of the set of equivalence classes.

2A. The Schröder-Bernstein theorem holds trivially in any finite $*$-ring. (A $*$-ring with unity is said to be *finite* [§15, Def. 3] if $e\sim 1$ implies $e=1$.)

3A. Let $\mathscr{H}$ be a separable, infinite-dimensional Hilbert space, with orthonormal basis $e_1, e_2, e_3, \ldots$. Let T be the operator such that $Te_n = e_{n+2}$ for all n; thus $T^*T=I$, $TT^*=E$, where E is the projection with range $[e_3, e_4, e_5, \ldots]$. Let F be the projection with range $[e_2, e_3, e_4, \ldots]$. Finally, let $\mathscr{A}$ be the $*$-ring generated by T and F.

The relations $T^*T=I$, $TT^*=E\leq F$ and $F\leq I$ show that $I\lesssim F$ and $F\lesssim I$ relative to the $*$-ring $\mathscr{A}$. Is $F\sim I$ relative to $\mathscr{A}$?

§ 13. The Parallelogram Law (P) and Related Matters

The law in question is reminiscent of the 'second isomorphism theorem' of abstract algebra:

Definition 1. A $*$-ring whose projections form a lattice is said to satisfy the *parallelogram law* if

(P) $$e - e\cap f \sim e\cup f - f$$

for every pair of projections e, f.

The projections of every weakly Rickart $*$-ring form a lattice [§5, Prop. 7], but even a Baer $*$-ring may fail to satisfy the parallelogram law (Exercise 1). Occasionally, the following variant of (P) is more convenient:

Proposition 1. *Let A be a $*$-ring with unity, whose projections form a lattice. The following conditions are equivalent:*

(a) *A satisfies the parallelogram law* (P);

(b) $e - e\cap(1-f) \sim f - (1-e)\cap f$ *for every pair of projections e, f.*

Proof. Replacement of f by $1-f$ in the relation (P) yields

$$e-e\cap(1-f)\sim[e\cup(1-f)]-(1-f)=f-[1-e\cup(1-f)]$$
$$=f-(1-e)\cap f. \quad \blacksquare$$

Proposition 1 may be interpreted as saying that, in the presence of (P), certain subprojections of e,f (indicated in (b)) are guaranteed to be equivalent; this conclusion reduces to the triviality $0\sim 0$ precisely when $e=e\cap(1-f)$, that is, when $ef=0$.

The projections that occur in (P) are familiar from [§5, Prop. 7]:

Proposition 2. *If A is a weakly Rickart $*$-ring such that* $\mathrm{LP}(x)\sim\mathrm{RP}(x)$ *for all $x\in A$, then A satisfies the parallelogram law* (P).

Proof. Apply the hypothesis to the element $x=e-ef$ [§5, Prop. 7]. ∎

An important application:

Corollary. *Every von Neumann algebra satisfies the parallelogram law* (P).

Proof. Let $\mathscr{A}$ be a von Neumann algebra of operators on a Hilbert space $\mathscr{H}$ [§4, Def. 5]. If T is any operator on $\mathscr{H}$, the 'canonical factorization' $T=WR$ is uniquely characterized by the following three properties: (i) $R\geq 0$, (ii) W is a partial isometry, and (iii) W^*W is the projection on the closure of the range of R, that is, $W^*W=\mathrm{LP}(R)$ as calculated in $\mathscr{L}(\mathscr{H})$. It follows that $W^*W=\mathrm{RP}(T)$, $WW^*=\mathrm{LP}(T)$, thus $\mathrm{LP}(T)\sim\mathrm{RP}(T)$ in $\mathscr{L}(\mathscr{H})$. The proof is concluded by observing that if $T\in\mathscr{A}$ then $W\in\mathscr{A}$ (therefore $\mathrm{LP}(T)\sim\mathrm{RP}(T)$ in $\mathscr{A}$).

Suppose $T\in\mathscr{A}$. If $U\in\mathscr{A}'$ is unitary, then $T=UTU^*=(UWU^*)(URU^*)$; since the properties (i), (ii), (iii) are satisfied by the positive operator URU^* and the partial isometry UWU^*, it follows from uniqueness that $UWU^*=W$, thus W commutes with U. Since $\mathscr{A}'$ is the linear span of its unitaries (as is any C^*-algebra with unity [cf. **23**, Ch. I, §1, No. 3, Prop. 3]), it follows that $W\in(\mathscr{A}')'=\mathscr{A}$. ∎

Later in the section it will be shown, more generally, that every AW^*-algebra—indeed, any weakly Rickart C^*-algebra—satisfies the parallelogram law (P). The proof will avoid the use of $\mathrm{LP}\sim\mathrm{RP}$ (known to hold in any AW^*-algebra [§20, Cor. of Th. 3], but of unknown status in Rickart C^*-algebras). The general strategy is to reduce the consideration of arbitrary pairs of projections e,f to pairs of projections in 'special position'; the following concept is central to such considerations:

Definition 2. Let A be a $*$-ring with unity, whose projections form a lattice (for example, A any Rickart $*$-ring). Projections e, f in A are said to be in *position p'* in case

$$e \cap (1-f) = (1-e) \cap f = 0 .$$

{Equivalenty, $e \cap (1-f) = 0$ and $e \cup (1-f) = 1$; that is, the projections e, $1-f$ are complementary.} The condition is obviously symmetric in e and f.

In Rickart $*$-rings, the concept has a useful reformulation:

Proposition 3. *In a Rickart $*$-ring, the following conditions on a pair of projections e, f imply one another:*
(a) *e, f are in position p';*
(b) $\mathrm{LP}(ef) = e$ *and* $\mathrm{RP}(ef) = f$.

Proof. Let $x = ef = e[1-(1-f)]$. Citing [§3, Prop. 7], we have

$$\mathrm{LP}(x) = e - e \cap (1-f), \qquad \mathrm{RP}(x) = e \cup (1-f) - (1-f);$$

thus, the conditions (b) are equivalent to $e \cap (1-f) = 0$ and $e \cup (1-f) = 1$. ∎

In a Rickart $*$-ring, the parallelogram law can be reformulated in terms of position p':

Proposition 4. *The following conditions on a Rickart $*$-ring A are equivalent:*
(a) *A satisfies the parallelogram law* (P);
(b) *if e, f are projections in position p', then $e \sim f$.*

Proof. (a) implies (b): If $e \cap (1-f) = (1-e) \cap f = 0$, then, in the presence of (P), $e \sim f$ by Proposition 1.

(b) implies (a): Let e, f be any pair of projections, and set $e' = \mathrm{LP}(ef)$, $f' = \mathrm{RP}(ef)$. Since $ef = e'(ef)f' = e'f'$, it follows from Proposition 3 that e', f' are in position p'; therefore, by hypothesis, $e' \sim f'$, that is,

$$e - e \cap (1-f) \sim e \cup (1-f) - (1-f) = f - (1-e) \cap f .$$

Since e, f are arbitrary, it follows from Proposition 1 that A satisfies (P). ∎

The proof of Proposition 4 yields a highly useful decomposition theorem:

Proposition 5. *Let A be a Rickart $*$-ring satisfying the parallelogram law* (P). *If e, f is any pair of projections in A, there exist orthogonal decompositions*

$$e = e' + e'', \qquad f = f' + f''$$

with e', f' in position p' (hence $e' \sim f'$ by Proposition 4) and $e''f = ef'' = 0$.

Proof. Let $e' = \mathrm{LP}(ef)$, $f' = \mathrm{RP}(ef)$; as noted in the proof of Proposition 4, e', f' are in position p'. Set $e'' = e - e'$, $f'' = f - f'$; obviously $e''(ef) = (ef)f'' = 0$, thus $e''f = ef'' = 0$. ∎

The rest of the section is concerned with developing sufficient conditions for (P) to hold. With an eye on Proposition 4, we seek conditions ensuring that projections in position p' are equivalent. For the most part, victory hinges on being able to analyze position p' considerations in terms of the following more stringent relation:

Definition 3. Let A be a $*$-ring with unity, whose projections form a lattice. Projections e, f in A are said to be in *position p* in case

$$e \cap f = e \cap (1-f) = (1-e) \cap f = (1-e) \cap (1-f) = 0.$$

{Equivalently, each of the pairs e, f and e, $1-f$ is in position p'.} The condition is obviously symmetric in e and f.

If e, f are in position p, then so is any pair g, h, where $g = e$ or $1-e$, and $h = f$ or $1-f$.

In Proposition 3, position p' is characterized in terms of the element ef; the characterization of position p involves both ef and its adjoint:

Proposition 6. *In a Rickart $*$-ring, the following conditions on a pair of projections e, f imply one another:*

(a) *e, f are in position p;*

(b) $\mathrm{RP}(ef - fe) = 1$.

Proof. (b) implies (a): Set $x = ef - fe$. Since $\mathrm{RP}(x) = 1$, the relations $e \cap f = 0$ and $e \cup f = 1$ are implied by the obvious computations

$$x(e \cap f) = 0, \qquad x(e \cup f) = x.$$

But $e(1-f) - (1-f)e = -x$ also has right projection 1, therefore $e \cap (1-f) = 0$ and $e \cup (1-f) = 1$. Thus $e \cap f = (1-e) \cap (1-f)$ $= e \cap (1-f) = (1-e) \cap f = 0$.

(a) implies (b): Let $x = ef - fe$, $g = \mathrm{RP}(x)$; assuming e, f are in position p, it is to be shown that $g = 1$. {For an insight on the success of the following strategem, compute $(ab - ba)^2$ for a pair of 2×2 matrices a, b over a commutative ring.} Set $z = x^* x = -x^2$; by direct computation,

$$\begin{aligned} z = -(ef - fe)^2 &= -efef + efe + fef - fefe \\ &= ef(1-e)fe + (1-e)fef(1-e) \\ &= fe(1-f)ef + (1-f)efe(1-f). \end{aligned}$$

From the last two formulas, it is clear that e and f commute with z. On the other hand,

$$g=\mathrm{RP}(x)=\mathrm{RP}(x^*x)=\mathrm{RP}(z)\in\{z\}''$$

[§3, Cor. 2 of Prop. 10], therefore g commutes with e and with f. Set $h=1-g$. Since $g=\mathrm{RP}(x)$ and since h commutes with e and f, we have

$$0=xh=(eh)(fh)-(fh)(eh),$$

thus eh, fh are commuting projections; citing [§1, Prop. 3], we have

$$(eh)(fh)=(eh)\cap(fh)\leq e\cap f=0,$$

thus

(1) $$(ef)h=0.$$

Since $e(1-f)-(1-f)e=-x$ also has right projection g, and since $e\cap(1-f)=0$ by hypothesis, the same reasoning yields

(2) $$[e(1-f)]h=0.$$

Adding (1) and (2), we have $eh=0$. Similarly $fh=0$. Thus $e\leq 1-h=g$ and $f\leq g$; since $e\cup f=1$, we conclude that $g=1$. ∎

An obvious way to fulfill condition (b) of Proposition 6 is to assume outright that $ef-fe$ is invertible; in the next proposition, it is shown that the invertibility of $ef-fe$ implies $e\sim f$, provided one also assumes a condition on the existence of 'square roots'. Historically, the first condition of this type, considered by I. Kaplansky ([**52**], [**54**]), was the following:

Definition 4. A $*$-ring is said to satisfy the *square-root axiom* (briefly, the (SR)-*axiom*) in case, for each element x, there exists $r\in\{x^*x\}''$ such that $r^*=r$ and $x^*x=r^2$.

Occasionally, the following weaker axiom suffices (later in the section, stronger axioms will be employed):

Definition 5. A $*$-ring is said to satisfy the *weak square-root axiom* (briefly, the (WSR)-*axiom*) in case, for each element x, there exists $r\in\{x^*x\}''$ (necessarily normal, but not necessarily self-adjoint) such that $x^*x=r^*r$ $(=rr^*)$.

A sample of the wholesome effect of square roots:

Lemma. *If A is a $*$-ring satisfying the* (WSR)-*axiom, and if the projections e, f are algebraically equivalent in the sense that $yx=e$ and $xy=f$ for suitable elements $x, y\in A$, then $e\sim f$.*

Proof. Replacing x and y by fxe and eyf, we can suppose $x \in fAe$, $y \in eAf$. We seek a partial isometry w such that $w^*w = e$, $ww^* = f$.

Choose $r \in \{y^*y\}''$ with $y^*y = r^*r = rr^*$, and set $w = rx$. Then

$$w^*w = x^*r^*rx = x^*y^*yx = (yx)^*(yx) = e^*e = e.$$

On the other hand, $ww^* = rxx^*r^*$; to proceed further, we show that r commutes with xx^*. Since $r \in \{y^*y\}''$, it suffices to note that $xx^* \in \{y^*y\}'$; indeed, xx^* and y^*y are self-adjoint elements whose product

$$(xx^*)(y^*y) = x(yx)^*y = xey = xy = f$$

is also self-adjoint. Thus $r \in \{x^*x\}'$, and

$$ww^* = rxx^*r^* = xx^*rr^* = (xx^*)(y^*y) = f. \quad \blacksquare$$

Armed with square roots, a considerable dent can be made on the parallelogram law problem:

Proposition 7. *If A is a $*$-ring with unity satisfying the* (WSR)*-axiom, and if e, f are projections in A such that $ef - fe$ is invertible, then $e \sim f \sim 1 - e \sim 1 - f$.*

Proof. Since the invertibility hypothesis for the pair e, f clearly holds also for the pairs e, $1-f$ and $1-e$, f, it is sufficient to show that $e \sim f$.

Let $z = (ef - fe)^*(ef - fe) = -(ef - fe)^2$ and write $B = \{z\}'$. As noted in the proof of Proposition 6, $e, f \in B$. Since $\{z\} \subset \{z\}'$, we have $B = \{z\}' \supset \{z\}'' = B'$, thus B has center $B \cap B' = B' = \{z\}''$. In particular, z is central in B.

We assert that efe is invertible in eBe. The proof begins by noting that $s = z^{-1}$ is also central in B; then $zs = sz = 1$ implies $(eze)(ese) = (ese)(eze) = e$, thus $eze = ez$ is invertible in eBe. From one of the formulas for z in the proof of Proposition 6, we have

$$ez = ef(1-e)fe = efe - (efe)^2 = efe(e - efe) = (e - efe)efe,$$

thus the invertibility of ez in eBe implies that of efe. Let $t \in eBe$ with $t(efe) = (efe)t = e$, that is,

$$(*) \qquad tfe = eft = e$$

(explicitly, $t = s(e - efe)$).

By the lemma, it will suffice to show that e and f are algebraically equivalent. To this end, define

$$x = ft, \qquad y = ef.$$

Obviously $x \in fAe$, $y \in eAf$, and $yx=(ef)(ft)=eft=e$ by (*). On the other hand, $xy=(ft)(ef)=ftf$; citing (*) at the appropriate step, we have

$$(ef-fe)xy=(ef-fe)ftf=eftf-feftf$$
$$=(eft)f-f(eft)f=ef-fef=(ef-fe)f,$$

therefore $xy=f$ by the invertibility of $ef-fe$. ∎

The technique of Proposition 7 suffices to establish the parallelogram law in the C^*-algebra case:

Theorem 1. *Every weakly Rickart C^*-algebra satisfies the parallelogram law* (P).

Proof. If A is a weakly Rickart C^*-algebra, then the projections of A form a lattice [§ 5, Prop. 7]. Let e, f be any pair of projections in A. To verify that e, f satisfy the relation (P), it suffices to work in the Rickart C^*-algebra $(e\cup f)A(e\cup f)$ [§ 5, Prop. 6]; dropping down, we can suppose without loss of generality that A has a unity element.

Set $z=(ef-fe)^*(ef-fe)=-(ef-fe)^2$ and consider the Rickart C^*-algebra $\{z\}'$ [§ 3, Prop. 10]. As noted in the proof of Proposition 7, $e, f \in \{z\}'$ and $\{z\}'$ has center $\{z\}''$. Dropping down to $\{z\}'$, we can suppose that z is in the center Z of A. (This will yield the sharper conclusion that the equivalence $e-e\cap f \sim e\cup f-f$ can be implemented by a partial isometry in $\{(ef-fe)^2\}'$.)

Write $Z=C(T)$, T a compact space with the properties noted in [§ 8, Prop. 1]. By C^*-algebra theory, we have $z \geq 0$ in Z (see the proof of [§ 7, Prop. 3]); setting

$$U=\{t\in T: z(t)>0\},$$

it follows that $\bar{U}$ is a clopen set whose characteristic function h is $\mathrm{RP}(z)$ [§ 8, Prop. 1].

If U is empty, that is, if $z=0$, then $ef-fe=0$ and the desired relation (P) reduces to the triviality $e-ef=(e+f-ef)-f$ [§ 1, Prop. 3].

Assuming U is nonempty, write $U=\bigcup P_n$, where P_n is a sequence (possibly finite) of disjoint, nonempty clopen sets (cf. the proof of [§ 8, Prop. 3]). Let h_n be the characteristic function of P_n; thus the h_n are orthogonal central projections with $\sup h_n=h$ (cf. [§ 7, Lemma to Prop. 1]).

Since z is bounded below on the compact-open set P_n, it follows that zh_n is invertible in h_nA; but

$$zh_n=-(ef-fe)^2h_n=-[(eh_n)(fh_n)-(fh_n)(eh_n)]^2,$$

therefore

(1) $$eh_n \sim h_n - fh_n$$

by Proposition 7 (note that every C^*-algebra satisfies the (SR)-axiom by easy spectral theory [cf. § 2, Example 5]). By Proposition 6, eh_n and fh_n are in position p in $h_n A$; in particular,

$$(eh_n) \cap (fh_n) = 0, \qquad (eh_n) \cup (fh_n) = h_n,$$

therefore (1) may be rewritten as

$$eh_n - (eh_n) \cap (fh_n) \sim (eh_n) \cup (fh_n) - fh_n.$$

Since h_n is central, the foregoing relation can, by lattice-theoretic trivia, be rewritten as

$$(1') \qquad (e - e \cap f)h_n \sim (e \cup f - f)h_n.$$

Since hA is the C^*-sum of the $h_n A$ [§ 10, Prop. 3], and since every partial isometry has norm ≤ 1, it follows from the relation (1′) that

$$(2) \qquad (e - e \cap f)h \sim (e \cup f - f)h.$$

What happens on $1-h$? Since $h = \mathrm{RP}(z)$, we have

$$0 = z(1-h) = (ef - fe)^*(ef - fe)(1-h),$$

therefore $(ef - fe)(1-h) = 0$, that is, $e(1-h)$ and $f(1-h)$ commute. Write $e' = e(1-h)$, $f' = f(1-h)$; as noted earlier, the relation

$$e' - e' \cap f' \sim e' \cup f' - f'$$

holds trivially, thus

$$(3) \qquad (e - e \cap f)(1-h) \sim (e \cup f - f)(1-h).$$

Adding (2) and (3), we arrive at (P). ∎

To proceed further, it is necessary to sharpen the conclusion of Proposition 7 (the price, of course, is a sharper hypothesis). As it stands, the relations $e \sim f$ and $1 - e \sim 1 - f$ obviously imply that e and f are unitarily equivalent, that is, $ueu^* = f$ for a suitable unitary element u; the sharper conclusion needed is that u can be taken to be a symmetry in the sense of the following definition:

Definition 6. In a $*$-ring with unity, a *symmetry* is a self-adjoint unitary ($u^* = u$, $u^2 = 1$).

In a $*$-ring with unity, the mapping $e \mapsto u = 2e - 1$ transforms projections e into symmetries u; if, in addition, 2 is invertible, then this mapping is *onto* the set of all symmetries, with inverse mapping $u \mapsto (\frac{1}{2})(1+u)$.

Definition 7. If e, f are projections such that $ueu = f$ for some symmetry u (hence also $ufu = e$), we say that e and f are *exchanged* by the symmetry u.

It can be shown that if, in Proposition 7, one assumes the (SR)-axiom, then the projections e, f can be exchanged by a symmetry (see Exercise 5). We content ourselves with a much simpler result (it is complicated enough) based on a stronger axiom. The stronger axiom depends on a general notion of positivity available in any $*$-ring (and therefore generally useless), consistent with the usual notion of positivity in C^*-algebras:

Definition 8. In any $*$-ring, an element x is called *positive*, written $x \geq 0$, in case $x = y_1^* y_1 + \cdots + y_n^* y_n$ for suitable elements $y_1, \ldots, y_n$.

The following properties are elementary: (1) if $x \geq 0$ then $x^* = x$; (2) if $x \geq 0$ then $y^* x y \geq 0$ for all y; (3) if $x \geq 0$ and $y \geq 0$, then $x + y \geq 0$. {Warning: $x \geq 0$ and $-x \geq 0$ is possible for nonzero x; equivalently, the relations $x \geq 0$, $y \geq 0$ and $x + y = 0$ need not imply $x = y = 0$.}

In particular, elements of the form $x^* x$ are positive; thus the following is an obvious strengthening of the (SR)-axiom:

Definition 9. A $*$-ring is said to satisfy the *positive square-root axiom* (briefly, the (PSR)-*axiom*) in case, for every $x \geq 0$, there exists $y \in \{x\}''$ with $y \geq 0$ and $x = y^2$.

The axiom we want is still stronger:

Definition 10. A $*$-ring is said to satisfy the *unique positive square-root axiom* (briefly, the (UPSR)-*axiom*) in case, for every $x \geq 0$, there exists a unique element y such that (1) $y \geq 0$, and (2) $x = y^2$; we assume, in addition, that (3) $y \in \{x\}''$ (but conditions (1) and (2) are already assumed to determine y uniquely).

Every C^*-algebra A satisfies the (UPSR)-axiom. {Proof: If $x \in A$, $x \geq 0$, there exists a unique $y \in A$ such that $y \geq 0$ and $x = y^2$; since $x \geq 0$ as an element of the C^*-algebra $\{x\}''$, it follows from uniqueness that $y \in \{x\}''$.}

The key to the rest of the section is the following result:

Proposition 8. *Let A be a $*$-ring with unity and proper involution, satisfying the* (UPSR)-*axiom. If e, f are projections such that $ef - fe$ is invertible, then e and f can be exchanged by a symmetry.*

Of course the pair $e, 1 - f$ also satisfies the hypothesis of Proposition 8, as do the pairs $1 - e, f$ and $1 - e, 1 - f$; the statement of the conclusion is confined to the pair e, f for simplicity. {Proposition 8 holds more generally with (UPSR) weakened to (SR), but with a considerably more complicated proof (Exercise 5).} To break up the rather long proof of Proposition 8, we separate out some of the earlier steps, which are valid under a weaker hypothesis, in the form of an admittedly ugly lemma:

Lemma. *Let A be a $*$-ring with unity and proper involution, satisfying the* (WSR)-*axiom, and suppose e, f are projections such that $ef-fe$ is invertible in A. Define $x=fe$. Then*

(1) $$x^*x=efe \quad \textit{is invertible in } eAe.$$

Let a be the inverse of efe in eAe; thus,

(2) $$a\in eAe, \quad a^*=a, \quad a(efe)=(efe)a=e \quad (\textit{that is, } afe=efa=e).$$

Choose $r\in\{x^*x\}''$ with $x^*x=r^*r$. Then

(3) $$r\in eAe,$$

(4) $$r \textit{ is invertible in } eAe, \textit{ with inverse } ar^*=r^*a,$$

(5) $$ar=ra.$$

Define $v=xar^*$. Then

(6) $$v^*v=e,$$

(7) $$x=vr,$$

(8) $$vv^*=f.$$

Proof. (1) See the proof of Proposition 7.

(2) The self-adjointness of a follows from that of efe.

(3) By the (WSR)-axiom, we may choose $r\in\{x^*x\}''=\{efe\}''$ such that $efe=r^*r=rr^*$. Since $e\in\{efe\}'$, it follows that $re=er$; a straightforward calculation then yields $(re-r)^*(re-r)=0$, therefore $re-r=0$ (the involution is assumed proper). Thus $r=re=er$, $r\in eAe$.

(4), (5) Since $r^*r=rr^*=efe$ is invertible in eAe, so is r; explicitly, the calculations

$$e=(efe)a=(rr^*)a=r(r^*a),$$

$$e=a(efe)=a(r^*r)=(ar^*)r$$

show that the inverse of r in eAe is $r^*a=ar^*$. Taking adjoints in the last equation, we have $ar=ra$.

(6) Setting $v=xar^*$, we have $v^*v=rax^*xar^*=ra[(efe)a]r^* =raer^*=(ra)r^*=(ar)r^*=a(efe)=e$ by (5) and (2).

(7) $vr=(xar^*)r=x[a(r^*r)]=x[a(efe)]=xe=x$.

(8) Writing $g=vv^*$, it remains to show that $g=f$. At any rate, g is a projection [§ 2, Prop. 2] and $fg=g$ (because $v=xar^*=(fe)ar^*\in fA$), thus $g\le f$. To show that $f-g=0$, it will suffice, by the invertibility of $ef-fe$, to show that

$$(ef-fe)(f-g)=0;$$

in fact, it will be shown that $ef(f-g)=fe(f-g)=0$. A straightforward computation yields $g=faf$, therefore

$$eg=e(faf)=(efa)f=ef$$

by (2); thus $e(f-g)=0$. On the one hand, this implies $fe(f-g)=0$; on the other hand, since $f-g\leq f$ we have also $ef(f-g)=e(f-g)=0$. ∎

Proof of Proposition 8. With notation as in the lemma, we assume, in addition, that $r\geq 0$.

Similarly, let $y=-(1-f)(1-e)$ (the minus sign is intentional) and consider $y^*y=(1-e)(1-f)(1-e)$. Since

$$(1-e)(1-f)-(1-f)(1-e)=ef-fe$$

is invertible, the lemma is again applicable, as follows.

(1′) $y^*y=(1-e)(1-f)(1-e)$ is invertible in $(1-e)A(1-e)$.

If b is the inverse of $(1-e)(1-f)(1-e)$ in $(1-e)A(1-e)$, then

(2′) $b^*=b, \quad b(1-f)(1-e)=(1-e)(1-f)b=1-e.$

Choosing $s\in\{y^*y\}''$ with $s\geq 0$ and $y^*y=s^2$, we have

(3′) $$s\in(1-e)A(1-e),$$

(4′) s is invertible in $(1-e)A(1-e)$, with inverse $bs=sb$

(recall that $s^*=s$; thus (5′) is redundant). Defining $w=ybs$, we have (the minus sign in the definition of y gives no trouble)

(6′) $$w^*w=1-e,$$

(7′) $$y=ws,$$

(8′) $$ww^*=1-f.$$

Define $u=v+w$. Obviously u is unitary and $ueu^*=f$; the proof will be concluded by showing that u is self-adjoint.

Set $t=r+s$. From (4) and (4′), it is clear that t is invertible in A (with $t^{-1}=ar+bs$). Since

$$ut=vr+vs+wr+ws=x+0+0+y,$$

and since

$$x+y=fe-(1-f)(1-e)=e+f-1,$$

we have $ut=e+f-1$. Thus, setting $z=e+f-1$, we have

(∗) $$z=ut,$$

where u and t are invertible and $z^*=z$. Since $t=r+s$, where $r\geq 0$ and $s\geq 0$, we have $t\geq 0$. Since, in addition, (∗) yields

$$z^2=z^*z=tu^*ut=t^2,$$

it follows from the (UPSR)-axiom that t is the unique positive square root of z^2, and in particular $t \in \{z^2\}''$; but $z \in \{z^2\}'$, therefore $tz = zt$, that is, $zt^{-1} = t^{-1}z$. Citing (*), we see that $u = zt^{-1} = t^{-1}z$ is the product of commuting self-adjoints, therefore $u^* = u$. ∎

In a Rickart *-ring, a condition weaker than the invertibility of $ef - fe$ is $\mathrm{RP}(ef - fe) = 1$, that is, position p (Proposition 6); still weaker is position p'. To arrive at the parallelogram law (P), we must show that projections in position p' are equivalent (Proposition 4); it would suffice to show that they can be exchanged by a symmetry. Thus, to establish the parallelogram law, it would suffice to prove the conclusion of Proposition 8 under the weaker hypothesis that e, f are in position p'; this is done in the next group of results (but the proofs require added axioms on A). It is convenient to separate out the intermediate case of position p as a lemma:

Lemma. *Let A be a Baer *-ring satisfying the* (EP)*-axiom and the* (UPSR)*-axiom. If e, f are projections in position p, then e and f can be exchanged by a symmetry (in particular, $e \sim f \sim 1 - e \sim 1 - f$).*

Proof. We show that e and f can be exchanged by a symmetry; it is then automatic that $1 - e \sim 1 - f$, and the parenthetical assertion of the lemma follows from the observation that $e, 1 - f$ are also in position p.

Let $x = ef - fe$, $z = x^*x = -(ef - fe)^2$, and write $B = \{z\}'$. As noted in the proof of Proposition 7, B has center $B' = \{z\}''$, and B contains e and f (hence also x).

By hypothesis, $\mathrm{RP}(z) = \mathrm{RP}(x) = 1$ (Proposition 6); we shall reduce matters to the situation of Proposition 8 by constructing a central partition of 1 in B such that z is invertible in each direct summand. Let (h_ι) be a maximal orthogonal family of nonzero projections in $\{z\}''$ such that, for each ι, zh_ι is invertible in $h_\iota B$ (the Zorn's lemma argument is launched by an application of the (EP)-axiom). We assert that $\sup h_\iota = 1$ (recall that suprema in B are unambiguous [§ 4, Prop. 7]). Writing $h = \sup h_\iota$, it is to be shown that $1 - h = 0$; since $\mathrm{RP}(z) = 1$, it will suffice to show that $z(1 - h) = 0$, equivalently, $x(1 - h) = 0$. Assume to the contrary. Then, by the (EP)-axiom, there exists an element

$$y \in \{(1-h)x^*x(1-h)\}'' = \{z(1-h)\}'' \subset \{z\}''$$

such that $z(1-h)y = k$, k a nonzero projection. Obviously $k \in \{z\}''$, $k \leq 1 - h$, and zk is invertible in kB, contradicting maximality of the family (h_ι).

We propose to apply Proposition 8 in each $h_\iota B$; to this end, we note that the (UPSR)-axiom is satisfied by B (Exercise 2) and therefore by $h_\iota B = h_\iota B h_\iota$ (Exercise 3). Since

$$(eh_\iota)(fh_\iota) - (fh_\iota)(eh_\iota) = xh_\iota$$

is invertible in $h_\iota B$ (because $(xh_\iota)(xh_\iota)^* = xx^*h_\iota = (xh_\iota)^*(xh_\iota) = zh_\iota$ is invertible in $h_\iota B$), it follows from Proposition 8 that there exists a symmetry u_ι in $h_\iota B$ such that

$$(*) \qquad u_\iota(eh_\iota)u_\iota = fh_\iota.$$

It remains to join the u_ι into a symmetry u exchanging e and f. {If A were an AW^*-algebra, the C^*-sum technique would do the trick; in a Baer $*$-ring, we must be more deft.} The strategy is to express the symmetry u_ι in terms of a projection g_ι of $h_\iota A$ (see the remarks following Definition 6), take the supremum g of the g_ι, and define $u = 2g - 1$. Part of the conclusion of Proposition 8 is $eh_\iota \sim h_\iota - eh_\iota$; therefore $2h_\iota$ has an inverse a_ι in $h_\iota B$ [§ 11, Lemma 2], thus $g_\iota = a_\iota(h_\iota + u_\iota)$ is a projection in $h_\iota B$, such that $u_\iota = 2g_\iota - h_\iota$. Define $g = \sup g_\iota$, $u = 2g - 1$. Since $gh_\iota = g_\iota$ for all ι [§ 11, Lemma 1], it follows that

$$uh_\iota = 2gh_\iota - h_\iota = 2g_\iota - h_\iota = u_\iota;$$

thus (*) yields $(ueu - f)h_\iota = 0$ for all ι, and $ueu - f = 0$ results from $\sup h_\iota = 1$. ∎

The above proof actually yields information for an arbitrary pair of projections:

Theorem 2. *Let A be a Baer $*$-ring satisfying the* (EP)-*axiom and the* (UPSR)-*axiom. If e, f is any pair of projections in A, there exists a projection h, central in the subring $B = \{-(ef - fe)^2\}'$, such that* (1) *eh and fh are in position p in Bh (hence may be exchanged by a symmetry in Bh), and* (2) *$e(1-h)$ and $f(1-h)$ commute. Explicitly, $h = \mathrm{RP}(ef - fe)$.*

Proof. With notation as in the proof of the lemma (but with the hypothesis $\mathrm{RP}(x) = 1$ suppressed), set $h = \sup h_\iota$; the argument given there shows that $h = \mathrm{RP}(x)$. On the one hand, $x(1-h) = 0$ shows that $e(1-h)$ and $f(1-h)$ commute. On the other hand, $(eh)(fh) - (fh)(eh) = xh = x$ has right projection h, therefore eh and fh are in position p in Bh (Proposition 6). ∎

We now advance to position p':

Lemma. *Notation as in Theorem 2. If, in addition, e, f are in position p', then $e(1-h) = f(1-h)$.*

Proof. Write $k = 1 - h$ and set $e'' = ek$, $f'' = fk$; we know from Theorem 2 that e'' and f'' commute. By hypothesis,

$$e \cap (1-f) = (1-e) \cap f = 0;$$

since k is central in B, it follows that

$$e'' \cap (k - f'') = (k - e'') \cap f'' = 0,$$

that is, in view of the commutativity of e'' and f'',

$$e''(k-f'')=(k-e'')f''=0.$$

Thus $e''=e''f''=f''$. ∎

Theorem 3. *Let A be a Baer $*$-ring satisfying the* (EP)*-axiom and the* (UPSR)*-axiom. If e,f are projections in position p', then e and f can be exchanged by a symmetry $2g-1$, g a projection.*

Proof. With notation as in the proof of Theorem 2, set $e'=eh$, $e''=e(1-h)$, $f'=fh$, $f''=f(1-h)$; thus

$$e=e'+e'', \qquad f=f'+f''.$$

By Theorem 2, e' and f' are in position p in Bh, and there exists a symmetry u' in Bh such that $u'e'u'=f'$; by the lemma, $e''=f''$. Then $u=u'+(1-h)$ is a symmetry in B (hence in A) and it is straightforward to check that $ueu=f$. A second look at the proof of Theorem 2 (rather, its lemma) shows that $u'=2g'-h$ for a suitable projection g', thus $u=2g-1$, where $g=g'+(1-h)$. ∎

Combining Theorem 3 with Proposition 4, we arrive at the climax of the section (see also Exercise 7):

Theorem 4. *The parallelogram law* (P) *holds in any Baer $*$-ring satisfying the* (EP)*-axiom and the* (UPSR)*-axiom.*

Theorems 3 and 4, combined with Proposition 5, yield an important decomposition theorem (see also Exercise 8):

Theorem 5. *Let A be a Baer $*$-ring satisfying the* (EP)*-axiom and the* (UPSR)*-axiom. If e,f is any pair of projections in A, there exist orthogonal decompositions*

$$e=e'+e'', \qquad f=f'+f''$$

such that $e'\sim f'$ and $e''f=ef''=0$. Explicitly, $e'=\mathrm{LP}(ef)$, $f'=\mathrm{RP}(ef)$, $e''=e-e'$, $f''=f-f'$; e' and f' are in position p', and can be exchanged by a symmetry.

Exercises

1A. In the Baer $*$-ring of all 2×2 matrices over the field of three elements [§1, Exer. 17], the parallelogram law (P) fails; so does the (SR)-axiom; so does the (EP)-axiom.

2A. Let A be a $*$-ring, B a $*$-subring such that $B=B''$. If A satisfies the (WSR)-axiom [(SR)-axiom, (PSR)-axiom, (UPSR)-axiom] then so does B.

3A. Let A be a $*$-ring with proper involution, and let e be a projection in A. If A satisfies the (WSR)-axiom [(SR)-axiom, (PSR)-axiom, (UPSR)-axiom] then so does eAe.

4A. If A is a weakly Rickart $*$-ring satisfying the (WSR)-axiom, and if e,f are projections such that $ef-fe$ is invertible in $(e\cup f)A(e\cup f)$, then $e\sim f\sim e\cup f-e \sim e\cup f-f$.

5C. Let A be a $*$-ring with unity and proper involution, satisfying the (SR)-axiom. If e,f are projections such that $ef-fe$ is invertible, then e and f can be exchanged by a symmetry. (This generalizes Proposition 8.)

6C. Let A be a Baer $*$-ring satisfying the (EP)-axiom and the (SR)-axiom. If e,f are projections in position p', then e and f can be exchanged by a symmetry. (This generalizes Theorem 3.)

7C. The parallelogram law (P) holds in every Baer $*$-ring satisfying the (EP)-axiom and the (SR)-axiom. (This generalizes Theorem 4.)

8C. Let A be a Baer $*$-ring satisfying the (EP)-axiom and the (SR)-axiom. If e,f is any pair of projections in A, there exist orthogonal decompositions $e=e'+e''$, $f=f'+f''$ with e',f' in position p' and $e''f=ef''=0$; in particular, $e'\sim f'$, indeed, e' and f' can be exchanged by a symmetry. (This generalizes Theorem 5.)

9A. The following conditions on a $*$-ring are equivalent: (a) the involution is proper, and the relations $x\geq 0$, $y\geq 0$, $x+y=0$ imply $x=y=0$; (b) $\sum_1^n x_i{}^*x_i=0$ implies $x_1=\dots=x_n=0$ (n arbitrary).

10A. In a $*$-ring satisfying the conditions of Exercise 9, the (PSR)-axiom and the (UPSR)-axiom are equivalent.

11A. Let A be a $*$-ring with proper involution, satisfying the following *strong square-root axiom* (SSR): If $x\in A$, $x\geq 0$, then there exists $y\in\{x\}''$ with $y^*=y$ and $x=y^2$. (The (SR)-axiom provides such a y only for positives x of the special form $x=t^*t$.) Assume, in addition, that (1) A has a central element i such that $i^2=-1$ and $i^*=-i$, and (2) $2x=0$ implies $x=0$. Then $\sum_1^n x_k^* x_k=0$ implies $x_1=\dots=x_n=0$ (cf. Exercise 9).

12A. Let A be a $*$-ring with proper involution, satisfying the conditions (1), (2) of Exercise 11. In such a $*$-ring, the (PSR)-axiom and the (UPSR)-axiom are equivalent.

13A. If A is a $*$-ring satisfying the (WEP)-axiom and the (SR)-axiom, then A satisfies the (EP)-axiom.

14A. Let A be a Baer $*$-ring, let $(e_\iota)_{\iota\in I}$ and $(f_\iota)_{\iota\in I}$ be equipotent families of orthogonal projections such that $e_\iota\sim f_\iota$ for all $\iota\in I$, and let $e=\sup e_\iota$, $f=\sup f_\iota$. We know that if $ef=0$ then $e\sim f$ [§11, Th. 1]. If A satisfies the parallelogram law (P), then the weaker condition $e\cap f=0$ also implies $e\sim f$.

15A. If e,f are projections in a $*$-ring A, such that $e\sim f$ and $ef=0$, then e and f can be exchanged by a symmetry in $(e+f)A(e+f)$.

16A. Theorems 2–5 hold in any Rickart C^*-algebra; in particular, any pair of projections in position p' can be exchanged by a symmetry.

17A. Suppose A is a Rickart $*$-ring in which every pair of projections in position p' can be exchanged by a symmetry. If e, f is any pair of projections in A, there exists a symmetry u such that $u(ef)u=fe$.

18A. In an arbitrary Baer $*$-ring, projections in position p need not be equivalent.

19C. In a von Neumann algebra A, projections e, f are in position p' (relative to A) if and only if $e \sim f$ relative to the von Neumann algebra generated by e and f.

§ 14. Generalized Comparability

Projections e, f in a $*$-ring A are said to be *comparable* if either $e \precsim f$ or $f \precsim e$. Rings in which any two projections are comparable are of interest in the same way that simply ordered sets are interesting examples of partially ordered sets [cf. § 12, Exer. 1]. In general, the concept of comparability is of limited use. (For example, if A contains a central projection h different from 0 and 1, and if e, f are nonzero projections such that $e \le h$ and $f \le 1-h$, then e and f cannot be comparable.) The pertinent concept in general $*$-rings is as follows:

Definition 1. Projections e, f in a $*$-ring A are said to be *generalized comparable* if there exists a central projection h such that

$$he \precsim hf, \qquad (1-h)f \precsim (1-h)e.$$

(When A has no unity element, the use of 1 is formal and the condition need not be symmetric in e and f.) We say that A has *generalized comparability* (briefly, A has GC) if every pair of projections is generalized comparable.

Generalized comparability may be reformulated in terms of the following concept, which generalizes, and is consistent with, an earlier definition [§ 6, Def. 2]:

Definition 2. Projections e, f in a $*$-ring A are said to be *very orthogonal* if there exists a central projection h such that $he = e$ and $hf = 0$. (That is, $e \le h$ and $f \le 1-h$, where 1 is used formally when A has no unity element—in which case, the relation need not be symmetric in e and f.)

If e, f are projections in a Baer $*$-ring A, then the following conditions are equivalent: (a) e, f are very orthogonal; (b) $C(e)C(f) = 0$; (c) $eAf = 0$ [§ 6, Cor. 1 of Prop. 3].

The relevance of very orthogonality to generalized comparability is as follows:

Proposition 1. *If e, f are projections in a $*$-ring, the following conditions are equivalent:*

(a) *e, f are generalized comparable;*

(b) *there exist orthogonal decompositions $e = e_1 + e_2$, $f = f_1 + f_2$ with $e_1 \sim f_1$ and f_2, e_2 very orthogonal.*

Proof. (a) implies (b): Choose h as in Definition 1, say

$$he \sim f_1' \leq hf, \quad (1-h)f \sim e_1'' \leq (1-h)e.$$

Writing $e_1' = he$ and $f_1'' = (1-h)f$, we have

$$(*) \qquad e_1' \sim f_1', \quad e_1'' \sim f_1''.$$

Obviously $e_1' e_1'' = 0$ and $f_1' f_1'' = 0$; setting

$$e_1 = e_1' + e_1'', \quad f_1 = f_1' + f_1'',$$

it follows from (*) that $e_1 \sim f_1$ [§1, Prop. 8]. Since $e_1 \leq e$ and $f_1 \leq f$, we may define $e_2 = e - e_1, f_2 = f - f_1$; it is routine to check that $he_2 = 0$ and $hf_2 = f_2$.

(b) implies (a): Assuming there exists such a decomposition, let h be a central projection such that $hf_2 = f_2$ and $he_2 = 0$. Then $he = he_1 \sim hf_1 \leq hf$ [§1, Prop. 7], thus $he \precsim hf$, and similarly $(1-h)f \precsim (1-h)e$. ∎

If e,f are generalized comparable, but are not very orthogonal, then Proposition 1 shows that e,f have nonzero subprojections e_1, f_1 such that $e_1 \sim f_1$; this is a phenomenon worth formalizing:

Definition 3. Projections e,f in a $*$-ring A are said to be *partially comparable* if there exist nonzero subprojections $e_0 \leq e$, $f_0 \leq f$ such that $e_0 \sim f_0$. We say that A has *partial comparability* (briefly, A has PC) if $eAf \neq 0$ implies e,f are partially comparable.

GC is stronger than PC:

Proposition 2. *If A is a $*$-ring with* GC, *then A has* PC.

Proof. Assuming e,f are projections that are not partially comparable, it is to be shown that $eAf = 0$. Write $e = e_1 + e_2$, $f = f_1 + f_2$ as in Proposition 1. By the hypothesis on e,f, necessarily $e_1 = f_1 = 0$, thus f,e are very orthogonal; if h is a central projection with $hf = f$ and $he = 0$, then $eAf = eAhf = ehAf = 0$. ∎

PC is implied by axioms of 'existence of projections' type; for instance:

Proposition 3. *If A is a $*$-ring satisfying the* (VWEP)-*axiom, then A has* PC.

Proof. Suppose e,f are projections such that $eAf \neq 0$, equivalently, $fAe \neq 0$. Let $x \in fAe$, $x \neq 0$. By hypothesis, there exists an element $y \in \{x^*x\}'$ with $(y^*y)(x^*x) = e_0$, e_0 a nonzero projection [§7, Def. 3], thus $e_0 = y^*(x^*x)y = (xy)^*(xy)$. Writing $w = xy$, we have $w^*w = e_0$; since the involution of A is proper [§2, Exer. 6], w is a partial isometry [§2, Prop. 2]. Set $f_0 = ww^*$. Since $x \in fAe$, the formula $e_0 = (y^*y)(x^*x)$ shows that $e_0 \leq e$, and $f_0 = ww^* = (xy)w^*$ shows that $f_0 \leq f$. ∎

In Baer $*$-rings, generalized comparability is intimately related to additivity of equivalence [§ 11, Def. 1]; in fact, a Baer $*$-ring has GC if and only if it has PC and equivalence is additive [§ 20, Th. 2]. The "only if" part appears to be fairly difficult—the proof we give in Section 20 involves most of the structure theory discussed in Part 2. The "if" part is easy:

Proposition 4. *If A is a Baer $*$-ring with* PC *and if equivalence in A is additive, then A has* GC.

Proof. Let e, f be any pair of projections in A. If $eAf = 0$ then e, f are very orthogonal and the generalized comparability of e and f is trivial. Assuming $eAf \neq 0$, let $(e_\iota)_{\iota \in I}$, $(f_\iota)_{\iota \in I}$ be a maximal pair of orthogonal families of nonzero projections such that $e_\iota \leq e$, $f_\iota \leq f$ and $e_\iota \sim f_\iota$ for all $\iota \in I$ (an application of PC starts the Zorn's lemma argument). Set $e' = \sup e_\iota$, $f' = \sup f_\iota$, $e'' = e - e'$, $f'' = f - f'$. On the one hand, $e' \sim f'$ by the assumed additivity of equivalence. On the other hand, $e'' A f'' = 0$ (if not, an application of PC would contradict maximality), therefore e'', f'' are very orthogonal. In view of Proposition 1, the decompositions $e = e' + e''$, $f = f' + f''$ show that e, f are generalized comparable. ∎

It is a corollary that every von Neumann algebra A has GC; for, it is easy to see that partial isometries in A are addable (e. g., they can be summed in the strong operator topology), and the validity of the (EP)-axiom [§ 7, Cor. of Prop. 3] ensures, via Proposition 3, that A has PC. For AW^*-algebras, essentially the same argument may be employed (except that the proof of addability is harder—see Section 20), but an alternative proof will shortly be given.

Proposition 4, and the fact that equivalence is orthogonally additive in any Baer $*$-ring [§ 11, Th. 1], naturally suggest the following definition:

Definition 4. We say that a $*$-ring has *orthogonal* GC if every pair of orthogonal projections is generalized comparable.

This condition is automatically fulfilled in a Baer $*$-ring with PC:

Proposition 5. *If A is a Baer $*$-ring with* PC, *then A has orthogonal* GC.

Proof. Let e, f be projections with $ef = 0$. The proofs proceeds as for Proposition 4, except that $e' \sim f'$ results from a theorem [§ 11, Th. 1] rather than an assumption. ∎

In the presence of the parallelogram law, GC and orthogonal GC are equivalent hypotheses:

Proposition 6. *If A is a Rickart $*$-ring with orthogonal* GC, *and if A satisfies the parallelogram law* (P), *then A has* GC.

Proof. Let e, f be any pair of projections in A. By the parallelogram law, write

$$e = e' + e'', \quad f = f' + f''$$

with $e' \sim f'$ and $e''f = ef'' = 0$ [§ 13, Prop. 5]. Since, by hypothesis, the orthogonal projections e'', f'' are generalized comparable, Proposition 1 yields decompositions

$$e'' = e_1 + e_2, \quad f'' = f_1 + f_2$$

with $e_1 \sim f_1$ and e_2, f_2 very orthogonal. Then

$$e = (e' + e_1) + e_2, \quad f = (f' + f_1) + f_2,$$

where $e' + e_1 \sim f' + f_1$ and e_2, f_2 are very orthogonal, therefore e, f are generalized comparable by Proposition 1. ∎

In a Baer $*$-ring satisfying the parallelogram law, the concepts PC, GC and orthogonal GC merge:

Proposition 7. *If A is a Baer $*$-ring satisfying the parallelogram law* (P), *then the following conditions on A are equivalent:* (a) *A has* PC; (b) *A has orthogonal* GC; (c) *A has* GC.

Proof. (a) implies (b) by Proposition 5; in the presence of (P), (b) implies (c) by Proposition 6; and (c) implies (a) by Proposition 2. ∎

Corollary 1. *Every AW^*-algebra has* GC.

Proof. An AW^*-algebra A satisfies the parallelogram law (P) [§ 13, Th. 1]; since A satisfies the (EP)-axiom [§ 7, Cor. of Prop. 3], and therefore has PC (Proposition 3), it follows from Proposition 7 that A has GC. ∎

Corollary 2. *If A is a Baer $*$-ring such that* $\mathrm{LP}(x) \sim \mathrm{RP}(x)$ *for all x in A, then A has* GC *and satisfies the parallelogram law* (P).

Proof. Since A satisfies (P) [§ 13, Prop. 2], by Proposition 7 it suffices to show that A has PC. Suppose e, f are projections such that $eAf \neq 0$, say $x = eaf \neq 0$; then $e_0 = \mathrm{LP}(x)$, $f_0 = \mathrm{RP}(x)$ are nonzero subprojections of e, f such that $e_0 \sim f_0$. ∎

The parallelogram law is not the most natural of hypotheses. Some ways of achieving it were shown in Section 13; an application (see also Exercise 5):

Theorem 1. *If A is a Baer $*$-ring satisfying the* (EP)-*axiom and the* (UPSR)-*axiom, then A has* GC *and satisfies the parallelogram law* (P).

Proof. A satisfies (P) [§ 13, Th. 4] and has PC (Proposition 3), therefore A has GC by Proposition 7. ∎

Incidentally, Theorem 1 provides a second proof of the AW^* case (Corollary 1 of Proposition 7).

We close the section with two items for later application. The first is for application in Section 17 [§ 17, Th. 2]:

Proposition 8. *Let A be a Rickart $*$-ring with* GC, *satisfying the parallelogram law* (P). *If e, f is any pair of projections in A, there exists a central projection h such that*

$$he \precsim hf,$$

$$(1-h)(1-e) \precsim (1-h)(1-f).$$

Proof. Apply GC to the pair $e \cap (1-f)$, $(1-e) \cap f$: there exists a central projection h such that

$$h[e \cap (1-f)] \precsim h[(1-e) \cap f], \tag{1}$$

$$(1-h)[(1-e) \cap f] \precsim (1-h)[e \cap (1-f)]. \tag{2}$$

It follows from the parallelogram law (see [§ 13, Prop. 1]) that

$$e - e \cap (1-f) \sim f - (1-e) \cap f$$

and (replacing e, f by $1-e, 1-f$)

$$(1-e) - (1-e) \cap f \sim (1-f) - e \cap (1-f),$$

therefore

$$h[e - e \cap (1-f)] \sim h[f - (1-e) \cap f], \tag{3}$$

$$(1-h)[(1-e) - (1-e) \cap f] \sim (1-h)[(1-f) - e \cap (1-f)]. \tag{4}$$

Adding (1) and (3) yields $he \precsim hf$, while (2) and (4) yield $(1-h)(1-e) \precsim (1-h)(1-f)$. ∎

The final proposition is for application in [§ 18, Prop. 5]:

Proposition 9. *Let A be a Baer $*$-ring with* PC, *and suppose $(e_\iota)_{\iota \in I}$ is a family of projections in A with the following property: for every nonzero central projection h, the set of indices*

$$\{\iota \in I : he_\iota \neq 0\}$$

is infinite; in other words, there exists no direct summand of A (other than 0) on which all but finitely many of the e_ι vanish.

Then, given any positive integer n, there exist n distinct indices $\iota_1, \dots, \iota_n$, and nonzero projections $g_\nu \leq e_{\iota_\nu}$ $(\nu = 1, \dots, n)$, such that

$$g_1 \sim g_2 \sim \cdots \sim g_n.$$

Proof. The proof is by induction on n. The case $n=1$ is trivial: the set $\{\iota: 1e_\iota \neq 0\}$ is infinite, and any of its members will serve as ι_1, with $g_1 = e_{\iota_1}$.

Assume inductively that all is well with $n-1$, and consider n. By assumption, there exist distinct indices $\iota_1, \ldots, \iota_{n-1}$ and nonzero projections $f_1, \ldots, f_{n-1}$ such that $f_\nu \leq e_{\iota_\nu}$ $(\nu = 1, \ldots, n-1)$ and $f_1 \sim \cdots \sim f_{n-1}$.

Since $C(f_1) \neq 0$, it is clear from the hypothesis that there exists an index ι_n distinct from $\iota_1, \ldots, \iota_{n-1}$ such that $C(f_1)e_{\iota_n} \neq 0$. Then $C(f_1)C(e_{\iota_n}) \neq 0$, thus $f_1 A e_{\iota_n} \neq 0$ [§ 6, Cor. 1 of Prop. 3]; citing PC, there exist nonzero subprojections $g_1 \leq f_1$ and $g_n \leq e_{\iota_n}$ such that $g_1 \sim g_n$. For $\nu = 2, \ldots, n-1$, the equivalence $f_1 \sim f_\nu$ transforms g_1 into a subprojection $g_\nu \leq f_\nu$ with $g_1 \sim g_\nu$. Thus $g_n \sim g_1 \sim g_\nu$ $(\nu = 2, \ldots, n-1)$.

{The proof shows that the indices for n may be obtained by augmenting the indices for $n-1$; but as n increases, the projections g_ν will in general shrink.} ∎

Exercises

1A. A Baer $*$-ring with orthogonal GC, but without PC (hence without GC): the ring of all 2×2 matrices over the field of three elements [§1, Exer. 17].

2B. A Baer $*$-ring A has GC if and only if (i) A has PC, and (ii) equivalence in A is additive.

3A. In a Baer $*$-ring with finitely many elements, PC and GC are equivalent.

4B. In a properly infinite Baer $*$-ring [§15, Def. 3], PC and GC are equivalent.

5C. If A is a Baer $*$-ring satisfying the (EP)-axiom and the (SR)-axiom, then A has GC and satisfies the parallelogram law (P). (This generalizes Theorem 1.)

6A. (i) If A is a $*$-ring with GC and if g is any projection in A, then gAg has GC.
(ii) If A is a Baer $*$-ring, if g is a projection in A, and if e, f are projections in gAg that are generalized comparable in gAg, then e, f are generalized comparable in A.

7A. If e, f are partially comparable projections in a Baer $*$-ring, then $C(e)C(f) \neq 0$.

8A. If A is a Rickart $*$-ring satisfying the parallelogram law (P), and if e, f are projections in A such that $ef \neq 0$, then e, f are partially comparable.

9A. Let A be a Baer $*$-ring satisfying the parallelogram law (P). If A satisfies any of the following conditions, then A has GC:
(1) For every projection e, $C(e) = \sup\{e' : e' \sim e\}$ [cf. §6, Exer. 7].
(2) If e, f are projections such that $eAf \neq 0$, then there exists a unitary u such that $euf \neq 0$.
(3) If e, f are projections such that $eAf \neq 0$, then there exists a projection g such that $e(2g)f \neq 0$.

10A. The following conditions on a $*$-ring A are equivalent: (a) A has GC; (b) A has orthogonal GC and, for every pair of projections e, f, there exist orthogonal decompositions $e = e' + e''$, $f = f' + f''$ with $e' \sim f'$ and $e''f'' = f''e''$.

11A. Let A be a Rickart $*$-ring in which every sequence of orthogonal projections has a supremum. As in [§12, Exer. 1], write $[e] = \{f : f \sim e\}$ and define $[e] \leq [f]$

iff $e \precsim f$. If A has GC then the set of equivalence classes is a lattice with respect to this ordering.

12A. Let A be a Baer $*$-ring satisfying the (EP)-axiom and the (SR)-axiom (or let A be a Rickart C^*-algebra with GC). If e, f is any pair of projections in A, there exist orthogonal decompositions $e = e_1 + e_2$, $f = f_1 + f_2$ such that e_1 and e_2 are exchangeable by a symmetry and e_2, f_2 are very orthogonal.

13B. If A is a Baer $*$-ring satisfying the (WEP)-axiom, then the following conditions are equivalent: (a) A has GC; (b) $\mathrm{LP}(x) \sim \mathrm{RP}(x)$ for all $x \in A$; (c) A satisfies the parallelogram law (P).

14B. Let A be a Baer $*$-ring with GC, and let e, f be any pair of projections in A. Either (1) $f \precsim e$, or (2) there exists a central projection h with the following property: for a central projection k, $ke \precsim kf$ iff $k \le h$. In case (2), such a projection h is unique, $h \ge 1 - C(e)$, and $(1-h)f \precsim (1-h)e$.

15A. If A is a Baer $*$-ring with PC, the following conditions on a pair of projections e, f imply one another: (a) $C(e) \le C(f)$; (b) $e = \sup e_\iota$ with (e_ι) an orthogonal family of projections such that $e_\iota \precsim f$ for all ι; (c) $e = \sup e_\iota$ with (e_ι) a family of projections such that $e_\iota \precsim f$ for all ι.

16A. Let A be a Rickart $*$-ring with GC, let n be a positive integer, and suppose that the $n \times n$ matrix ring A_n is a Rickart $*$-ring satisfying the parallelogram law (P). Then A_n has GC.

17C. Let A be a Rickart $*$-ring with orthogonal GC (e.g., let A be a Baer $*$-ring with PC) and let e be a projection in A. The following conditions on e are equivalent: (a) e is central in A; (b) e commutes with every projection in A (that is, e is central in the reduced ring A°); (c) e has a unique complement.

18A. (i) If A is a Baer $*$-ring with PC, then a projection in A is central iff it commutes with every projection of A (thus a projection is central in A iff it is central in the reduced ring A° [§ 3, Exer. 18]).

(ii) The converse of (i) is false: there exists a Baer $*$-ring A such that $A^\circ = A$ but A does not have PC.

19A. Let A be a $*$-ring with unity. A partial isometry u in A is said to be *extremal* if the projections $1 - u^*u$ and $1 - uu^*$ are very orthogonal in the sense of Definition 2. {The terminology is motivated by the fact that if A is an AW^*-algebra, then the closed unit ball of A is a convex set whose extremal points are precisely the extremal partial isometries.} For example, if u is an isometry ($u^*u = 1$) or a co-isometry ($uu^* = 1$) then u is an extremal partial isometry; when A is factorial, there are no others [§ 6, Def. 3].

If A has GC and if w is any partial isometry in A, then there exists an extremal partial isometry u that 'extends' w, in the sense that $u(w^*w) = w$.

20D. *Problem:* If A is a Baer $*$-ring with PC, does it follow that A has GC?

21D. *Problem:* If A is a Baer $*$-ring satisfying the parallelogram law (P), does it follow that A has PC?

22D. *Problem:* If A is a Baer $*$-ring with PC, does it follow that A satisfies the parallelogram law (P)?

Part 2: Structure Theory

Chapter 3

Structure Theory of Baer $*$-Rings

Part 1 of the book dealt with more or less general Baer $*$-rings, liberally seasoned with such axioms of a general nature as are needed to make the arguments work. From here on, most arguments entail qualitative distinctions between Baer $*$-rings; a particular argument will generally apply only to certain kinds of Baer $*$-rings (with or without extra axioms—usually with). {Some analogous qualitative considerations in group theory: commutativity, finiteness, solvability, decomposability, simplicity, etc.} Such qualitative distinctions are the basis of structure theory.

By structure theory we mean the description of general Baer $*$-rings in terms of simpler ones. {The best-loved model of a successful structure theory describes finitely generated abelian groups in terms of cyclic ones.} When we say that a ring—or a class of rings—is simpler, we mean, vaguely, that less can happen in it. An inventory of the things that can happen in a Baer $*$-ring will lead off with annihilation, commutativity, projection lattice operations, and equivalence of projections; for structure theory, the most important happenings are *commutativity* and *equivalence*. {These are, in a sense, opposite sides of the same coin; equivalence is interesting only when there exist partial isometries w for which w^*w and ww^* are different.}

Structure theory comes in two grades, fine and coarse; in both cases the center of the ring, aptly, plays a central role. In fine structure theory, we seek to describe general Baer $*$-rings in terms of factorial ones (i. e., Baer $*$-rings in which 0 and 1 are the only central projections), accepting whatever equivalence behavior the factors may exhibit. In the coarse structure theory, we accept general centers (i. e., we do not insist on factors) but seek direct sum decompositions into summands in which equivalence behavior is limited in various ways.

Following in the wake of a complete structure theory is a classification theory, i. e., a full listing of the various kinds of objects that can occur, with a specification of when two objects are isomorphic. By these standards, the structure theory of von Neumann algebras, despite intensive cultivation for nearly four decades, remains incomplete even for separable Hilbert spaces; for Baer $*$-rings, there is barely a beginning.

Chapter 7 is devoted to the fine structure of one special class of Baer $*$-rings (namely, the Baer $*$-rings in which $u^*u=1$ implies $uu^*=1$, and $\mathrm{LP}(x)\sim\mathrm{RP}(x)$ for all x); this isn't much, but it's all there is at the present state of the subject.

The present chapter is devoted to the coarse structure theory of Baer $*$-rings. As far as it goes, this theory goes remarkably smoothly; the general Baer $*$-ring theory is not essentially harder than the special case of von Neumann algebras.

{*Remarks.* However, we should not give the impression that practically all coarse structure theory carries over from von Neumann algebras to Baer $*$-rings. We cite here three examples to the contrary. (1) The coarse structure theory of von Neumann algebras, when applied simultaneously to an algebra and its commutant, leads to spatial isomorphism invariants for the algebra [cf. **23**, Ch. III, § 6, No. 4, Prop. 10]; the lack of an appropriate notion of 'spatial' for Baer $*$-rings (or even for AW^*-algebras) is a natural boundary for the theory. (2) The coarse structure theory of von Neumann algebras of 'Type I' leads to a complete system of $*$-isomorphism invariants for such an algebra, the invariants consisting in a set of cardinal numbers together with a corresponding set of commutative von Neumann algebras (cf. [**23**, Ch. III, § 3, Prop. 2], [**81**, Part II, Th. 10]); for AW^*-algebras of Type I there is a partial theory of this sort ([**48**], [**49**]), fully satisfactory in the 'finite' case, but for 'infinite' algebras the theory bogs down in unresolved questions of cardinal uniqueness [cf. **48**, Th. 4]; for Baer $*$-rings, only a few wisps of such a theory are yet in hand (cf. Section 18). (3) A final example concerns the problem of describing the automorphisms and derivations of a ring; this is a large topic in von Neumann algebras [cf. **23**, Ch. III, § 9], a small topic in AW^*-algebras [cf. **49**], and a non-topic in Baer $*$-rings.

For von Neumann algebras, there is a highly developed fine structure theory, called reduction theory [cf. **23**, Ch. II], that is applicable to algebras of arbitrary 'type'; there are limitations to the theory (the reduction is uncanonical and is largely limited to algebras acting on separable Hilbert spaces), but nothing like it exists for general AW^*-algebras, let alone Baer $*$-rings.}

§ 15. Decomposition into Types

Throughout this section, A is a Baer $*$-ring; no additional axioms need be imposed on A. {In fact, there exists an involution-free version: with projections replaced by idempotents, and lattice operations by deft strategies, the results were originally proved by I. Kaplansky for arbitrary Baer rings [§ 4, Exer. 4] (see [**54**] for an exposition of this theory).} All of the definitions, and many of the propositions (but none of the theorems), can be formulated in an arbitrary $*$-ring with unity; it is when suprema must be taken that the Baer $*$-ring condition is required. (In this connection, see the remarks at the end of the section.)

The coarse structure theory is cast in terms of the concepts of 'finite projection' and 'abelian projection'.

Definition 1. A is said to be *finite* if $x^*x=1$ implies $xx^*=1$. We agree to regard the ring $\{0\}$ as finite. If A is not finite, it is called *infinite*. A projection $e\in A$ is said to be *finite* (relative to A) if the $*$-ring eAe is finite in the foregoing sense. (In particular, 0 is a finite projection.)

So to speak, A is finite iff every isometry in A is unitary. Another formulation is that A is finite iff $e\sim 1$ implies $e=1$; the key point is that if $x^*x=1$, then $(xx^*)(xx^*)=x(x^*x)x^*=xx^*$ shows that $e=xx^*$ is a projection with $e\sim 1$.

Proposition 1. *Let e, f be projections with $f\leq e$. Then f is finite relative to A iff it is finite relative to eAe.*

Proof. $fAf=f(eAe)f$. ∎

A projection is finite iff it cannot be 'deformed' into a proper subprojection of itself via a partial isometry in the ring:

Proposition 2. *A projection e is finite iff $e\sim f\leq e$ implies $f=e$.*

Proof. The projections $f\leq e$ are precisely the projections of eAe. The condition $e\sim f\leq e$ means that there exists a partial isometry w such that $w^*w=e$ and $ww^*=f\leq e$ [§ 1, Prop. 6]; such an element w satisfies $fw=w=we$ [§ 1, Prop. 5], therefore $w\in eAe$, thus the equivalence $e\sim f$ is implemented in eAe. The proposition now follows at once from Definition 1. ∎

All projections dominated by a finite projection are finite:

Proposition 3. *If e is a finite projection in A and if $f\precsim e$, then f is also finite.*

Proof. Say $f\sim f'\leq e$. Since fAf is $*$-isomorphic to $f'Af'$ [§ 1, Prop. 9], it is no loss of generality to suppose that $f\leq e$.

Assuming $f\sim g\leq f$, it is to be shown that $g=f$. Let v be a partial isometry such that $v^*v=f$, $vv^*=g\leq f$. Setting $w=v+(e-f)$, we have $w^*w=e$ and $ww^*=g+(e-f)=e-(f-g)\leq e$; thus $e\sim e-(f-g)\leq e$, therefore $e-(f-g)=e$ (Proposition 2), that is, $f-g=0$. ∎

If A is finite, we may put $e=1$ in Proposition 3:

Corollary. *If A is finite, then every projection in A is finite.*

We now parallel the foregoing with 'abelian' in place of 'finite':

Definition 2. A is said to be *abelian* if every projection in A is central. A projection $e \in A$ is said to be *abelian* (relative to A) if the $*$-ring eAe is abelian in the foregoing sense.

An abelian ring need not be commutative, but for AW^*-algebras there is no ambiguity:

Examples. 1. Every division ring with involution is trivially an abelian Baer $*$-ring.

2. An AW^*-algebra is abelian if and only if it is commutative. {Proof: An AW^*-algebra is the closed linear span of its projections [cf. § 8, Prop. 1]. See also Exercise 3.}

If Baer $*$-rings are approached through the more general Baer rings (as in [**52**], [**54**]), it is appropriate to define a Baer $*$-ring to be abelian if all of its idempotents are central [**54**, p. 10] or if all of its idempotents commute with each other [**52**, p. 5]; these definitions are equivalent to Definition 2 (see Exercise 2).

Every abelian projection is finite; this follows from the fact that in an abelian ring, equivalence collapses to equality:

Proposition 4. (i) *In an abelian ring, $e \sim f$ implies $e = f$.* (ii) *Every abelian ring is finite.* (iii) *Every abelian projection is finite.*

Proof. It is clearly sufficient to prove (i). Suppose $e \sim f$ in an abelian ring, and let w be a partial isometry such that $w^* w = e$, $w w^* = f$. Since f is central, $wf = fw = w$, therefore $e \leq f$ [§ 1, Prop. 5]. Similarly $f \leq e$. ∎

Paralleling Proposition 1, we have (with identical proof):

Proposition 5. *Let e, f be projections with $f \leq e$. Then f is abelian relative to A iff it is abelian relative to eAe.*

Paralleling Proposition 2, abelianness may be characterized as follows:

Proposition 6. *The following conditions on a projection e in A are equivalent:*

(a) *e is abelian;*

(b) *$f \leq e$ implies $f = eC(f)$.*

(c) *$f \leq e$ implies $f = eh$ for some central projection h in A.*

Proof. Immediate from [§ 6, Prop. 4] and Definition 2. ∎

Paralleling Proposition 3:

Proposition 7. *If e is an abelian projection in A and if $f \precsim e$, then f is also abelian.*

Proof. We suppose, as in the proof of Proposition 3, that $f \leq e$; then $fAf \subset eAe$. If $g \leq f$ then g is central in eAe, therefore also in fAf. ∎

The structure theorems depend on exhaustion arguments whose essence is the following proposition, to the effect that finiteness and abelianness are cumulative, provided that the projections being combined are very orthogonal:

Proposition 8. *If $(e_\iota)_{\iota \in I}$ is a very orthogonal family of finite [abelian] projections and if $e = \sup e_\iota$, then e is also finite [abelian].*

Proof. Write $h_\iota = C(e_\iota)$, $h = \sup h_\iota$. By hypothesis, the h_ι are orthogonal, therefore $h_\iota e = e_\iota$ [§ 11, Lemma 1]. Moreover, $h = C(e)$ [§ 6, Prop. 1, (iv)].

Suppose the e_ι are finite. Assuming $e \sim f \leq e$, it is to be shown that $f = e$. For each ι, $h_\iota e \sim h_\iota f \leq h_\iota e$, that is, $e_\iota \sim h_\iota f \leq e_\iota$; since e_ι is finite, $h_\iota f = e_\iota = h_\iota e$, thus $(e-f)h_\iota = 0$ for all ι, therefore $(e-f)h = 0$. But $e - f \leq e \leq C(e) = h$, thus $e - f = (e-f)h = 0$.

Now suppose the e_ι are abelian. Assuming $f \leq e$, it will suffice to show that $f = eC(f)$ (Proposition 6). For each ι, $h_\iota f \leq h_\iota e = e_\iota$; since e_ι is abelian,
$$h_\iota f = e_\iota C(h_\iota f) = e_\iota h_\iota C(f) = h_\iota e C(f),$$
thus $(eC(f) - f)h_\iota = 0$ for all ι; therefore $(eC(f) - f)h = 0$, and, since $h = C(e)$, this yields $eC(f) - f = 0$. ∎

The direct summands in the coarse structure theory are chosen so as to satisfy one or several of the following conditions:

Definition 3. (1a) Repeating Definition 1, A is said to be *finite* if $x^*x = 1$ implies $xx^* = 1$.

(1b) A is said to be *properly infinite* if the only finite central projection is 0; in other words, A has no finite direct summand other than $\{0\}$ [§ 3, Exer. 4]. We agree to regard the ring $\{0\}$ as properly infinite (thus $\{0\}$ is the only ring that is both finite and properly infinite).

(2a) Repeating Definition 2, A is said to be *abelian* if all of its projections are central.

(2b) We say that A is *properly nonabelian* if the only abelian central projection is 0 (in other words, the only abelian direct summand of A is $\{0\}$). We agree to regard the ring $\{0\}$ as properly nonabelian (thus $\{0\}$ is the only ring that is both abelian and properly nonabelian).

(3a) A is said to be *semifinite* if it has a faithful finite projection, that is, a finite projection e such that $C(e) = 1$.

(3b) A is said to be *purely infinite* if it contains no finite projection other than 0. Abusing the notation slightly, we agree to regard the ring $\{0\}$ as both semifinite and purely infinite; no other ring can be both.

(4a) A is said to be *discrete* if it has a faithful abelian projection, that is, an abelian projection e such that $C(e)=1$.

(4b) A is said to be *continuous* if it contains no abelian projection other than 0. We agree to regard the ring $\{0\}$ as both discrete and continuous; no other ring can be both.

A central projection h in A is said to be finite [properly infinite, etc.] if the direct summand hA is finite [properly infinite, etc.].

Remarks. 1. We follow here the terminology of J. Dixmier [**19**], except for the term 'properly nonabelian', which is ad hoc (but rounds things out nicely). Some authors use the term 'purely infinite' for the condition (1b), and 'Type III' for the condition (3b); the 'type' terminology is explained later in the section.

2. Since every abelian projection is finite, the following implications are immediate from the definitions: properly infinite $\Rightarrow$ properly nonabelian; discrete $\Rightarrow$ semifinite; purely infinite $\Rightarrow$ continuous. Obviously, abelian $\Rightarrow$ discrete, and finite $\Rightarrow$ semifinite.

3. If a Baer $*$-factor contains a nonzero finite [abelian] projection, then it is semifinite [discrete].

4. Every finite-dimensional von Neumann algebra is finite [cf. §17, Prop. 1]. If $\mathscr{H}$ is an infinite-dimensional Hilbert space, then $\mathscr{L}(\mathscr{H})$ is properly infinite and discrete (a projection with one-dimensional range is abelian). It is harder to produce examples of (i) rings that are semifinite but not finite or discrete, (ii) purely infinite rings, and (iii) continuous rings that are not purely infinite (see [**23**]).

Each of the eight classes of rings described in Definition 3 is 'pure' in the following sense:

Proposition 9. *If A is finite [properly infinite, etc. as in Definition 3] and h is a central projection in A, then hA is also finite [properly infinite, etc.].*

Proof. For 'finite' and 'abelian' see Propositions 3 and 7. The assertions for 'properly infinite', 'properly nonabelian', 'purely infinite' and 'continuous' follow at once from the fact that the central projections of hA are central in A (cf. [§6, Exer. 4 or Prop. 4]).

Suppose A is semifinite [discrete], and let e be a faithful finite [abelian] projection in A; then he is finite [abelian] and $C(he)=hC(e)=h$ shows that he is faithful in hA, thus hA is semifinite [discrete]. ∎

The coarse structure theorems now follow easily from Proposition 8 and the definitions:

Theorem 1. *If A is any Baer $*$-ring, there exist unique central projections h_1, h_2, h_3, h_4 such that*

(1) $h_1 A$ is finite and $(1-h_1)A$ is properly infinite;

(2) $h_2 A$ is abelian and $(1-h_2)A$ is properly nonabelian;

(3) $h_3 A$ is semifinite and $(1-h_3)A$ is purely infinite;

(4) $h_4 A$ is discrete and $(1-h_4)A$ is continuous.

A central projection k is finite iff $k \le h_1$, and properly infinite iff $k \le 1-h_1$; abelian iff $k \le h_2$, and properly nonabelian iff $k \le 1-h_2$; semifinite iff $k \le h_3$, and purely infinite iff $k \le 1-h_3$; discrete iff $k \le h_4$, and continuous iff $k \le 1-h_4$.

Proof. (1) Let (h_ι) be a maximal orthogonal family of nonzero, finite central projections, and set $h = \sup h_\iota$ (if there are no such projections, set $h=0$). Then hA is finite by Proposition 8, and $(1-h)A$ is properly infinite by maximality. This proves the existence of a central projection h_1 satisfying (1).

Let k be a central projection. If $k \le h_1$ then k is finite (Proposition 9). Conversely, if k is finite then $k(1-h_1)$ is both finite and properly infinite (Proposition 9), therefore $k(1-h_1)=0$, that is, $k \le h_1$. Similarly, k is properly infinite iff $k \le 1-h_1$. In particular, it is clear that h_1 is uniquely determined by the properties (1).

(2) Let (h_ι) be a maximal orthogonal family of nonzero, abelian central projections, and set $h = \sup h_\iota$. Then hA is abelian by Proposition 8, and $(1-h)A$ is properly nonabelian by maximality. The argument continues in the format of (1).

(3) Let (h_ι) be a maximal orthogonal family of nonzero, semifinite central projections, and set $h = \sup h_\iota$. For each ι let e_ι be a finite projection such that $C(e_\iota) = h_\iota$, and set $e = \sup e_\iota$; then e is finite (Proposition 8) and $C(e) = \sup C(e_\iota) = h$, thus hA is semifinite. By maximality, $(1-h)A$ is purely infinite. The rest follows the format of (1).

(4) Same as (3), with 'finite', 'semifinite' and 'purely infinite' replaced by 'abelian', 'discrete' and 'continuous'. ∎

There is another standard decomposition of Baer $*$-rings, obtained by meshing certain of the decompositions in Theorem 1; this is described in the next two theorems.

Definition 4. A is said to be of *Type I* if it is discrete; *Type II* if it is continuous and semifinite; and *Type III* if it is purely infinite. (Thus, A is of Type I if it has a faithful abelian projection; Type II if it has a faithful finite projection, but no abelian projections other than 0; and Type III if it has no finite projections other than 0.) We allow the ring $\{0\}$ to be of all three types; no other ring can be of more than one type. A central projection h in A is said to be of Type ν ($\nu =$ I, II, III) if the direct summand hA is of Type ν.

Every Baer $*$-ring may be decomposed uniquely according to type:

Theorem 2. *If A is any Baer $*$-ring, there exist unique orthogonal central projections h_{I}, h_{II}, h_{III} such that $h_\nu A$ is of Type ν ($\nu=\mathrm{I}, \mathrm{II}, \mathrm{III}$) and $h_{\mathrm{I}}+h_{\mathrm{II}}+h_{\mathrm{III}}=1$.*

A central projection k is of Type ν iff $k \le h_\nu$.

In the notation of Theorem 1, $h_{\mathrm{I}}=h_4$, $h_{\mathrm{II}}=h_3(1-h_4)$, $h_{\mathrm{III}}=1-h_3$.

Proof. Define h_ν ($\nu=\mathrm{I}, \mathrm{II}, \mathrm{III}$) by the indicated formulas. Clearly h_ν is of Type ν (for $\nu=\mathrm{II}$, cf. Proposition 9). Since $h_4 \le h_3$ (cf. Remark 2 following Definition 3), we have $h_{\mathrm{II}}=h_3-h_4$, thus the h_ν are orthogonal with sum 1.

Let k be a central projection. If k is of Type ν for $\nu=\mathrm{I}$ or III, then $k \le h_{\mathrm{I}}$ or $k \le h_{\mathrm{III}}$, respectively (Theorem 1). If k is of Type II, that is, if k is semifinite and continuous, then $k \le h_3$ and $k \le 1-h_4$ by Theorem 1, therefore $k \le h_3(1-h_4)=h_{\mathrm{II}}$.

Conversely, every direct summand of a ring of Type ν is also of Type ν (cf. Proposition 9).

It remains to prove uniqueness. Suppose k_{I}, k_{II}, k_{III} are central projections such that k_ν is of Type ν and $k_{\mathrm{I}}+k_{\mathrm{II}}+k_{\mathrm{III}}=1$; since $k_\nu \le h_\nu$, the relation

$$(h_{\mathrm{I}}-k_{\mathrm{I}})+(h_{\mathrm{II}}-k_{\mathrm{II}})+(h_{\mathrm{III}}-k_{\mathrm{III}})=1-1=0$$

shows that $h_\nu-k_\nu=0$ ($\nu=\mathrm{I}, \mathrm{II}, \mathrm{III}$). ∎

For Type I Baer $*$-rings with PC, there is a further decomposition into 'homogeneous' rings (see Section 18); this, too, qualifies as coarse structure theory (but an extra hypothesis is needed, and there are unresolved questions of uniqueness).

Meshing Theorem 2 with the decomposition (1) of Theorem 1, we have:

Theorem 3. *If A is any Baer $*$-ring, there exist unique orthogonal central projections*

$$h_{\mathrm{I_{fin}}},\ h_{\mathrm{I_{inf}}},\ h_{\mathrm{II_{fin}}},\ h_{\mathrm{II_{inf}}},\ h_{\mathrm{III}}$$

such that:

(i) $h_{\mathrm{I_{fin}}}$ *is Type* I *and finite,*

(ii) $h_{\mathrm{I_{inf}}}$ *is Type* I *and properly infinite,*

(iii) $h_{\mathrm{II_{fin}}}$ *is Type* II *and finite,*

(iv) $h_{\mathrm{II_{inf}}}$ *is Type* II *and properly infinite,*

(v) h_{III} *is Type* III,

(vi) *their sum is* 1.

Each of these five projections is maximal in its stated property, that is, it contains every central projections k having that property.

Proof. In the notations of Theorems 1 and 2, define

$$h_{\mathrm{I_{fin}}} = h_{\mathrm{I}} h_1,$$
$$h_{\mathrm{I_{inf}}} = h_{\mathrm{I}}(1-h_1),$$
$$h_{\mathrm{II_{fin}}} = h_{\mathrm{II}} h_1,$$
$$h_{\mathrm{II_{inf}}} = h_{\mathrm{II}}(1-h_1).$$

The final assertion is clear from Theorems 1 and 2. ∎

Remarks. 1. Why the peculiar definition of 'abelian'? One could have proceeded otherwise, as follows. Call a projection e *commutative* if eAe is commutative. The analogues of Propositions 5 and 7 hold: if $f \le e$ then f is commutative relative to A iff it is commutative relative to eAe; if e is commutative and $f \precsim e$, then f is commutative. The analogue of the key Proposition 8 also holds: If (e_ι) is a very orthogonal family of commutative projections, then $e = \sup e_\iota$ is also commutative. {Proof: Write $h_\iota = C(e_\iota)$, $h = \sup h_\iota = C(e)$. Assuming $x, y \in eAe$, it is to be shown that $xy = yx$. For each ι, $h_\iota x = h_\iota x h_\iota = h_\iota(exe)h_\iota = e_\iota x e_\iota$ and $h_\iota y = e_\iota y e_\iota$; since $e_\iota x e_\iota$ and $e_\iota y e_\iota$ commute, we have $h_\iota(xy - yx) = 0$ for all ι, therefore $h(xy - yx) = 0$, that is, $xy - yx = 0$.} This paves the way for the analogue of Theorem 1: There exists a unique central projection h such that hA is commutative and $(1-h)A$ has no commutative summands; also, there exists a unique central projection k such that kA has a faithful commutative projection and $(1-k)A$ has no commutative projections other than 0.

To repeat the question, why the peculiar definition of 'abelian'? A capsule answer is that it keeps division rings together with fields (Example 1 following Definition 2). A fuller answer is to be found in an inventory of examples of Type I rings. Historically, the term 'Type I' was first assigned by Murray and von Neumann **[67]** to von Neumann algebras isomorphic to $\mathscr{L}(\mathscr{H})$, the algebra of all continuous linear mappings on a space with an inner product (complete) taking values in the complex field; algebras of other types appear as subrings of Type I factors [§ 4, Def. 5]. Subsequently, Kaplansky **[49]** showed that every Type I AW^*-algebra may be represented as the algebra of all continuous 'operators' on a space with an inner product taking values in a commutative AW^*-algebra. (The question of which AW^*-algebras may be suitably embedded in a Type I algebra is not yet fully answered; see, e. g. **[26]**, **[38]**, **[79]**, **[90]**.) A purely algebraic example in this vein: if D is an involutive division ring, n is a positive integer, $\mathscr{E}$ is the left vector space of $n$ples $x = (\lambda_i)$, $y = (\mu_i)$ over D, and if the inner product $[x, y] = \sum_1^n \lambda_i \mu_i^*$ is definite (that is, $[x, x] = 0$ implies $x = 0$), then the ring $\mathscr{L}(\mathscr{E})$ of all linear mappings on $\mathscr{E}$ is a (factorial) Baer $*$-ring [§ 4, Exer. 20] of Type I (viewing $\mathscr{L}(\mathscr{E})$ as the matrix ring D_n, the matrix e with 1 in the northwest corner and zeros elsewhere is abelian—eD_ne is $*$-isomorphic with D—and faithful).

In summary, the peculiar definition of 'abelian' (1) keeps together division rings and fields, and therefore (2) includes among the Type I rings the full matrix rings D_n just described; a motivation for the definition is the desire, consistent with the classical usage, to ascribe Type I behavior to such matrix rings. Another way to put the matter is as follows. The von Neumann factors of Type I are characterized by the property of having a minimal projection. If P is a minimal projection in the von Neumann factor $\mathscr{A}$, then $P\mathscr{A}P$ is the field of complex numbers. However, if e is a minimal projection in a Baer $*$-factor A, then eAe may be a noncommutative division ring—the ring $A=D_n$ described above is an example. {In this connection, see [**54**, p. 19, Exer. 12; p. 25, Exer. 7; and p. 98, Exer. 1].}

2. Let A be any $*$-ring with unity. Definitions 1 and 2 are meaningful for A. So is Definition 3, provided 'e faithful' is interpreted to mean that the only central projection h such that $he=e$ is $h=1$ [cf. § 6, Def. 1]. Propositions 1–5 and 7 are valid for A. So is Proposition 9, the key point being that if e is faithful in A, then he is faithful in hA [cf. § 6, Exer. 4].

Exercises

1A. (i) Every minimal projection in a $*$-ring is abelian (see [§ 4, Exer. 3]).

(ii) In a $*$-ring with finitely many elements, every nonzero projection is the sum of orthogonal minimal projections.

(iii) If A is a Baer $*$-ring with finitely many elements, then A is finite, of Type I, and is the direct sum of factors.

2A. The following conditions on a Rickart $*$-ring A are equivalent: (a) A is abelian (in the sense of Definition 2); (b) every idempotent in A is central; (c) all idempotents of A commute with each other.

3A. A Baer $*$-ring is commutative if and only if it is abelian and is generated by its projections in the sense that $(\tilde{A})''=A$.

4A. If (h_ι) is a family of (not necessarily orthogonal) central projections in a Baer $*$-ring, such that each h_ι is finite [properly infinite, etc. as in Definition 3], then $\sup h_\iota$ is also finite [properly infinite, etc.].

5A. In an abelian Rickart $*$-ring, $\mathrm{LP}(x)=\mathrm{RP}(x)$ for all x.

6A. If A is a Baer $*$-ring, e is an abelian projection in A, and f, g are subprojections of e such that $fg=0$, then $C(f)C(g)=0$.

7A. If e is any projection in a Baer $*$-ring, there exists a central projection h such that, on setting $e'=he$ and $e''=(1-h)e$, $e'Ae'$ is finite and $e''Ae''$ is properly infinite.

8A. In the notation of Theorem 1, $e \leq h_4$ for all abelian projections e, and $f \leq h_3$ for all finite projections f. If, in addition, A satisfies one of the conditions (i), (ii), (iii) of [§ 6, Exer. 7], then $h_4 = \sup\{e: e \text{ abelian}\}$, $h_3 = \sup\{f: f \text{ finite}\}$.

9C. (i) A von Neumann algebra is of Type I if and only if it is $*$-isomorphic to a von Neumann algebra (on a suitable Hilbert space) whose commutant is abelian (i.e., commutative).

(ii) A von Neumann algebra is semifinite if and only if it is $*$-isomorphic to a von Neumann algebra (on a suitable Hilbert space) whose commutant is finite.

10C. The decomposition into types can be formulated axiomatically in certain complete lattices.

11C. A von Neumann algebra A with center Z is finite if and only if there exists a function $\mathrm{tr}: A \to Z$ such that (1) $\mathrm{tr}(a+b) = \mathrm{tr}(a) + \mathrm{tr}(b)$ for all $a, b \in A$, (ii) $\mathrm{tr}(z) = z$ for all $z \in Z$, and (iii) $\mathrm{tr}(ab) = \mathrm{tr}(ba)$ for all $a, b \in A$. Such a function is unique (it is called the center-valued normalized *trace function* of the finite algebra A) and has, in addition, the following properties: (iv) $\mathrm{tr}(za) = z\,\mathrm{tr}(a)$ for all $z \in Z$, $a \in A$; (v) $\mathrm{tr}(a^*) = (\mathrm{tr}(a))^*$ for all $a \in A$; (vi) $a \geq 0$ implies $\mathrm{tr}(a) \geq 0$; (vii) if $a \geq 0$ and $\mathrm{tr}(a) = 0$, then $a = 0$; (viii) tr is continuous for the ultraweak topology.

12B. Let A be a Baer $*$-ring, A° the reduced ring of A [§ 3, Exer. 18].

(i) A projection in A is finite relative to A iff it is finite relative to A°.

(ii) If A has PC, then a projection in A is abelian relative to A iff it is abelian relative to A°.

(iii) If A has PC, then the central projections described in Theorems 1–3 are the same for A° as for A (so to speak, A and A° have the 'same' coarse structure).

13C. (i) If A is an AW^*-algebra, then $A = A^\circ$ iff the abelian summand of A is 0.

(ii) If A is any AW^*-algebra, and if e, f are projections in A, then $e \sim f$ iff $e \overset{*}{\sim} f$.

(iii) A von Neumann algebra coincides with the algebra generated by its projections iff its abelian summand is finite-dimensional.

14A. If A is a Baer $*$-factor and e is a nonzero abelian projection in A (in particular, A is of Type I), then e is minimal and eAe is a $*$-ring without divisors of zero.

§ 16. Matrices

The elementary material on matrices presented here could just as easily have been included in Part 1—it does not depend on type—but the main role of matrices is in structure theory. Matrix rings have already occurred in several of the exercises; for the record, here is the official definition:

Definition 1. If B is a ring and n is a positive integer, we write B_n for the set of all $n \times n$ matrices $x = (x_{ij})$, $y = (y_{ij})$ with entries in B, equipped with the usual ring structure:

$$x + y = (x_{ij} + y_{ij}), \qquad xy = \left(\sum_{k=1}^{n} x_{ik} y_{kj}\right).$$

If, in addition, B has an involution, we regard B_n as a $*$-ring, with $*$-transposition as the involution: $x^* = (z_{ij})$, where $z_{ij} = x_{ji}^*$.

It is generally quite hard to show that a matrix ring is a Baer $*$-ring (see Chapter 9). The question of when a Baer $*$-ring is a matrix ring is (at least, piecewise) easier; for example, it is shown in Section 17 that if A is a properly infinite Baer $*$-ring with orthogonal GC, then, for every positive integer n, A is $*$-isomorphic with A_n; in Section 18, every finite Baer $*$-ring of Type I with PC is 'decomposed' into matrix rings over abelian rings; in Section 19, it is shown that every continuous Baer $*$-ring A with PC is, for each positive integer n, $*$-isomorphic to B_n for a suitable Baer $*$-subring B of A. We now discuss the elementary algebra underlying such matrix representations.

Suppose B is a $*$-ring with unity, n is a positive integer, and $A=B_n$ as in Definition 1. For $i=1,\dots,n$, let e_i be the ith diagonal unit, that is, the matrix with 1 in the (i, i) position and zeros elsewhere. The original ring B may be recaptured by 'compression': B is $*$-isomorphic to e_iAe_i, via the correspondence

$$b \leftrightarrow \operatorname{diag}(0,\dots,0,b,0,\dots,0),$$

under which $b\in B$ is paired with the matrix having b in the (i, i) position and zeros elsewhere. Observe that $e_i \sim e_j$ (via the partial isometry w_{ji} having 1 in the (j, i) position and zeros elsewhere). Thus $e_1,\dots,e_n$ are orthogonal, equivalent projections in A with sum 1 (the identity matrix). Important for structure theory is the converse:

Proposition 1. *Let A be a $*$-ring with unity, let n be a positive integer, and suppose that A contains n orthogonal, equivalent projections $e_1,\dots,e_n$ with sum 1. Then A is $*$-isomorphic with the ring $(e_1Ae_1)_n$ of all $n\times n$ matrices over e_1Ae_1.*

Explicitly, for $i=1,\dots,n$ let w_i be a partial isometry such that $w_i^* w_i=e_1$ and $w_i w_i^*=e_i$. Then the mapping φ: $A\to(e_1Ae_1)_n$ defined by

$$\varphi(a)=(w_i^* a w_j) \qquad (a\in A)$$

is a $*$-isomorphism.

Proof. If $a\in A$ then $w_i^* a w_j \in e_1Ae_1$ for all i, j, thus $\varphi(a)\in(e_1Ae_1)_n$. Clearly $\varphi(a+b)=\varphi(a)+\varphi(b)$ and $\varphi(a^*)=\varphi(a)^*$.

φ is injective: if $w_i^* a w_j=0$ for all i, j, then $e_i a e_j=w_i(w_i^* a w_j)w_j^*=0$ for all i, j, thus $a=1a1=\left(\sum_i e_i\right)a\left(\sum_j e_j\right)=\sum_{i,j} e_i a e_j=0$.

φ is surjective: if $(a_{ij})\in(e_1 Ae_1)_n$, define $a=\sum_{i,j} w_i a_{ij} w_j^*$; noting that $w_i^* w_j=0$ when $i\neq j$, the calculation

$$w_i^* a w_j=\sum_{k,l} w_i^* w_k a_{kl} w_l^* w_j = w_i^* w_i a_{ij} w_j^* w_j = e_1 a_{ij} e_1 = a_{ij}$$

shows that $\varphi(a)=(a_{ij})$.

Finally, $\varphi(ab)=\varphi(a)\varphi(b)$ results from the calculation

$$\sum_k (w_i^* a w_k)(w_k^* b w_j) = (w_i^* a)\Big(\sum_k w_k w_k^*\Big)(b w_j)$$
$$= (w_i^* a)\Big(\sum_k e_k\Big)(b w_j) = w_i^*(ab) w_j. \quad \blacksquare$$

As an application, we explain in matricial terms a calculation made in the proof of [§ 11, Lemma 3]:

Proposition 2. *Let A be a weakly Rickart $*$-ring and let e, f be projections in A with $ef=0$. Then $e \sim f$ if ond only if there exists a projection g such that*

$$2ege = e, \quad 2fgf = f, \quad 2geg = 2gfg = g.$$

Proof. If such a projection g exists then, defining $w = 2fge$, we have

$$w^* w = 4egfge = 2e(2gfg)e = 2ege = e,$$

and similarly $ww^* = f$, thus $e \sim f$ (the hypothesis $ef=0$ is not needed here).

Conversely, suppose $e \sim f$. Dropping down to $(e+f)A(e+f)$, we can suppose that A is a Rickart $*$-ring and $e+f=1$ (see [§ 1, Prop. 4], [§ 5, Prop. 6]). Let w be a partial isometry such that $w^* w = e$, $ww^* = f$ $= 1-e$. Define $w_1 = e$, $w_2 = w$ and consider the $*$-isomorphism

$$\varphi: A \to (eAe)_2$$

given by $\varphi(a) = (w_i^* a w_j)$ as in Proposition 1. For $a \in A$ write $a_{ij} = w_i^* a w_j$ $(i, j = 1, 2)$. Since $w_{11} = ewe = 0$, $w_{12} = eww^* = 0$, $w_{21} = w^* we = e$, $w_{22} = w^* ww = ew = 0$, we have

$$\varphi(w) = \begin{pmatrix} 0 & 0 \\ e & 0 \end{pmatrix}.$$

It follows that (or similar computations yield)

$$\varphi(e) = \begin{pmatrix} e & 0 \\ 0 & 0 \end{pmatrix}, \quad \varphi(f) = \begin{pmatrix} 0 & 0 \\ 0 & e \end{pmatrix}.$$

It is more suggestive to write 1 for the unity element of eAe and identify A with $(eAe)_2$; then

$$e = \begin{pmatrix} 1 & 0 \\ 0 & 0 \end{pmatrix}, \quad f = \begin{pmatrix} 0 & 0 \\ 0 & 1 \end{pmatrix}, \quad w = \begin{pmatrix} 0 & 0 \\ 1 & 0 \end{pmatrix}.$$

Note that if $a = (a_{ij}) \in A$ then

$$eae = \begin{pmatrix} a_{11} & 0 \\ 0 & 0 \end{pmatrix}, \quad faf = \begin{pmatrix} 0 & 0 \\ 0 & a_{22} \end{pmatrix}.$$

Since 2 is invertible in A [§ 11, Lemma 2], we may define

$$g=\begin{pmatrix}\frac{1}{2} & \frac{1}{2}\\ \frac{1}{2} & \frac{1}{2}\end{pmatrix}=\frac{1}{2}\begin{pmatrix}1 & 1\\ 1 & 1\end{pmatrix}.$$

Evidently g is a projection, and direct calculation yields the desired relations $2ege=e$, etc. {Note, incidentally, that $2g=e+f+w+w^*$ (cf. the proof of [§ 11, Lemma 3]).} ∎

Exercises

1A. If A is a Rickart *-ring and if, for some positive integer n, A is *-isomorphic with the *-ring B_n of $n\times n$ matrices over a *-ring B with unity, then $n1$ is invertible in A. In fact, $(n!)1$ is invertible in A.

2A. Let A be a ring with unity and let n be a positive integer. If I is a subset of A, write I_n for the subset of A_n consisting of those matrices whose entries are all in I. Then $I\mapsto I_n$ maps the set of all (two-sided) ideals of A bijectively onto the set of all ideals of A_n, and $A_n/I_n\cong(A/I)_n$ for every ideal I. In particular, I is maximal in A iff I_n is maximal in A_n; A is simple iff A_n is simple. If A is a *-ring, then $I\mapsto I_n$ pairs the *-ideals of A with those of A_n.

3A. Let B be a ring with unity, let n be a positive integer, and let B_n be the ring of all $n\times n$ matrices over B. For $b\in B$ write

$$\operatorname{diag}(b)=\operatorname{diag}(b,b,\dots,b)=(\delta_{ij}b)$$

for the 'scalar' matrix with b on the diagonal and zeros elsewhere. For any subset S of B, write

$$D(S)=\{\operatorname{diag}(s): s\in S\};$$

also, write S_n for the set of all $n\times n$ matrices with entries in S.

(i) If S is any subset of A containing 0 and 1, then $(S_n)'=D(S')$.

(ii) Let A be a subring of B such that $A=A''$. Then $A_n=(D(A'))'$; in particular, A_n is a subring of B_n such that $A_n=A_n''$. Moreover, $D(A')=(A_n)'$, thus A_n and $D(A')$ are each others' commutants in B_n. Similarly $(A')_n=(D(A))'$, $D(A)=((A')_n)'$.

(iii) Assume, in addition, that B is a *-ring, and let A be a *-subring of B such that $A=A''$. Then each of A_n, $(A')_n$, $D(A)$, $D(A')$ is a *-subring of B_n that coincides with its bicommutant in B_n. If B_n is a Rickart *-ring, then A_n and $(A')_n$ are also Rickart *-rings (with unambiguous RP's and LP's); if B_n is a Baer *-ring, then A_n and $(A')_n$ are Baer *-subrings of B_n.

(iv) Related remarks: If B is a von Neumann algebra on a Hilbert space $\mathscr{H}$, it is easy to show that B_n may be regarded as a von Neumann algebra on the direct sum of n copies of $\mathscr{H}$. If B is an AW^*-algebra, then so is B_n [§ 62, Cor. 1 of Th. 1]. For B a Baer *-ring, the situation is complicated, and largely unknown (see Section 55).

4A. Let B be a Baer *-ring, A a *-subring of B such that $A=A''$, $Z=A\cap A'$ the center of A, and n a positive integer. Assume that B_n is also a Baer *-ring (cf. Exercise 3, (iv)).

If $x\in A_n$, $x'\in(A')_n$, the following conditions are equivalent: (a) $xx'=0$; (b) there exists a projection $c\in Z_n$ such that $xc=0$ and $cx'=x'$; (c) there exists $c\in Z_n$ such that $xc=0$ and $cx'=x'$; (d) $xD(Z')x'=0$, where $D(Z')$ is the set of all matrices $\operatorname{diag}(z')$ with z' in $Z'=(A\cap A')'=(A\cup A')''$ (see Exercise 3).

5A, C. (i) If λ is a complex number with $|\lambda|\leq\frac{1}{2}$, then there exists a real number α, $\frac{1}{2}\leq\alpha\leq 1$, such that the matrix

$$\begin{pmatrix}\alpha & \lambda \\ \lambda^* & 1-\alpha\end{pmatrix}$$

is a projection.

(ii) If $A=B_n$, where $n\geq 2$ and B is a complex $*$-algebra with unity, then the reduced ring A° [§ 3, Exer. 18] is a $*$-subalgebra of A. If, in particular, B is the field of complex numbers (or even a C^*-algebra), then $A^\circ=A$.

§ 17. Finite and Infinite Projections

Throughout this section, A denotes a $*$-ring with unity (often, but not always, a Baer $*$-ring).

Recall that A is said to be finite if $x^*x=1$ implies $xx^*=1$, and a projection $e\in A$ is called finite if the ring eAe is finite [§ 15, Def. 1]. Nonfiniteness may be characterized as follows:

Proposition 1. (i) *If A is an infinite $*$-ring with unity, then there exists a sequence f_n of orthogonal, mutually equivalent, nonzero projections.*

(ii) *Conversely, if A is a Rickart $*$-ring that contains such a sequence, and if every sequence of orthogonal projections in A has a supremum, then A is infinite.*

Proof. (i) By hypothesis, there exists a projection $e\neq 1$ such that $1\sim e$ (see the remark following [§ 15, Def. 1]). Let w be a partial isometry such that $w^*w=1$, $ww^*=e$, and let $\varphi\colon A\to eAe$ be the $*$-isomorphism defined by $\varphi(x)=wxw^*$. In particular, φ is an order-preserving bijection of the projections of A onto the projections of eAe. Define $e_1=1$ and, inductively, $e_{n+1}=\varphi(e_n)$ for $n=1,2,3,\dots$ In particular, $e_2=\varphi(1)=e$. Since $e_1\geq e_2$ and $e_1\neq e_2$, an application of φ to the inequality yields $e_2\geq e_3$, $e_2\neq e_3$. Continuing inductively, we see that the sequence of projections e_n is strictly decreasing. Defining $f_n=e_n-e_{n+1}$ $(n=1,2,3,\dots)$, we have an orthogonal sequence of nonzero projections; moreover,

$$\varphi(f_n)=\varphi(e_n)-\varphi(e_{n+1})=e_{n+1}-e_{n+2}=f_{n+1}$$

shows that $f_n\sim f_{n+1}$ [§ 1, Prop. 9].

(ii) Suppose f_n is an orthogonal sequence of nonzero projections such that $f_1\sim f_2\sim f_3\sim\cdots$. By hypothesis, we may define $e=\sup\{f_n\colon n\geq 1\}$ and $f=\sup\{f_n\colon n\geq 2\}$. Then $f_1f=0$, and $e=f_1+f$ by the associativity of suprema. We have $e\sim f$ [§ 11, Prop. 1], where $f\leq e$ and $e-f=f_1\neq 0$, thus e is not a finite projection [§ 15, Prop. 2]; it follows that A is not finite [§ 15, Cor. of Prop. 3]. ∎

It is useful to have some terminology to describe orthogonal families such as those occurring in Proposition 1:

Definition 1. An orthogonal family of nonzero projections (e_ι) is called a *partition*, with *terms* e_ι; if $e=\sup e_\iota$ exists, it is a *partition of* e. Two equipotent partitions $(e_\iota)_{\iota\in I}$, $(f_\iota)_{\iota\in I}$ are *equivalent* if $e_\iota \sim f_\iota$ for all $\iota\in I$. A partition (e_ι) is *homogeneous* if its terms are mutually equivalent. {Warning: The word 'homogeneous' has another meaning in the context of Type I rings (see the next section).} A homogeneous partition (e_ι) is called *maximal* if it cannot be enlarged, that is, there does not exist a projection e such that $e\sim e_\iota$ and $ee_\iota=0$ for all ι.

Remarks. 1. We admit partitions with finitely many terms; but the message of Proposition 1 is that infinite homogeneous partitions correlate with nonfiniteness of the ring.

2. If $e\sim f$ and (e_ι) is a partition of e, then there exists a partition (f_ι) of f that is equivalent to (e_ι) [§1, Prop. 9]. If, in addition, (e_ι) is homogeneous, then so is (f_ι).

3. Every homogeneous partition can be enlarged to a maximal one (a routine application of Zorn's lemma). We remark that if (e_ι) is a homogeneous partition of e, then (e_ι) can be enlarged if and only if $e_\iota \lesssim 1-e$; if (e_ι) is maximal, the leftover scrap $1-e$ can sometimes be absorbed by dropping down to a direct summand:

Proposition 2. *If A is a Baer $*$-ring with orthogonal* GC, *and if (e_ι) is a maximal homogeneous partition in A with infinitely many terms, then there exists a nonzero central projection h and a homogeneous partition (f_ι) of h that is equivalent to (he_ι).*

Proof. Fix an index ι_0; for simplicity, write $\iota_0=1$. Set $e=\sup e_\iota$. Since $1-e$ and e_1 are orthogonal, by hypothesis there exists a central projection h such that

$$(*) \qquad h(1-e)\lesssim he_1$$

and $(1-h)e_1 \lesssim (1-h)(1-e)$. Necessarily $he_1\neq 0$; for, $he_1=0$ would imply $e_1=(1-h)e_1\lesssim(1-h)(1-e)\leq 1-e$, contrary to the maximality of (e_ι).

Since (he_ι) is a homogeneous partition of he, it will suffice, by Remark 2 above, to show that $he\sim h$. In view of the Schröder-Bernstein theorem ([§1, Th. 1] or [§12, Prop. 1]), it is enough to show that $h\lesssim he$. Let $f=\sup\{e_\iota\colon \iota\neq 1\}$; thus $e=f+e_1$. Since the family (e_ι) is infinite, we have $e\sim f$ [§11, Prop. 1], therefore

$$(**) \qquad he\sim hf;$$

adding (∗) and (∗∗), we have

$$h = he + h(1-e) \precsim hf + he_1 = h(f+e_1) = he. \quad \blacksquare$$

The hypotheses of Proposition 2 can be fulfilled in any infinite Baer ∗-ring with orthogonal GC (apply Remark 3 above to the sequence f_n whose existence is ensured by Proposition 1), leading to a direct summand with a certain property. When A is properly infinite, the argument can be pursued to exhaustion:

Proposition 3. *If A is a properly infinite Baer ∗-ring with orthogonal* GC, *then there exists a sequence e_n that is a homogeneous partition of* 1.

Proof. By Proposition 1 there exists a homogeneous partition with infinitely many terms, which we can suppose to be maximal (Remark 3 following Definition 1). Invoking Proposition 2, there exists a nonzero central projection h that possesses a homogeneous partition $(f_\iota)_{\iota \in I}$ with infinitely many terms. Since $\aleph_0 \operatorname{card} I = \operatorname{card} I$, the index set I can be written as the union of a disjoint sequence of equipotent sets I_n,

$$I = I_1 \cup I_2 \cup I_3 \cup \cdots.$$

Defining $f_n = \sup\{f_\iota : \iota \in I_n\}$ $(n = 1, 2, 3, \ldots)$, we have $f_m \sim f_n$ for all m, n by orthogonal additivity of equivalence [§11, Th. 1], and $\sup f_n = h$ by the associativity of suprema. Summarizing, there exists (in every infinite Baer ∗-ring with orthogonal GC) a nonzero central projection h, and a sequence f_n that is a homogeneous partition of h.

Let $(h_\alpha)_{\alpha \in \Lambda}$ be a maximal family of orthogonal, nonzero central projections such that, for each $\alpha \in \Lambda$, there exists a sequence $e_{\alpha n}$ that is a homogeneous partition of h_α. Defining

$$e_n = \sup\{e_{\alpha n} : \alpha \in \Lambda\} \qquad (n = 1, 2, 3, \ldots),$$

we have $e_m \sim e_n$ for all m, n, and $\sup e_n = \sup h_\alpha$ (for the reasons cited above). It will suffice to show that $\sup h_\alpha = 1$. Assume to the contrary that $1 - \sup h_\alpha \neq 0$. Then $(1 - \sup h_\alpha)A$ is infinite (because A is properly infinite); by the first part of the proof, it contains a nonzero central projection k and a sequence f_n that is a homogeneous partition of k. This contradicts the maximality of the family $(h_\alpha)_{\alpha \in \Lambda}$. ∎

In the conclusion of the above proposition, the e_n are mutually equivalent; this can be achieved by making them equivalent to 1:

Theorem 1. *Let A be a properly infinite Baer ∗-ring with orthogonal* GC.

(1) *There exists an orthogonal sequence of projections f_n such that $\sup f_n = 1$ and $f_n \sim 1$ for all n.*

(2) *For each positive integer* m, *there exist orthogonal projections* $g_1,\dots,g_m$ *such that* $g_1+\dots+g_m=1$ *and* $g_i\sim 1$ *for all* i.

Proof. By Proposition 3, there exists a sequence e_n that is a homogeneous partition of 1.

(1) Write the index set $I=\{1,2,3,\dots\}$ as the union of a disjoint sequence of infinite subsets,

$$I=I_1\cup I_2\cup I_3\cup\cdots,$$

and define $f_n=\sup\{e_i\colon i\in I_n\}$. The f_n are mutually orthogonal, $\sup f_n=1$, and $f_n\sim 1$ [§ 11, Prop. 1].

(2) The proof is similar, based on a partition of I into infinite subsets $I_1,\dots,I_m$. ∎

A matricial interpretation of (2):

Corollary. *If* A *is a properly infinite Baer* $*$*-ring with orthogonal* GC, *then, for each positive integer* m, A *is* $*$*-isomorphic to the ring* A_m *of all* $m\times m$ *matrices over* A.

Proof. By the theorem, there exists a partition $e_1,\dots,e_m$ of 1 such that $e_i\sim 1$ for all i. Thus A is $*$-isomorphic to $(e_1Ae_1)_m$ [§ 16, Prop. 1]. The equivalence $e_1\sim 1$ induces a $*$-isomorphism of e_1Ae_1 with A [§ 1, Prop. 9], which in turn induces a $*$-isomorphism of $(e_1Ae_1)_m$ with A_m. ∎

Another application of Theorem 1 (see Theorem 3 for a more general result):

Theorem 2. *Let* A *be a Baer* $*$*-ring with* GC, *satisfying the parallelogram law* (P). *If* e,f *are finite projections in* A, *then* $e\cup f$ *is also finite (thus the finite projections in* A *form a lattice).*

Proof. {We remark that in a Baer $*$-ring satisfying the parallelogram law, the hypotheses GC, PC, and orthogonal GC are equivalent [§ 14, Prop. 7].}

The subring $(e\cup f)A(e\cup f)$ is also a Baer $*$-ring with GC and satisfying (P), in which e,f are finite projections; dropping down to it, we can suppose $e\cup f=1$.

Citing (P), we have $1-e=e\cup f-e\sim f-e\cap f\le f$, thus $1-e\precsim f$; since f is finite, so is $1-e$ [§ 15, Prop. 3]. Thus $1=e+(1-e)$ is the sum of orthogonal finite projections. Changing notation, we can suppose $ef=0$, $e+f=1$.

Assume to the contrary that A is not finite. Then, by structure theory, there exists a nonzero central projection k such that kA is properly infinite [§ 15, Th. 1]. Clearly kA has GC and satisfies (P), and

ke, kf are orthogonal finite projections with sum k. Dropping down to kA, we seek a contradiction from the following conditions: A is a properly infinite Baer $*$-ring with GC and satisfying (P), and e, f are orthogonal finite projections such that $e+f=1$. By Theorem 1 (with $m=2$) there exists a projection g such that $g \sim 1 \sim 1-g$. Apply [§ 14, Prop. 8] to the pair g, e: there exists a central projection h such that

$$hg \precsim he, \qquad (1-h)(1-g) \precsim (1-h)(1-e);$$

since e and $1-e$ are finite, it follows from the relations

$$hg \precsim he \leq e$$

$$(1-h)g \sim (1-h)(1-g) \precsim (1-h)(1-e) \leq 1-e,$$

that hg and $(1-h)g$ are also finite. Since hg and $(1-h)g$ are very orthogonal finite projections, their sum g is also finite [§ 15, Prop. 8]. Then $1 \sim g$ shows that 1 is finite—that is, A is finite—contrary to supposition. ∎

Remarks. 1. Theorem 2 applies to any Baer $*$-ring satisfying the (EP)-axiom and the (UPSR)-axiom [§ 14, Th. 1]—in particular, to any AW^*-algebra. More generally, it applies to any Baer $*$-ring satisfying the (EP)-axiom and the (SR)-axiom [§ 14, Exer. 5].

2. The parallelism between finite and abelian projections is evident in Section 15. Theorem 2 marks a parting of the ways: if e and f are finite [abelian] projections, then $e \cup f$ often [hardly ever] has the same property.

We now consider some conditions, on a $*$-ring A with unity, closely related to (and usually equivalent to) finiteness; these conditions are sometimes more convenient to apply than the definition of finiteness. The following list is not exhaustive (see, e. g., Exercises 4–6 and 10–12):

(1) A does not contain an infinite homogeneous partition.

(2a) $e \sim f$ implies e, f are unitarily equivalent, that is, $ueu^* = f$ for a suitable unitary u.

(2b) $e \sim f$ implies $1-e \sim 1-f$.

(3a) A is finite (that is, $x^*x=1$ implies $xx^*=1$).

(3b) $e \sim 1$ implies $e=1$.

(3c) $e \sim f \leq e$ implies $f=e$.

Proposition 4. *Let A be a $*$-ring with unity.*

(i) *The implications* (2a) $\Leftrightarrow$ (2b) $\Rightarrow$ (3a) $\Leftrightarrow$ (3b) $\Leftrightarrow$ (3c) *and* (1) $\Rightarrow$ (3a, b, c) *hold.*

(ii) *If A has* GC *then* (3c) $\Rightarrow$ (2b), *thus the conditions* (2), (3) *are equivalent and* (1) *implies them.*

(iii) *If A is a Rickart $*$-ring in which every orthogonal sequence of projections has a supremum, then* (3a) $\Rightarrow$ (1).

(iv) *If A is a Rickart $*$-ring with* GC, *in which every orthogonal sequence of projections has a supremum, then the six conditions* (1)–(3c) *are equivalent.*

Proof. (i): (3a)$\Leftrightarrow$(3b) is remarked following [§ 15, Def. 1].

(3c)$\Rightarrow$(3b) is obvious.

(3a)$\Rightarrow$(3c): Suppose A is finite and $e \sim f \le e$. Since e is finite [§15, Cor. of Prop. 3], it follows that $f = e$ [§ 15, Prop. 2].

Thus (3a)$\Leftrightarrow$(3b)$\Leftrightarrow$(3c).

(2a)$\Rightarrow$(2b): Suppose $e \sim f$. By hypothesis, there exists a unitary u such that $ueu^* = f$. Then $w = u(1-e)$ satisfies $w^*w = 1-e$, $ww^* = u(1-e)u^* = uu^* - ueu^* = 1-f$, thus $1-e \sim 1-f$.

(2b)$\Rightarrow$(2a): Suppose $e \sim f$. By hypothesis, $1-e \sim 1-f$. Let w, v be partial isometries such that $w^*w = e$, $ww^* = f$ and $v^*v = 1-e$, $vv^* = 1-f$. Since $w \in fAe$ and $v \in (1-f)A(1-e)$, it is routine to verify that $u = w+v$ is unitary and $ueu^* = f$.

Thus (2a)$\Leftrightarrow$(2b).

(2a)$\Rightarrow$(3b) is obvious.

(1)$\Rightarrow$(3a) is covered by Proposition 1, (i).

(ii) Assuming A has GC, suppose (3c) holds. Given $e \sim f$, we are to show that $1-e \sim 1-f$. Applying GC to the pair $1-e$, $1-f$, there exists a central projection h such that

$$(*) \qquad h(1-e) \sim f' \le h(1-f),$$

$$(**) \qquad (1-h)(1-f) \sim e' \le (1-h)(1-e).$$

Since $he \sim hf$, addition with $(*)$ yields

$$h = he + h(1-e) \sim hf + f' \le h,$$

therefore $hf + f' = h$ by (3c), that is, $f' = h(1-f)$. Thus $(*)$ reads $h(1-e) \sim h(1-f)$, and $(1-h)(1-f) \sim (1-h)(1-e)$ follows similarly from $(**)$; adding these equivalences, we have $1-e \sim 1-f$.

(iii) Quote Proposition 1, (ii).

(iv) Immediate from (i)–(iii). ∎

An application (more generally, see Exercise 8):

Proposition 5. *Let A be a Baer $*$-ring with* GC, *satisfying the parallelogram law* (P). *If e, f are finite projections in A such that $e \sim f$, then $1-e \sim 1-f$.*

Proof. The subring $B = (e \cup f)A(e \cup f)$ is also a Baer $*$-ring with GC and satisfies (P), and it is finite by Theorem 2. Applying Proposition 4, (ii) in B, there exists a unitary v in B such that $vev^* = f$; then $u = v + (1 - e \cup f)$ is unitary in A and $ueu^* = f$, therefore $1-e \sim 1-f$ (Proposition 4, (i)). ∎

The final result, a generalization of Theorem 2, is for application in Section 58; it can be omitted by the reader who is omitting Chapter 9:

Theorem 3. *Let A be a Rickart ∗-ring with* GC, *satisfying the parallelogram law* (P), *such that every sequence of orthogonal projections in A has a supremum. If* e, f *are finite projections in A, then* $e \cup f$ *is also finite (thus the finite projections in A form a lattice).*

Proof. As argued in the proof of Theorem 2, we can suppose $ef = 0$, $e + f = 1$.

Assume to the contrary that A is not finite, and let g_n be a sequence of orthogonal, equivalent, nonzero projections (Proposition 1). Define

$$g = \sup\{g_n : n = 1, 2, 3, \dots\},$$

$$g' = \sup\{g_n : n \text{ odd}\},$$

$$g'' = \sup\{g_n : n \text{ even}\}.$$

Then $g'g'' = 0$, $g = g' + g''$, and $g' \sim g \sim g''$ [§ 11, Prop. 1].

Applying GC to the pair $g' \cap e$, $g'' \cap f$, there exists a central projection h such that

(i) $$h(g' \cap e) \precsim h(g'' \cap f),$$

(ii) $$(1-h)(g'' \cap f) \precsim (1-h)(g' \cap e).$$

Citing (P), we have

(iii) $$h(g' - g' \cap e) \sim h(g' \cup e - e).$$

The left sides of (i) and (iii) are obviously orthogonal; prior to adding them, we check that the right sides are orthogonal too. Indeed, $g'' \cap f$ and $g' \cup e$ are orthogonal; for, $g'' \cap f$ is orthogonal to g' (because $g'g'' = 0$) and to e (because $fe = 0$), hence to $g' \cup e$. Adding (i) and (iii), we have

$$hg' = h(g' \cap e) + h(g' - g' \cap e) \precsim h(g'' \cap f) + h(g' \cup e - e);$$

since $g'' \cap f \leq f$ and $g' \cup e - e \leq 1 - e = f$, this yields $hg' \precsim f$, and since $hg \sim hg'$ we have

(∗) $$hg \precsim f.$$

Again citing (P),

(iv) $$(1-h)(g'' - g'' \cap f) \sim (1-h)(g'' \cup f - f),$$

and (ii) and (iv) yield, by a similar argument,

(∗∗) $$(1-h)g \precsim e.$$

From (∗) and (∗∗) we see that hg and $(1-h)g$ are finite (by [§ 15, Prop. 3], which holds in any ∗-ring). But hg is the supremum of the sequence

hg_n of orthogonal, equivalent projections; it follows from Proposition 1, (ii) that $hg_n=0$ for all n, thus $hg=0$. Similarly $(1-h)g=0$. Then $g=hg+(1-h)g=0$, a contradiction. {One could also argue that $g=hg+(1-h)g$ is finite [cf. § 15, Prop. 8], contradicting the relations $g\sim g'\leq g$, $g'\neq g$.} ∎

Theorem 3 applies, in particular, to any Rickart C^*-algebra with GC ([§ 8, Lemma 3], [§ 13, Th. 1]).

Exercises

1A. If a $*$-ring with unity has finitely many elements, then it is finite (in the sense of [§ 15, Def. 1]).

2A. If h is a finite central projection in a $*$-ring, and if $h\lesssim e$, then $h\leq e$. (Cf. [§ 1, Exer. 15].)

3A. Let A be a $*$-ring with unity and GC. If e, f are finite projections such that $e\sim f$, and if $e'\leq e$, $f'\leq f$ are subprojections such that $e'\sim f'$, then $e-e'\sim f-f'$.

4A. Let A be a symmetric $*$-ring with unity [§ 1, Exer. 7] satisfying the (WSR)-axiom (for example, let A be a C^*-algebra with unity). If A is finite, then $yx=1$ implies $xy=1$.

5A. Let A be a Rickart $*$-ring satisfying the (WSR)-axiom. If A is finite, then $yx=1$ implies $xy=1$.

6A. Let A be a Rickart $*$-ring such that $\mathrm{LP}(x)\sim\mathrm{RP}(x)$ for all x in A. If A is finite, then $yx=1$ implies $xy=1$.

7A. If A is a properly infinite Baer $*$-ring with orthogonal GC, then $n1$ is invertible for every positive integer n, thus A may be regarded as an algebra over the field of rational numbers.

8A. Let A be a Rickart $*$-ring satisfying the conditions of Theorem 3. If e, f are finite projections in A such that $e\sim f$, then $1-e\sim 1-f$.

9A. Theorem 3 remains true with 'Rickart $*$-ring' replaced by 'weakly Rickart $*$-ring'. In particular, the finite projections in a weakly Rickart C^*-algebra with GC form a lattice.

10A. (i) If A is a finite Rickart $*$-ring satisfying the parallelogram law (P), then the projection lattice of A is *modular*, that is, if e, f, g are projections with $e\leq g$, then $(e\cup f)\cap g=e\cup(f\cap g)$. (For a result in the converse direction, see Exercise 11.)

(ii) The finite projections form a modular lattice in a weakly Rickart $*$-ring satisfying the conditions of Theorem 3.

11C. If A is an AW^*-algebra whose projection lattice is modular (see Exercise 10), then A is finite.

12A, C, D. Let A be a Rickart $*$-ring. Projections e, f in A are said to be *perspective* if they admit a common complement, that is, there exists a projection g such that $e\cup g=1$, $e\cap g=0$ and $f\cup g=1$, $f\cap g=0$.

(i) Let h be a projection in A, and let e, f be projections in hAh. If e, f are perspective in hAh, then they are perspective in A; the converse is true if h is a central projection of A.

(ii) Let h be a projection in A, and let e, f be projections in hAh. Suppose that the projection lattice of A is modular (cf. Exercises 10, 11). If e, f are perspective in A then they are perspective in hAh.

(iii) If e, f are in position p', then they are perspective.

(iv) Suppose A satisfies the parallelogram law (P). If e, f are perspective, then they are unitarily equivalent.

(v) Suppose A satisfies the parallelogram law (P). If $e \sim f$ implies e, f are perspective, then A is finite.

(vi) If $e \sim f$ and $ef = 0$ then e, f are perspective. (In fact, it suffices to assume that the orthogonal projections e, f are algebraically equivalent in the sense of [§ 1, Exer. 6].)

(vii) Suppose A satisfies the conditions of Theorem 3. If e, f are finite projections in A such that $e \sim f$, then e, f are perspective. (In view of (iv), this yields another proof of Exercise 8.)

(viii) Here are some applications of (vii). Consider the following conditions on a pair of finite projections e, f: (a) e, f are perspective; (b) e, f are unitarily equivalent; (c) $e \sim f$.

Then (a)$\Leftrightarrow$(b)$\Leftrightarrow$(c) under any one of the following four hypotheses: (1) A satisfies the conditions of Theorem 3. (2) A is a Baer $*$-ring with GC, satisfying the parallelogram law (P). (3) A is a Baer $*$-ring satisfying the (EP)-axiom and the (SR)-axiom. (4) A is an AW^*-algebra.

(ix) Suppose A is either a Rickart C^*-algebra with GC, or a Baer $*$-ring satisfying the (EP)-axiom and the (SR)-axiom. If e, f are finite projections in A such that $e \sim f$, then there exists a symmetry exchanging e and f.

(x) If A is a von Neumann algebra, and e, f are projections in A (not necessarily finite), then e, f are perspective iff they are unitarily equivalent.

(xi) *Problem:* Does (x) generalize to AW^*-algebras?

13A. Let A be a finite Rickart $*$-ring with GC, and let n be a positive integer. Then, under either of the following two hypotheses, the matrix ring A_n is also finite:

(1) A_n is a Rickart $*$-ring, satisfying the parallelogram law (P), in which every sequence of orthogonal projections has a supremum.

(2) A is a Rickart C^*-algebra and A_n is a Rickart $*$-ring (hence also a Rickart C^*-algebra).

14A. Let A be a $*$-ring with unity and GC. Suppose that e, f are projections in A, n is a positive integer, $(e_i)_{1 \le i \le n}$ and $(f_i)_{1 \le i \le n}$ are homogeneous partitions of e and f, respectively; suppose, in addition, that $e \sim f$ and that e is finite (hence f, e_i, f_i are also finite). Then $e_i \sim f_i$. (Cf. Exercise 15.)

15A. Let A be a Baer $*$-ring with GC, satisfying the parallelogram law (P). Let e be a projection in A, and suppose $e = e_1 + e_2 = f_1 + f_2$ are orthogonal decompositions of e with $e_1 \sim e_2$ and $f_1 \sim f_2$. Then $e_i \sim f_i$. (Cf. Exercise 14.)

16A. Let A be a Rickart $*$-ring satisfying the conditions of Theorem 3, and assume that A is properly infinite [§ 15, Def. 3]. If e, f are finite projections in A, then $e \precsim 1 - f$. In particular, $e \precsim 1 - e$ for every finite projection e. (Cf. [§ 18, Prop. 6].)

17A. Let A be a Baer $*$-ring with GC and let e, f be projections in A such that $e \sim f$.

(i) There exist orthogonal decompositions $e=e_1+e_2$, $f=f_1+f_2$ such that e_i and f_i are unitarily equivalent $(i=1,2)$.

(ii) When A is a von Neumann algebra, it follows that e_i, f_i are perspective $(i=1,2)$.

18A. Let A be a properly infinite Baer $*$-ring with GC, satisfying the parallelogram law (P), and suppose A contains a nonzero finite projection g (thus A is not purely infinite). Then there exist a nonzero central projection h and a homogeneous partition $(f_\iota)_{\iota\in I}$ of h with $hg\sim f_\iota$ (I is necessarily infinite).

19A. Let A be a semifinite Baer $*$-ring with PC, and let e be any nonzero projection in A. Then (i) e is the supremum of an orthogonal family of finite projections; (ii) eAe is semifinite; (iii) if e is faithful, then e contains a faithful finite projection. (Cf. [§ 18, Exer. 2].)

20C, D. (i) Let A be a Baer $*$-ring satisfying the (EP)-axiom and the (SR)-axiom, and let A° be the reduced ring of A [§3, Exer. 18]. If e, f are finite projections in A, then the following conditions are equivalent: (a) $e\sim f$; (b) e, f can be exchanged by a symmetry $2g-1$, g a projection; (c) $e \overset{\circ}{\sim} f$. In particular, in a finite Baer $*$-ring satisfying the indicated axioms, the relations $e\sim f$ and $e\overset{\circ}{\sim} f$ are equivalent.

(ii) If A is an AW^*-algebra, and e, f are any projections in A, then $e\sim f$ iff $e\overset{\circ}{\sim} f$.

(iii) *Problem:* Does the conclusion of (ii) hold for the rings of (i)?

21B. If A is a properly infinite Baer $*$-ring with PC, then (i) partial isometries in A are addable, and (ii) A has GC.

22C, D. (i) In a properly infinite von Neumann algebra, every unitary element is the product of four symmetries.

(ii) *Problem:* Does (i) generalize to AW^*-algebras?

23A. Let A be a finite Baer $*$-ring with GC. If e is a projection in A such that, for every positive integer n, A contains n orthogonal projections equivalent to e, then $e=0$.

24D. *Problem:* If A is a Baer $*$-ring with GC and if e, f are finite projections in A, is $e\cup f$ finite?

§ 18. Rings of Type I; Homogeneous Rings

Recall that a Baer $*$-ring is said to be of Type I (or discrete) if it contains a faithful abelian projection, that is, an abelian projection e such that $C(e)=1$ [§15, Def. 3].

Consider, for example, the Baer $*$-ring $\mathscr{L}(\mathscr{H})$ of all bounded linear operators on a Hilbert space $\mathscr{H}$ [§4, Prop. 3]. Let $(\xi_\iota)_{\iota\in I}$ be an orthonormal basis of $\mathscr{H}$ and let $(E_\iota)_{\iota\in I}$ be the corresponding family of one-dimensional projections, that is, $E_\iota\xi=(\xi|\xi_\iota)\xi_\iota$ for all $\xi\in\mathscr{H}$. If $\iota, \varkappa\in I$ then E_ι and $E_\varkappa$ are unitarily equivalent via the obvious permutation operator, thus $(E_\iota)_{\iota\in I}$ is a homogeneous partition of the identity operator, in the sense of [§17, Def. 1]; moreover, $E_\iota\mathscr{L}(\mathscr{H})E_\iota$ is the set of scalar multiples of E_ι, thus E_ι is an abelian projection in $\mathscr{L}(\mathscr{H})$. This situation is a special case of the following:

Definition 1. A Baer $*$-ring A is said to be *homogeneous* if there exists a homogeneous partition of 1 whose terms are abelian projections; that is, there exists an orthogonal family $(e_\iota)_{\iota \in I}$ of abelian projections such that $\sup e_\iota = 1$ and $e_\iota \sim e_\varkappa$ for all ι and $\varkappa$. Though $\operatorname{card} I$ is in general not known to be unique, we speak of A as being homogeneous of *order* $\operatorname{card} I$.

Remarks. 1. A homogeneous Baer $*$-ring is of Type I; for, in the notation of Definition 1, we have [§6, Prop. 1] $C(e_\iota) = C(e_\varkappa)$ for all ι and $\varkappa$, therefore $1 = C(1) = C(\sup e_\iota) = \sup C(e_\iota) = C(e_\varkappa)$ for any $\varkappa$.

2. In the wake of the operatorial example given above, it should be emphasized that the rings $e_\iota A e_\iota$ in Definition 1 are assumed to be abelian, but need not be commutative (cf. the remarks at the end of Section 15).

3. For a homogeneous von Neumann algebra, the order is unique; more generally, the order is unique for a homogeneous AW^*-algebra whose center satisfies a certain decomposability condition (see Exercise 10), but the question of uniqueness in general is open. However, when the order is finite, the tendency to uniqueness is strong (see Proposition 2 below).

4. A Baer $*$-ring A is homogeneous of order n, n a positive integer, if and only if A is $*$-isomorphic to the ring B_n of $n \times n$ matrices over an abelian Baer $*$-ring B; this is immediate from [§16, Prop. 1] and the discussion preceding it. In such a ring, 1 is the sum of n finite (even abelian) orthogonal projections; we do not know if A must be finite (Exercise 12), but A is finite if it has GC and satisfies the parallelogram law (P) [§ 17, Th. 2]—in particular, when A satisfies the (EP)-axiom and the (SR)-axiom [§14, Exer. 5].

5. In a Baer $*$-ring with PC, the conditions $e_\iota \sim e_\varkappa$ in Definition 1 may be replaced by $C(e_\iota) = C(e_\varkappa)$ $(=1)$, in view of the following proposition (cf. Exercise 1):

Proposition 1. *Let e, f be abelian projections in a Baer $*$-ring A, such that $C(e) = C(f)$. Then $e \sim f$ under either of the following hypotheses:* (i) *A has* GC, *or* (ii) *$ef = 0$ and A has* PC.

Proof. Under either hypothesis, e and f are generalized comparable (see [§14, Prop. 5]). Let h be a central projection such that $he \precsim hf$ and $(1-h)f \precsim (1-h)e$. Say $he \sim f' \le hf$. Citing [§6, Prop. 1], we have

$$hC(e) = C(he) = C(f') \le C(hf) = hC(f) = hC(e),$$

thus $C(f') = hC(f)$. But $f' = fC(f')$ since f is abelian [§15, Prop. 6], thus

$$f' = f[hC(f)] = hf\,C(f) = hf,$$

therefore $he \sim hf$. Similarly $(1-h)e \sim (1-h)f$. ∎

Proposition 1 implies a slight generalization of itself:

Corollary. *Let e, f be projections in a Baer $*$-ring A, such that $C(e) \le C(f)$, and suppose e is abelian. Then $e \precsim f$ under either of the following hypotheses:* (i) *A has* GC, *or* (ii) *$ef=0$ and A has* PC.

Proof. As in the proof of Proposition 1, let h be a central projection such that $he \precsim hf$ and $(1-h)f \precsim (1-h)e$. It clearly suffices to show that $(1-h)e \sim (1-h)f$; since both projections are abelian [§15, Prop. 7], and their central covers are equal by the calculation

$$(1-h)C(f)=C[(1-h)f] \le C[(1-h)e]=(1-h)C(e) \le (1-h)C(f),$$

they are equivalent by Proposition 1. ∎

If a finite Baer $*$-ring is homogeneous, then the order is necessarily finite [§17, Prop. 1]; in the presence of generalized comparability, we can show that it is unique:

Proposition 2. *Let A be a finite Baer $*$-ring with* GC, *and suppose A is homogeneous of order n (n a positive integer). Say $1=e_1+\cdots+e_n$, where the e_i are orthogonal, equivalent, abelian projections. Then:*

(1) *Every faithful abelian projection in A is equivalent to the e_i.*

(2) *If $(f_\iota)_{\iota\in I}$ is any orthogonal family of faithful abelian projections, then* $\operatorname{card} I \le n$.

(3) *n is unique.*

(4) *The abelian ring e_1Ae_1 is unique up to $*$-isomorphism.*

Thus, if A is a finite, homogeneous Baer $*$-ring with GC, *then A is $*$-isomorphic to an $n\times n$ matrix ring B_n, where B is an abelian Baer $*$-ring; these conditions determine n uniquely, and B uniquely up to $*$-isomorphism.*

Proof. (1) Quote Proposition 1, (i).

(2) Assume to the contrary that there exist $n+1$ orthogonal, faithful abelian projections $f_1,\ldots,f_n,f_{n+1}$. By (1), $e_i \sim f_i$ $(i=1,\ldots,n)$, thus

$$1=e_1+\cdots+e_n \sim f_1+\cdots+f_n \ne 1,$$

contrary to the finiteness of A.

(3) Immediate from (2).

(4) Immediate from (1) [§1, Prop. 9]. ∎

Note that the following definition is limited to n finite and A finite:

Definition 2. Let n be a positive integer. A Baer $*$-ring A is said to be of *Type* I_n if (i) A is finite, and (ii) A is homogeneous (hence Type I) of order n.

Remarks. 1. A Baer $*$-ring A is of Type I_n (n a positive integer) iff A is finite and is $*$-isomorphic to the ring B_n of $n\times n$ matrices over an abelian Baer $*$-ring B (Remark 4 following Definition 1). If A has GC, then n is unique and B is unique up to $*$-isomorphism (Proposition 2). When A has GC and satisfies the parallelogram law (P), the assumption that A is finite is redundant (Remark 4 following Definition 1).

2. For $\aleph$ an infinite cardinal, 'Type $I_\aleph$' is defined in Definition 3; for the present, we proceed without this concept.

Warning. 'Type I_n' (n a positive integer) means 'finite and homogeneous of order n'; it is conceivable that a homogeneous ring of order n may fail to be of Type I_n (i.e., it might fail to be finite). {*Problem*: Can this happen?}

A homogeneous Baer $*$-ring is of Type I (Remark 1 following Definition 1); it follows that if a Baer $*$-ring A possesses an orthogonal family (h_α) of central projections such that $\sup h_\alpha = 1$ and $h_\alpha A$ is homogeneous for all α, then A is of Type I [§ 15, Th. 1, (4)]. In the converse direction, we show in Theorem 1 below that every Type I Baer$*$-ring with PC possesses such a family (h_α); the incisive result in this direction is as follows:

Proposition 3. *Let A be a Baer $*$-ring with* PC, *and suppose A contains a nonzero abelian projection e (that is, A is not continuous). Then A has a homogeneous direct summand. More precisely, there exists a nonzero central projection h and a homogeneous partition of h having he as one of its terms (thus hA is homogeneous).*

For clarity, we separate out two lemmas.

Lemma 1. *If A is a Baer $*$-ring of Type* I *with* PC, *then every nonzero projection in A contains a nonzero abelian projection.*

Proof. Let f be a nonzero projection, e a faithful abelian projection. Then $C(e)C(f) = C(f) \neq 0$, therefore $eAf \neq 0$ [§ 6, Cor. 1 of Prop. 3]; in view of PC, there exist nonzero subprojections $e_0 \le e$, $f_0 \le f$ such that $e_0 \sim f_0$ [§ 14, Def. 3]. Since $f_0 \precsim e$, f_0 is abelian [§ 15, Prop. 7]. ∎

Lemma 2. *If A is a Baer $*$-ring of Type* I *with* PC, *and f is any faithful projection in A, then f contains a faithful abelian projection.*

Proof. Let $(f_\iota)_{\iota \in I}$ be a maximal family of nonzero abelian projections such that $f_\iota \le f$ and $C(f_\iota)C(f_\varkappa) = 0$ when $\iota \neq \varkappa$ (get started by Lemma 1). Set $f' = \sup f_\iota$; f' is abelian [§ 15, Prop. 8], thus it will suffice to show that $C(f') = 1$. Set $h = 1 - C(f')$ and assume to the contrary that $h \neq 0$. Since f is faithful, $hf \neq 0$. By Lemma 1, hf contains a nonzero abelian projection f_0; since $C(f_0) \le C(hf) \le h$, $C(f_0)$ is orthogonal to every $C(f_\iota)$, contrary to maximality. ∎

Proof of Proposition 3. The direct summand $C(e)A$ obviously has PC, and e is a faithful abelian projection in it; dropping down to $C(e)A$, we can suppose A is of Type I and $C(e)=1$.

Set $e_1=e$ and expand $\{e_1\}$ to a maximal orthogonal family $(e_\iota)_{\iota\in I}$ of faithful abelian projections. Set $f=\sup e_\iota$. By maximality, $1-f$ does not contain a faithful abelian projection; by Lemma 2, $1-f$ is not faithful. Setting $h=1-C(1-f)$, we have $h\neq 0$ and $h(1-f)=0$. Thus

$$(*) \qquad h=hf=h\sup e_\iota=\sup he_\iota .$$

If $\iota\neq\varkappa$ then, since $C(e_\iota)=C(e_\varkappa)=1$, we have $e_\iota\sim e_\varkappa$ by Proposition 1, (ii), therefore $he_\iota\sim he_\varkappa$; then $(*)$ shows that $(he_\iota)_{\iota\in I}$ is a homogeneous partition of h with $he_1=he$ abelian. ∎

In a ring of Type I, Proposition 3 can be pursued to exhaustion:

Theorem 1. *If A is a Baer $*$-ring of Type* I *with* PC, *there exists an orthogonal family (h_α) of nonzero central projections such that* $\sup h_\alpha=1$ *and every $h_\alpha A$ is homogeneous.*

Proof. Let (h_α) be a maximal orthogonal family of nonzero central projections such that every $h_\alpha A$ is homogeneous (get started by Proposition 3). Set $h=1-\sup h_\alpha$ and assume to the contrary that $h\neq 0$; then hA is also a Baer $*$-ring of Type I with PC, and an application of Proposition 3 to it contradicts maximality. ∎

Attached to each homogeneous summand $h_\alpha A$ there is a cardinal number, namely, the cardinal number of the index set of a homogeneous partition of h_α with abelian terms. As remarked in Definition 1, no uniqueness is claimed for this cardinal number; nevertheless, homogeneous summands with same cardinal number can be lumped together:

Lemma. *If A is a Baer $*$-ring and $(h_\alpha)_{\alpha\in\Lambda}$ is an orthogonal family of central projections such that every $h_\alpha A$ is homogeneous of order $\aleph$, then* $(\sup h_\alpha)A$ *is also homogeneous of order $\aleph$.*

Proof. For each α, let $(e_{\alpha\iota})_{\iota\in I}$ be a homogeneous partition of h_α with the $e_{\alpha\iota}$ abelian and $\operatorname{card} I=\aleph$. For each $\iota\in I$, define

$$e_\iota=\sup\{e_{\alpha\iota}\colon \alpha\in\Lambda\}.$$

The e_ι are abelian [§15, Prop. 8] and mutually orthogonal, $e_\iota\sim e_\varkappa$ for all ι and $\varkappa$ [§11, Th. 1], and

$$C(e_\iota)=\sup_\alpha C(e_{\alpha\iota})=\sup_\alpha h_\alpha . \quad ∎$$

We spread out the applications so as to avoid the pitfall indicated following Definition 2.

Theorem 2. *If A is a finite Baer $*$-ring of Type* I *with* PC, *there exists a sequence h_n of central projections such that* (i) *the h_n are orthogonal,* (ii) $\sup h_n = 1$, *and* (iii) *$h_n A$ is either* 0 *or of Type* I_n. *If, in addition, A has* GC, *then condition* (i) *is redundant and the h_n are unique.*

Proof. If hA is a homogeneous direct summand of A, then the order of hA must be finite [§17, Prop. 1]; since hA is a finite Baer $*$-ring, the conditions of Definition 2 are met, that is, hA is of Type I_n for some positive integer n. An application of Theorem 1, and a lumping together of the various orders via the lemma, produces the desired sequence h_n.

Suppose, in addition, that A has GC. Let h be any central projection such that hA is homogeneous, say of order m. We assert that $h \le h_m$. In other words, if $n \ne m$ we assert that $h h_n = 0$; if, on the contrary, $h h_n$ were nonzero, then the summand $h h_n A$ would be homogeneous of both orders m and n, contrary to Proposition 2. {Thus, h_m may be described invariantly as the largest central projection h such that hA is homogeneous of order m (modulo the possibility that $h=0$).} The final assertion of the theorem follows at once. ∎

Remark. Let A be a Baer $*$-ring of Type I with PC, and write A as the direct sum of a finite ring and a properly infinite ring [§15, Th. 1]. Theorem 2 may be applied to the finite summand. The properly infinite summand can be decomposed via Theorem 1; as warned following Definition 2, finite orders could conceivably occur here—but not when A has GC and satisfies the parallelogram law (P).

Theorem 3. *Let A be a Baer $*$-ring of Type* I *with* PC. *Consider cardinal numbers* $\aleph \le \operatorname{card} A$.

There exists an orthogonal family $(h_\aleph)_{\aleph \le \operatorname{card} A}$ of central projections such that (i) $\sup h_\aleph = 1$, *and* (ii) *$h_\aleph A$ is either* 0 *or homogeneous of order* $\aleph$. *If, in addition, A satisfies the parallelogram law* (P), *then, for $\aleph$ finite, the $h_\aleph$ are uniquely determined.*

Proof. {The choice of card A is not critical; it is simply a convenient upper bound for the cardinal numbers that can occur.}

The existence of a family $(h_\aleph)$ satisfying (i) and (ii) is immediate from Theorem 1 and the lemma to Theorem 2.

If, in addition, A satisfies (P), then A has GC [§ 14, Prop. 7]. It follows that if $\aleph$ is finite, then $h_\aleph A$ is finite (Remark 4 following Definition 1); thus, setting $h = \sup\{h_\aleph : \aleph \text{ finite}\}$, we conclude that hA is finite [§ 15, Exer. 4]. On the other hand, if $\aleph$ is infinite then $h_\aleph A$ is properly infinite; for, if $(e_\iota)_{\iota \in I}$ is a homogeneous partition of $h_\aleph$, with the e_ι abelian and $\operatorname{card} I = \aleph$, and if k is any nonzero central projection in $h_\aleph A$, then $(k e_\iota)_{\iota \in I}$ is a homogeneous partition of k with infinitely many terms, therefore $k(h_{\aleph}A)=kA$ is infinite [§ 17, Prop. 1]. Since $1-h=\sup\{h_{\aleph}:\aleph$ infinite$\}$, it follows that $(1-h)A$ is properly infinite [§ 15, Exer. 4]. Thus

$$A=hA+(1-h)A$$

is the unique central decomposition of A into finite and properly infinite summands described in [§ 15, Th. 1]. The final assertion of the theorem now follows from Theorem 2 applied to hA. ∎

We cautiously augment the notation of Definition 2:

Definition 3. If A is a homogeneous Baer $*$-ring of order $\aleph$, $\aleph$ infinite, we say that A is of *Type* $\mathrm{I}_{\aleph}$ (but we do not imply that $\aleph$ is uniquely determined by A). {Such a ring is of Type I (Remark 1 following Definition 1) and is properly infinite (see the proof of Theorem 3).}

Remarks. 1. Suppose A is a homogeneous Baer $*$-ring. If A is finite, then it can only have finite order [§ 17, Prop. 1]. If A is infinite, it is conceivable that A has both finite order and infinite order (see the warning following Definition 2); to put it another way, it is conceivable that A is not of Type $\mathrm{I}_{\aleph}$ for any $\aleph$, finite or infinite. However, it is a feature of Definition 2 that A can't be both Type I_n (n finite) and Type $\mathrm{I}_{\aleph}$ ($\aleph$ infinite).

2. If A is a Baer $*$-ring of Type I, with GC and satisfying the parallelogram law (P), Theorem 3 provides a 'decomposition' of A into rings of Type $\mathrm{I}_{\aleph}$ ($\aleph\le\operatorname{card}A$). This result is applicable to a Baer $*$-ring of Type I satisfying the (EP)-axiom and the (UPSR)-axiom [§ 14, Th. 1] (see also [§ 14, Exer. 5]), in particular to an AW^*-algebra of Type I. Better yet, for AW^*-algebras the 'central direct sum' alluded to in Theorem 3 is an honest C^*-sum [§ 10, Prop. 2], and we have proved the following theorem:

Theorem 4. *If A is an AW^*-algebra of Type* I, *then A is the C^*-sum of a family of homogeneous AW^*-algebras. In detail, there exists an orthogonal family* $(h_{\aleph})_{\aleph\le\operatorname{card}A}$ *of central projections such that* (i) $\sup h_{\aleph}=1$, *and* (ii) $h_{\aleph}A$ *is either* 0 *or of Type* $\mathrm{I}_{\aleph}$*; thus A is the C^*-sum of homogeneous algebras $h_{\aleph}A$. For $\aleph$ finite, the $h_{\aleph}$ are uniquely determined.*

We conclude the section with miscellaneous results for later application. The first is a sharpening of Proposition 2, part (2):

Proposition 4. *If A is a Baer $*$-ring of Type* I_n *(n finite), with* PC, *then A does not contain $n+1$ orthogonal, equivalent, nonzero projections.*

Proof. By hypothesis, A is finite and homogeneous of order n (Definition 2). Let $e_1,\dots,e_n$ be a homogeneous partition of 1 into abelian projections.

Assuming $f_1, \dots, f_{n+1}$ are orthogonal, equivalent projections, it is to be shown that $f_1=0$. It will suffice to show that $e_1 A f_1 = 0$. (This will imply that $0 = C(e_1)C(f_1) = C(f_1)$, therefore $f_1 = 0$.)

Assume to the contrary that $e_1 A f_1 \neq 0$. By PC, there exist nonzero subprojections $e_0 \le e_1$, $f_0 \le f_1$ such that $e_0 \sim f_0$. Since e_1 is abelian, $e_0 = h e_1$ for a suitable central projection h [§ 15, Prop. 6]. Then, for each $i = 1, \dots, n$, we have

$$h e_i \sim h e_1 = e_0 \sim f_0 \le f_1 \sim f_i,$$

thus $h e_i \precsim f_i$ and so $h e_i \precsim h f_i$. Then

$$h = h e_1 + \cdots + h e_n \precsim h f_1 + \cdots + h f_n \le h;$$

in view of finiteness, this implies $h = h f_1 + \cdots + h f_n$, therefore $h f_{n+1} = 0$. Then $h f_1 \sim \cdots \sim h f_n \sim h f_{n+1} = 0$, therefore $h = 0$, a contradiction. ∎

The next proposition is a powerful tool for reducing infinite considerations to finite ones; there are key applications to additivity of equivalence [§ 20, Prop. 5] and dimension in finite rings ([§ 28, Prop. 1], [§ 29, Lemma 4, Prop. 1], [§ 33, Th. 4]), and, in modified form, to matrix rings over finite Baer $*$-rings [§ 61, Lemma 3]:

Proposition 5. *Let A be a finite Baer $*$-ring of Type* I, *with* PC, *and let $(e_\iota)_{\iota \in I}$ be an orthogonal family of projections in A. There exists an orthogonal family $(h_\alpha)_{\alpha \in \Lambda}$ of nonzero central projections with* $\sup h_\alpha = 1$, *such that, for each $\alpha \in \Lambda$, the set $\{\iota \in I : h_\alpha e_\iota \neq 0\}$ is finite.*

Proof. By an obvious exhaustion argument, it suffices to find a nonzero central projection h such that $h e_\iota = 0$ for all but finitely many ι. In view of Theorem 2, we can suppose A to be of Type I_n, n a positive integer.

Assume to the contrary that no such h exists. Applying [§ 14, Prop. 9], we may construct $n+1$ nonzero, equivalent, orthogonal projections (recall that the e_ι are orthogonal); this contradicts Proposition 4. ∎

The final result is for application in [§ 20, Prop. 5] (see also Exercise 5):

Proposition 6. *Let A be a Baer $*$-ring of Type* I, *with* GC *and satisfying the parallelogram law* (P), *and assume that A has no abelian direct summand. If e is any abelian projection in A, then $e \precsim 1 - e$.*

Proof. By Theorem 1, there exists an orthogonal family (h_α) of nonzero central projections such that $\sup h_\alpha = 1$ and each $h_\alpha A$ is homogeneous; moreover, by hypothesis, the order of $h_\alpha A$ is ≥ 2 (possibly infinite). It will suffice to show that $h_\alpha e \precsim h_\alpha(1-e)$ for all α [§ 11, Prop. 2]. Dropping down, we can suppose that A is homogeneous of order ≥ 2.

Let f be a faithful abelian projection, $f \precsim 1-f$ (Definition 1 and Remark 1 following it). Since e is abelian and f is faithful, we have $e \precsim f$ by the corollary of Proposition 1. Say $e \sim g \leq f$; since abelian projections are finite [§ 15, Prop. 4], it follows that $1-e \sim 1-g$ [§ 17, Prop. 5]. Then

$$e \precsim f \precsim 1-f \leq 1-g \sim 1-e,$$

thus $e \precsim 1-e$. {Note that when A is finite, the equivalence $1-e \sim 1-g$ is a consequence of GC [§ 17, Prop. 4]; in other words, (P) can be omitted when A is finite.} ∎

Exercises

1A. Let A be the Baer $*$-ring of all 2×2 matrices over the field of three elements, and let e, f be the projections described in [§1, Exer. 17]. Then e, f are faithful, abelian (even minimal) projections—in particular, A is a finite Baer $*$-factor of Type I_2—but e is not equivalent to f. (Cf. Proposition 1.)

2A. If A is a Baer $*$-ring of Type I with PC, and e is any nonzero projection in A, then (i) e is the supremum of an orthogonal family of abelian projections, therefore (ii) eAe is of Type I. (Cf. [§17, Exer. 19].)

3A. Let A be a Baer $*$-ring with PC, and suppose e is an abelian projection in A. If f is any projection such that $C(f) \geq C(e)$, then f contains an abelian projection f_0 such that $C(f_0) = C(e)$. (Cf. the sharper conclusion of the corollary of Proposition 1 when A has GC.)

4A. Let A be a Baer $*$-ring of Type I with PC. If e is any projection in A, there exists an abelian projection f such that $f \precsim e$ and $C(f) = C(e)$; if, in addition, A has GC, then f is unique up to equivalence.

5A. Let A be a Baer $*$-ring satisfying the conditions of Proposition 6. If e, f are abelian projections in A, then $e \precsim 1-f$.

6A. Let A be a $*$-ring such that every nonzero right ideal contains a nonzero projection. (For example, the (VWEP)-axiom [§7, Def. 3] is sufficient.) A nonzero projection e in A is minimal if and only if eAe is a division ring.

7C. If A is an AW^*-algebra of Type I whose center Z is $*$-isomorphic to a von Neumann algebra, then A is also $*$-isomorphic to a von Neumann algebra.

8C. If A is an AW^*-algebra of Type I, with center Z, and if B is a Type I AW^*-subalgebra of A such that $B \supset Z$, then $B = B''$.

9C. Every AW^*-algebra of Type I may be represented as the algebra of all bounded module operators on a suitable 'AW^*-module' (the latter being a generalized Hilbert space whose inner product takes values in a commutative AW^*-algebra).

10C, D. A $*$-ring B is said to be *orthoseparable* if every orthogonal family of nonzero projections in B is countable; it is *locally orthoseparable* if there exists an orthogonal family (e_ι) of projections with $\sup e_\iota = 1$, such that every $e_\iota B e_\iota$ is orthoseparable.

(i) Every von Neumann algebra is locally orthoseparable.

(ii) If A is an AW^*-algebra whose center Z is locally orthoseparable (equivalently, Z is the C^*-sum of a family of orthoseparable commutative AW^*-algebras), and if A is homogeneous of order $\aleph$, then $\aleph$ is unique.

(iii) If a von Neumann algebra is homogeneous of order $\aleph$, then $\aleph$ is unique. Thus the term 'Type $\mathrm{I}_\aleph$' is completely unambiguous in the context of von Neumann algebras.

(iv) *Problem:* Can local orthoseparability be omitted in (ii)?

11C. Let A be a finite AW^*-algebra of Type I, having no abelian summand. If x is any self-adjoint (or skew-adjoint) element of A, there exist orthogonal projections e, f, g in $\{x\}'$ such that $e+f+g=1$, $e\sim f$ and $g\precsim e$.

12D. *Problem:* If A is a homogeneous Baer $*$-ring of order n, n a positive integer, is A finite?

13A. Let A be a finite Baer $*$-factor of Type I, with PC.

(i) All nonzero abelian projections of A are minimal and are mutually equivalent.

(ii) Every orthogonal family of nonzero projections in A is finite.

(iii) Every nonzero projection in A is the sum of finitely many orthogonal, minimal projections.

(iv) Any two projections e, f in A are comparable ($e\precsim f$ or $f\precsim e$).

(v) A is of Type I_n, n unique; explicitly, if e is any minimal projection in A, then eAe is a $*$-ring without divisors of zero, and A is $*$-isomorphic to $(eAe)_n$.

14A. (i) If A is a Baer $*$-factor with finitely many elements and with 'orthogonal comparability' ($ef=0$ implies $e\precsim f$ or $f\precsim e$), then $A=K_n$ for a suitable positive integer n and a finite involutive field K (i.e., a Galois field with a distinguished automorphism $k\mapsto k^*$ such that $\sum_1^n k_i^* k_i=0$ implies $k_1=\cdots=k_n=0$). {Included is the ring of Exercise 1.}

(ii) If A is a Baer $*$-ring with finitely many elements and with orthogonal GC, then A is the direct sum of finitely many matrix rings of the sort described in (i). In particular, A is semisimple (even $*$-regular).

15A. If A is a Baer $*$-ring of Type I_n (n finite) with PC, then n is unique.

16A, D. Let A be a Baer $*$-ring of Type I with PC, and let e, f be faithful abelian projections in A.

(i) There exist equivalent partitions of e, f. More precisely, there exists an orthogonal family (h_α) of central projections with $\sup h_\alpha=1$, such that $h_\alpha e\sim h_\alpha f$ for all α.

(ii) *Problem:* Is $e\sim f$? Does it help to assume that A is of Type I_n (n finite)?

17A. (i) If B is a $*$-regular ring [§ 3, Exer. 6] and e is a minimal projection in B, then eBe is a division ring.

(ii) If B is a regular Baer $*$-factor of Type I_n, then B is $*$-isomorphic to D_n for a suitable involutive division ring D.

§ 19. Divisibility of Projections in Continuous Rings

Recall that a Baer $*$-ring is said to be continuous if it contains no abelian projection other than 0. By 'divisibility' of projections, we mean the following:

Theorem 1. *Let A be a continuous Baer $*$-ring with* PC. *If e is any nonzero projection in A and n is any positive integer, there exists a homogeneous partition of e of length n; that is, there exist orthogonal projections $e_1, \ldots, e_n$ such that $e = e_1 + \cdots + e_n$ and $e_1 \sim e_2 \sim \cdots \sim e_n$.*

The theorem has key applications to additivity of equivalence [§ 20, Prop. 4], dimension in finite rings [§ 26, Prop. 15], reduction of finite rings (cf. [§ 36, Prop. 1], [§ 37], [§ 41, Th. 1]), and matrix rings over finite rings (cf. [§ 59], [§ 60, Lemma 1]).

We approach the proof through two lemmas, the first of which reformulates nonabelianness:

Lemma 1. *Let A be a Baer $*$-ring with* PC, *and let e be a projection in A. The following conditions on e are equivalent:* (a) *e is not abelian;* (b) *there exists a pair e', e'' of orthogonal, nonzero subprojections of e such that $e' \sim e''$.*

Proof. (a) implies (b): Since eAe is not abelian, there exists a projection $f \le e$ such that $f \ne eC(f)$ [§ 15, Prop. 6], thus $C(f)(e-f) \ne 0$; then $C(f)C(e-f) \ne 0$, therefore $fA(e-f) \ne 0$ [§ 6, Cor. 1 of Prop. 3]. In view of PC, there exist nonzero subprojections $e' \le f$, $e'' \le e-f$ such that $e' \sim e''$.

(b) implies (a) in any $*$-ring A. For, suppose e', e'' are orthogonal subprojections of e such that $e' \sim e''$ (in A). Then $e' \sim e''$ in eAe [§ 1, Prop. 4]. If eAe is abelian, then e', e'' are central projections in eAe, therefore $e' = e''$ [§ 1, Exer. 15]; since e', e'' are orthogonal, it follows that $e' = e'' = 0$. ∎

When e contains no abelian projection other than 0, we may continue inductively:

Lemma 2. *Let A be a Baer $*$-ring with* PC, *and suppose e is a nonzero projection whose only abelian subprojection is* 0. *Then, given any positive integer n, there exists a homogeneous partition of length n whose terms are $\le e$.*

Proof. It is sufficient to consider $n = 2^m$ $(m = 0, 1, 2, 3, \ldots)$. For $m = 0$ there is nothing to do. Assume all is well with $m-1$. By Lemma 1, there exist orthogonal nonzero projections e', e'' such that $e' + e'' \le e$ and $e' \sim e''$. Since e' contains no abelian projection other than 0, the induction hypothesis applied to e' yields 2^{m-1} orthogonal, equivalent, nonzero subprojections of e'; the equivalence $e' \sim e''$ induces an equivalent partition with 2^{m-1} terms $\le e''$. ∎

Proof of Theorem 1. Consider $n$ples of equipotent families

$$(*) \qquad (e_{1\iota})_{\iota \in I}, \ldots, (e_{n\iota})_{\iota \in I}$$

of projections such that (a) the $e_{i\iota}$ are nonzero subprojections of e, (b) $e_{i\iota}e_{j\varkappa}=0$ if $i\neq j$ or $\iota\neq\varkappa$, and (c) for each $\iota\in I$, $e_{1\iota}\sim e_{2\iota}\sim\cdots\sim e_{n\iota}$. Lemma 2 ensures that such an $n$ple exists with I a singleton. We can suppose, by Zorn's lemma, that the family (∗) cannot be enlarged (by increasing the index set I), i.e., that it is maximal in the properties (a)–(c). Define

$$e_i=\sup\{e_{i\iota}\colon \iota\in I\} \qquad (i=1,\dots,n);$$

then $e_1,\dots,e_n$ are mutually orthogonal, nonzero subprojections of e, and $e_1\sim\cdots\sim e_n$ by orthogonal additivity of equivalence [§ 11, Th. 1]. Setting $f=e_1+\cdots+e_n$, it suffices to show that $f=e$; indeed, $e-f\neq 0$ would imply that $e-f$ contained some homogeneous partition of length n (Lemma 2), which, adjoined to (∗), would contradict maximality. ∎

Corollary. *If A is a continuous Baer ∗-ring with* PC, *and n is any positive integer, there exists a projection g in A such that A is ∗-isomorphic to the ring $(gAg)_n$ of all $n\times n$ matrices over gAg.*

Proof. Applying the theorem to $e=1$, the corollary follows at once from [§ 16, Prop. 1]. {We remark that gAg is also a continuous Baer ∗-ring [cf. § 15, Prop. 5] and obviously has PC.} ∎

Exercises

1A. If A is a finite Baer ∗-ring of Type II, with GC, then the projection g of the Corollary is uniquely determined by n up to equivalence, and the ring gAg is unique up to ∗-isomorphism.

2A. Let A be a Baer ∗-ring with PC, having no summand of Type $\mathrm{I}_{\mathrm{fin}}$ [cf. § 15, Th. 3].

(i) If n is any positive integer, there exists a projection e in A such that A is ∗-isomorphic to $(eAe)_n$.

(ii) If A is a complex ∗-algebra, then the reduced ring A° [§ 3, Exer. 18] is a ∗-subalgebra of A.

Chapter 4

Additivity of Equivalence

§ 20. General Additivity of Equivalence

We take up again the theme considered in Section 11.

Throughout this section, A is a Baer $*$-ring (satisfying various axioms as needed). We assume given a pair of equipotent families of projections $(e_\iota)_{\iota\in I}$, $(f_\iota)_{\iota\in I}$ such that (i) the e_ι are orthogonal, (ii) the f_ι are orthogonal, and (iii) $e_\iota \sim f_\iota$ for all $\iota\in I$. We write

$$e=\sup e_\iota, \qquad f=\sup f_\iota .$$

Thus, $(e_\iota)_{\iota\in I}$, $(f_\iota)_{\iota\in I}$ are equivalent partitions of e, f [§ 17, Def. 1]. For each $\iota\in I$, we denote by w_ι a fixed partial isometry such that $w_\iota^* w_\iota = e_\iota$, $w_\iota w_\iota^* = f_\iota$.

Our aim in this section is to prove the following results (Theorem 1): (1) If A has GC, then $e\sim f$; (2) If, in addition, A has no abelian direct summand, then the partial isometries w_ι are addable, in the sense that there exists a partial isometry w such that $w^*w=e$, $ww^*=f$ (hence $e\sim f$) and $we_\iota = w_\iota = f_\iota w$ for all $\iota\in I$ (cf. Exercise 4); (3) If A is any AW^*-algebra, then the partial isometries w_ι are addable.

The plan of attack is to use the structure theory developed in Sections 15–19 to reduce the problem to various special cases. The case that A is abelian is pathological: all projections are central, equivalent projections are equal (therefore $e=f$), and something drastic (e.g., spectral theory in a commutative AW^*-algebra) is needed to add up the partial isometries. In all other cases, equivalence counts for something, and the core result is the one proved in Section 11, which we repeat here for convenience:

Proposition 1. *If* $ef=0$, *then the partial isometries* w_ι *are addable.*

More generally, it suffices that e be equivalent to a subprojection of $1-f$:

Proposition 2. *If* $e \precsim 1-f$, *then the partial isometries* w_ι *are addable.*

Proof. Let u be a partial isometry such that $u^*u=e$, $uu^*=f'\le 1-f$. Defining $f_\iota'=ue_\iota u^*$, we have a partition $(f_\iota')_{\iota\in I}$ of f' [§ 1, Prop. 9], and $f_\iota\sim f_\iota'$ via the partial isometry $v_\iota=uw_\iota^*$. By Proposition 1, there exists a partial isometry v such that $v^*v=f$, $vv^*=f'$ and $vf_\iota=v_\iota=f_\iota'v$ for all ι. Then $w=v^*u$ is a partial isometry with $w^*w=e$, $ww^*=f$, and it is routine to check that $we_\iota=w_\iota=f_\iota w$. ∎

Proposition 3. *If A is a properly infinite Baer *-ring with orthogonal* GC, *then the partial isometries* w_ι *are addable.*

Proof. By [§ 17, Th. 1], there exists a projection g such that $g\sim 1\sim 1-g$. Since $1\sim g$ we have $e\sim e'\le g$ for suitable e'. Similarly, $1\sim 1-g$ yields $f\sim f'\le 1-g$ for suitable f'. Then $e\sim e'\le g\le 1-f'$ shows that $e\precsim 1-f'$.

The equivalence $f\sim f'$ induces a partition $(f_\iota')_{\iota\in I}$ of f' with $f_\iota\sim f_\iota'$. Then $e_\iota\sim f_\iota\sim f_\iota'$, and the composed equivalences $e_\iota\sim f_\iota'$ can be added via Proposition 2. Say $v^*v=e$, $vv^*=f'$, with $ve_\iota=f_\iota'v$ the partial isometry implementing $e_\iota\sim f_\iota'$. Composing v with the partial isometry implementing $f'\sim f$, we get a partial isometry w that 'adds' the w_ι; the routine details are similar to those in Proposition 2. ∎

A lemma preparatory to disposing of the continuous case:

Lemma 1. *If* $h_1,\dots,h_r$ *are orthogonal central projections with* $h_1+\cdots+h_r=1$, *such that for each j the partial isometries* $h_j w_\iota$ *are addable, then the* w_ι *are also addable.*

Proof. A routine application of finite addability [§ 1, Prop. 8]. ∎

Proposition 4. *If A is a continuous Baer *-ring with* GC, *then the partial isometries* w_ι *are addable.*

Proof. Since GC implies PC [§ 14, Prop. 2], projections in A are divisible in the sense of [§ 19, Th. 1]. For each $\iota\in I$ write

$$e_\iota=e_\iota'+e_\iota'' \quad \text{with } e_\iota'\sim e_\iota'' \tag{1}$$

(via a partial isometry unrelated to the given partial isometries w_ι). The partial isometry w_ι induces a decomposition

$$f_\iota=f_\iota'+f_\iota'' \quad \text{with } f_\iota'\sim f_\iota'', \tag{2}$$

where

$$e_\iota'\sim f_\iota' \quad \text{via } w_\iota e_\iota', \tag{3}$$

$$e_\iota''\sim f_\iota'' \quad \text{via } w_\iota e_\iota'', \tag{4}$$

(and $f_\iota' \sim f_\iota''$ is implemented by a partial isometry in whose geneaology we are not interested). Define

$$e' = \sup e_\iota', \quad e'' = \sup e_\iota''$$

$$f' = \sup f_\iota', \quad f'' = \sup f_\iota''.$$

Since $e' e'' = 0$, we infer from (1) and Proposition 1 that

(5) $$e' \sim e'',$$

and similarly

(6) $$f' \sim f''.$$

We assert that the partial isometries $w_\iota e_\iota'$ (see (3)) are addable. To this end, let h be a central projection such that

(7) $$h e' \precsim h f',$$

(8) $$(1-h) f' \precsim (1-h) e'.$$

From (7) and (6), we have

$$h e' \precsim h f' \sim h f'' \leq h(1-f') = h - h f' \leq 1 - h f',$$

thus $h e' \precsim 1 - h f'$, therefore the $h w_\iota e_\iota'$ are addable by Proposition 2. Similarly, by (8) and (5), the $(1-h) w_\iota e_\iota'$ are addable. It follows from Lemma 1 that the $w_\iota e_\iota'$ are addable. Let w' be a partial isometry such that

(9) $$w'^* w' = e', \quad w' w'^* = f', \quad w' e_\iota' = w_\iota e_\iota' = f_\iota' w'.$$

Similarly, using (5)–(8) in the proper combinations, there exists a partial isometry w'' that adds the $w_\iota e_\iota''$:

(10) $$w''^* w'' = e'', \quad w'' w''^* = f'', \quad w'' e_\iota'' = w_\iota e_\iota'' = f_\iota'' w''.$$

Then $w = w' + w''$ is a partial isometry implementing $e \sim f$, and it is straightforward to check that $w e_\iota = w_\iota = f_\iota w$. ∎

There remains the Type I (i.e., discrete) case. It suffices to consider separately the abelian and properly nonabelian cases ([§15, Th. 1] and Lemma 1). The next two lemmas are preparation for the properly nonabelian case; the gist of the first is that the problem of adding partial isometries can be solved if a refinement of it can be solved:

Lemma 2. *Suppose that for each* $\iota \in I$, $(e_{\iota\varkappa})_{\varkappa \in K_\iota}$ *is a partition of* e_ι. *Let* $(f_{\iota\varkappa})_{\varkappa \in K_\iota}$ *be the equivalent partition of* f_ι *induced by* w_ι, *that is,* $f_{\iota\varkappa} = w_\iota e_{\iota\varkappa} w_\iota^*$; *thus,* $e_{\iota\varkappa} \sim f_{\iota\varkappa}$ *via the partial isometry* $w_{\iota\varkappa} = w_\iota e_{\iota\varkappa}$. *If the doubly indexed family of partial isometries* $w_{\iota\varkappa}$ *is addable, then the* w_ι *are also addable.*

Proof. Let w be a partial isometry such that

$$w^* w = \sup_{\iota,\varkappa} e_{\iota\varkappa} = \sup_\iota (\sup_\varkappa e_{\iota\varkappa}) = \sup e_\iota = e,$$

$ww^* = \sup_{\iota,\varkappa} f_{\iota\varkappa} = f$, and $we_{\iota\varkappa} = w_{\iota\varkappa} = f_{\iota\varkappa} w$ for all $\iota, \varkappa$. It is routine to check that $we_\iota = w_\iota = f_\iota w$ for all ι; for example, $we_\iota = w_\iota$ results from the fact that $(we_\iota - w_\iota)e_{\iota\varkappa} = 0$ for all $\varkappa \in K_\iota$. ∎

Lemma 1 generalizes to arbitrary central partitions:

Lemma 3. *If $(h_\alpha)_{\alpha\in\Lambda}$ is an orthogonal family of central projections with $\sup h_\alpha = 1$, such that the doubly indexed family of partial isometries $w_{\alpha\iota} = h_\alpha w_\iota$ is addable, then the w_ι are also addable.*

Proof. Set $e_{\alpha\iota} = h_\alpha e_\iota$; for each $\iota \in I$, $(e_{\alpha\iota})_{\alpha\in\Lambda}$ is a partition of e_ι. Set $f_{\alpha\iota} = w_\iota e_{\alpha\iota} w_\iota^* = h_\alpha f_\iota$; thus $e_{\alpha\iota} \sim f_{\alpha\iota}$ via the partial isometry $w_\iota e_{\alpha\iota} = h_\alpha w_\iota = w_{\alpha\iota}$. By hypothesis, the $w_{\alpha\iota}$ are addable, therefore the w_ι are addable by Lemma 2. ∎

The most complicated case is properly nonabelian Type $\mathrm{I}_{\mathrm{fin}}$:

Proposition 5. *Let A be finite, of Type* I, *with no abelian direct summand, and assume that A has* GC. *Then the partial isometries w_ι are addable.*

Proof. Each e_ι is the supremum of an orthogonal family of abelian projections, by an obvious exhaustion argument based on [§ 18, Lemma 1 to Prop. 3]. In view of Lemma 2, we can suppose, without loss of generality, that the e_ι (hence the f_ι) are abelian. {We remark that this is a way of losing countability of the index set.}

Let $(h_\alpha)_{\alpha\in\Lambda}$ be an orthogonal family of nonzero central projections with $\sup h_\alpha = 1$, such that, for each α, the set

$$I_\alpha = \{\iota \in I : h_\alpha e_\iota \neq 0\} \; (= \{\iota \in I : h_\alpha f_\iota \neq 0\})$$

is finite [§ 18, Prop. 5]. By Lemma 3, it suffices to show that the doubly indexed family of partial isometries $w_{\alpha\iota} = h_\alpha w_\iota$ is addable. Only the nonzero $w_{\alpha\iota}$ need be considered.

For a fixed $\alpha \in \Lambda$, consider $I_\alpha = \{\iota_1, \ldots, \iota_n\}$; thus $h_\alpha e_{\iota_1}, \ldots, h_\alpha e_{\iota_n}$ are precisely the nonzero $h_\alpha e_\iota$ (for this α). By [§ 6, Prop. 5], there exists a finite central partition of h_α, say $k_1, \ldots, k_r$, such that for each pair of indices j, ν, either $k_j(h_\alpha e_{\iota_\nu}) = 0$ or $k_j(h_\alpha e_{\iota_\nu})$ is faithful in $k_j(h_\alpha A)$, that is (since $k_j \leq h_\alpha$), either $k_j e_{\iota_\nu} = 0$ or $k_j e_{\iota_\nu}$ is faithful in $k_j A$. We make mental note of the partial isometries $k w_\iota$ ($k = k_1, \ldots, k_r$; $\iota = \iota_1, \ldots, \iota_n$).

Do this for each $\alpha \in \Lambda$. In view of Lemma 3 (or 2), it is sufficient to show that the partial isometries $k w_\iota$ so obtained are addable. Revising the notation, we can suppose that the h_α already have the property that $h_\alpha e_\iota$ is either 0 or faithful in $h_\alpha A$.

Summarizing, we are in the following situation: the e_ι (and f_ι) are abelian; $(h_\alpha)_{\alpha\in\Lambda}$ is an orthogonal family of nonzero central projections with $\sup h_\alpha = 1$; for each α, the set $I_\alpha = \{\iota \in I : h_\alpha e_\iota \neq 0\}$ is finite; for each pair of indices α, ι, either $h_\alpha e_\iota = 0$ or $h_\alpha e_\iota$ is faithful in $h_\alpha A$. It is to be shown that the partial isometries $w_{\alpha\iota} = h_\alpha w_\iota$ are addable. Write $J = \{(\alpha, \iota) : \alpha \in \Lambda,\ \iota \in I_\alpha\}$. We will show that J can be partitioned into three disjoint subsets,

(*) $$J = J^1 \cup J^2 \cup J^3,$$

such that for each t $(t = 1, 2, 3)$ the family of partial isometries $w_{\alpha\iota}$ $((\alpha, \iota) \in J^t)$ is addable; in view of finite addability [§ 1, Prop. 8], this will complete the proof.

For each $\alpha \in \Lambda$, let $n(\alpha)$ be the number of elements in I_α, and write $I_\alpha = \{\iota_1, \ldots, \iota_{n(\alpha)}\}$. (Strictly speaking, ι_ν should be indexed to indicate dependence on α, but enough is enough.) We know that $h_\alpha e_{\iota_1}, \ldots, h_\alpha e_{\iota_{n(\alpha)}}$ are faithful abelian projections in $h_\alpha A$; citing [§ 18, Prop. 1], we have

(1) $$h_\alpha e_{\iota_1} \sim h_\alpha e_{\iota_2} \sim \cdots \sim h_\alpha e_{\iota_{n(\alpha)}}.$$

Similarly (or therefore)

(2) $$h_\alpha f_{\iota_1} \sim h_\alpha f_{\iota_2} \sim \cdots \sim h_\alpha f_{\iota_{n(\alpha)}}.$$

{Note that the partial isometries implementing (1), (2) have nothing to do with the given partial isometries w_ι.} Partition I_α into three disjoint subsets,

$$I_\alpha = I_\alpha^1 \cup I_\alpha^2 \cup I_\alpha^3,$$

as follows: if $n(\alpha) = 0$, that is, if $I_\alpha = \varnothing$, set $I_\alpha^1 = I_\alpha^2 = I_\alpha^3 = \varnothing$; if $n(\alpha)$ is an even integer, partition I_α into two subsets I_α^1, I_α^2 each with $n(\alpha)/2$ elements, and set $I_\alpha^3 = \varnothing$; if $n(\alpha)$ is an odd integer, select I_α^3 to be any singleton in I_α, and partition $I_\alpha - I_\alpha^3$ into two subsets I_α^1, I_α^2 each with $(n(\alpha) - 1)/2$ elements. In all cases,

$$\operatorname{card} I_\alpha^1 = \operatorname{card} I_\alpha^2 \quad \text{and} \quad \operatorname{card} I_\alpha^3 \le 1.$$

Define

$$e_\alpha^t = \sup\{h_\alpha e_\iota : \iota \in I_\alpha^t\} \qquad (t = 1, 2, 3),$$

with the understanding that $e_\alpha^t = 0$ when $I_\alpha^t = \varnothing$; the suprema are in fact finite sums, and the sum defining e_α^3 has at most one term. Similarly, define

$$f_\alpha^t = \sup\{h_\alpha e_\iota : \iota \in I_\alpha^t\} \qquad (t = 1, 2, 3).$$

Note that

(3) $$h_\alpha e = \sup\{h_\alpha e_\iota : \iota \in I\} = \sup\{h_\alpha e_\iota : \iota \in I_\alpha\} = e_\alpha^1 + e_\alpha^2 + e_\alpha^3,$$

and similarly

(4) $$h_\alpha f = f_\alpha^1 + f_\alpha^2 + f_\alpha^3.$$

From (1), and the equipotence of I_α^1, I_α^2, finite additivity yields

(5) $$e_\alpha^1 \sim e_\alpha^2;$$

similarly,

(6) $$f_\alpha^1 \sim f_\alpha^2.$$

{The equivalences (5), (6) have nothing to do with the w_ι.} Also, since $h_\alpha e_\iota \sim h_\alpha f_\iota$ (via $h_\alpha w_\iota$), finite addability yields

(7) $$e_\alpha^1 \sim f_\alpha^1 \quad (\text{via} \sum \{h_\alpha w_\iota : \iota \in I_\alpha^1\}),$$

(8) $$e_\alpha^2 \sim f_\alpha^2 \quad (\text{via} \sum \{h_\alpha w_\iota : \iota \in I_\alpha^2\}).$$

Define

$$e^1 = \sup_\alpha e_\alpha^1, \qquad e^2 = \sup_\alpha e_\alpha^2, \qquad e^3 = \sup_\alpha e_\alpha^3,$$

$$f^1 = \sup_\alpha f_\alpha^1, \qquad f^2 = \sup_\alpha f_\alpha^2, \qquad f^3 = \sup_\alpha f_\alpha^3.$$

From (3) and (4), we see that

$$e = e^1 + e^2 + e^3, \qquad f = f^1 + f^2 + f^3$$

by the associativity of suprema. By orthogonal additivity of equivalence (Proposition 1), (5) yields

(9) $$e^1 \sim e^2;$$

similarly, (6) yields

(10) $$f^1 \sim f^2.$$

{The equivalences (9), (10) have nothing to do with the w_ι.}

The desired partition (*) is obtained by defining

$$J^t = \{(\alpha, \iota) : \alpha \in \Lambda, \ \iota \in I_\alpha^t\} \qquad (t = 1, 2, 3).$$

Checking the definitions, we see that

(11) $$e^t = \sup\{h_\alpha e_\iota : (\alpha, \iota) \in J^t\},$$

(12) $$f^t = \sup\{h_\alpha f_\iota : (\alpha, \iota) \in J^t\},$$

for $t = 1, 2, 3$.

We assert that the partial isometries $w_{\alpha\iota}$ $((\alpha, \iota) \in J^1)$ are addable (hence $e^1 \sim f^1$). To this end, let h be a central projection such that

(13) $$he^1 \precsim hf^1,$$

(14) $$(1-h) f^1 \precsim (1-h) e^1.$$

From (13) and (10), we have

$$he^1 \precsim hf^1 \sim hf^2 \leq h(1-f^1) = h - hf^1 \leq 1 - hf^1,$$

thus $he^1 \precsim 1-hf^1$; noting the formulas (11), (12) with $t=1$, it follows from Proposition 2 that the partial isometries $hw_{\alpha\iota}$ ($(\alpha,\iota)\in J^1$) are addable. Similarly, it follows from (14) and (9) that the partial isometries $(1-h)w_{\alpha\iota}$ ($(\alpha,\iota)\in J^1$) are addable. In view of Lemma 1, the assertion is proved.

Similarly, the partial isometries $w_{\alpha\iota}$ ($(\alpha,\iota)\in J^2$) are addable.

It remains to consider J^3. Note that each e^3_γ is an abelian projection (recall that $\operatorname{card} I^3_\alpha = 0$ or 1), thus e^3 is the supremum of a very orthogonal family of abelian projections; it follows that e^3 is abelian [§ 15, Prop. 8], and similarly f^3 is abelian. We assert that

$$(15) \qquad e^3 \sim f^3 .$$

Since e^3, f^3 are abelian, it suffices to show that $C(e^3)=C(f^3)$ [§ 18, Prop. 1]. Indeed, for any $(\alpha,\iota)\in J$ we have $h_\alpha e_\iota \sim h_\alpha f_\iota$, hence $C(h_\alpha e_\iota) = C(h_\alpha f_\iota)$; in particular, $C(e^3_\alpha)=C(f^3_\alpha)$ for all α (recall that e^3_α is either 0 or one of the $h_\alpha e_\iota$), therefore

$$C(e^3)=C(\sup_\alpha e^3_\alpha)=\sup_\alpha C(e^3_\alpha)=\sup_\alpha C(f^3_\alpha)=C(f^3).$$

Thus (15) is verified (but the equivalence has nothing to do with the w_ι). On the other hand, since A has no abelian summand, the abelian projection f^3 satisfies $f^3 \precsim 1-f^3$ (see the remark at the end of the proof of [§ 18, Prop. 6]); combining this with (15), we have

$$(16) \qquad e^3 \precsim 1-f^3 .$$

The equivalence in (16) has nothing to do with the w_ι, but that does not matter; noting the formulas (11), (12) with $t=3$, it follows from Proposition 2 that the partial isometries $w_{\alpha\iota}$ ($(\alpha,\iota)\in J^3$) are addable. ∎

The abelian case is half trivial, half pathological:

Proposition 6. *If A is abelian then $e=f$, but the partial isometries w_ι may fail to be addable.*

Proof. Since all projections in A are central, $e_\iota \sim f_\iota$ reduces to $e_\iota = f_\iota$, thus

$$e=\sup e_\iota = \sup f_\iota = f .$$

An example of nonaddability is given in Exercise 4. ∎

An abelian AW^*-algebra is commutative (Remark 2 following [§ 15, Def. 2]), and addability is rescued by spectral theory:

Proposition 7. *If A is a commutative AW^*-algebra, then the partial isometries w_ι are addable.*

Proof. We know that $w_\iota^* w_\iota = w_\iota w_\iota^* = e_\iota = f_\iota$. Adjoining $e_0 = f_0 = w_0 = 1-e$ to the family, we can suppose $e=f=1$. We seek a unitary element w such that $we_\iota = w_\iota$ for all ι. Since $\|w_\iota\| \le 1$ for all $\iota \in I$, the existence of w follows from the fact that A is the C^*-sum of the $e_\iota A$ [§ 10, Prop. 2]. ∎

Let us gather up all the threads:

Theorem 1. *Equivalence is additive in a Baer $*$-ring A with* GC. *Explicitly, let each of $(e_\iota)_{\iota\in I}$, $(f_\iota)_{\iota\in I}$ be an orthogonal family of projections, such that $e_\iota \sim f_\iota$ for all $\iota \in I$, let $e = \sup e_\iota$, $f = \sup f_\iota$ and, for each $\iota \in I$, let w_ι be a partial isometry such that $w_\iota^* w_\iota = e_\iota$, $w_\iota w_\iota^* = f_\iota$. Then:*

(i) $e \sim f$.

(ii) *If, moreover, A has no abelian direct summand, then the partial isometries w_ι are addable.*

(iii) *If A is any AW^*-algebra, then the partial isometries w_ι are addable.*

Proof. The idea is to break up A into a finite number of manageable direct summands, using Lemma 1 to put the summands back together. The needed central projections are provided by successive applications of the coarse structure theory [§ 15, Th. 1], as follows.

Let h_1 be a central projection such that

(1) $$h_1 A \text{ is properly infinite}$$

and $(1-h_1)A$ is finite. Let $h_2 \le 1-h_1$ be a central projection such that

(2) $$h_2 A \text{ is continuous}$$

(and finite) and $(1-h_1-h_2)A$ is Type I and finite. Let $h_3 \le 1-h_1-h_2$ be a central projection such that

(3) $$h_3 A \text{ is Type I, finite, with no abelian summand,}$$

and such that, on setting $h_4 = 1-h_1-h_2-h_3$,

(4) $$h_4 A \text{ is abelian}.$$

The summands (1), (2), (3) are covered by Propositions 3, 4, 5, respectively; in these cases, the partial isometries w_ι are addable, which proves (ii). When the summand (4) is present we may apply Proposition 6 to it, concluding only that $e \sim f$, and this proves (i); but in the AW^* case, Proposition 7 is available, so the partial isometries are addable on the summand (4) also, and this proves (iii). ∎

An application (see also Exercise 6):

Corollary. *Parts* (i) *and* (ii) *of Theorem 1 hold when A is a Baer $*$-ring satisfying the* (EP)*-axiom and the* (UPSR)*-axiom.*

Proof. A has GC [§ 14, Th. 1]. ∎

The following result is a remarkable application of Theorem 1 to the comparative anatomy of the axioms (see also Exercise 12):

Theorem 2. *The following conditions on a Baer $*$-ring A are equivalent:* (a) *A has* GC; (b) *A has* PC *and equivalence in A is additive.*

Proof. (a) implies (b): Theorem 1 and [§ 14, Prop. 2].
(b) implies (a): This is [§ 14, Prop. 4]. ∎

Another useful consequence of Theorem 1(see also Exercise 2):

Theorem 3. *If A is a Baer $*$-ring with* GC *satisfying the* (WEP)-*axiom, then* $\mathrm{LP}(x) \sim \mathrm{RP}(x)$ *for all x in A.*

Proof. {We remark that in a von Neumann algebra, the conclusion comes free of charge with polar decomposition.}

We can suppose $x \neq 0$. Write $e = \mathrm{RP}(x)$, $f = \mathrm{LP}(x)$. Let $(e_\iota)_{\iota \in I}$ be a maximal family of orthogonal, nonzero projections such that, for each $\iota \in I$, there exists $y_\iota \in \{x^* x\}''$ with $x^* x y_\iota^* y_\iota = e_\iota$ (the routine Zorn's lemma argument is launched by the (WEP)-axiom). Obviously e_ι also belongs to the commutative ring $\{x^* x\}''$; replacing y_ι by $y_\iota e_\iota$, we can suppose that $y_\iota e_\iota = y_\iota$, that is, $\mathrm{RP}(y_\iota) \leq e_\iota$. Since $e_\iota \leq \mathrm{RP}(y_\iota)$ results from $x^* x y_\iota^* y_\iota = e_\iota$, we conclude that $\mathrm{RP}(y_\iota) = e_\iota$.

From $e x^* = x^*$, we see that $e_\iota \leq e$ for all ι. We assert that $\sup e_\iota = e$. Write $g = \sup e_\iota$; at any rate, $g \leq e$, and $g \in \{x^* x\}''$ [§ 4, Prop. 7]. To show that $e \leq g$ it will suffice to show that $xg = x$. Assume to the contrary that $x(1-g) \neq 0$. By the (WEP)-axiom, there exists

$$y \in \{[x(1-g)]^* [x(1-g)]\}'' = \{(1-g) x^* x\}'' \subset \{x^* x\}''$$

such that $(1-g) x^* x y^* y = h$, h a nonzero projection. Then $gh = 0$, thus $e_\iota h = 0$ for all $\iota \in I$, contradicting maximality.

Define $w_\iota = x y_\iota$ $(\iota \in I)$; then $w_\iota^* w_\iota = y_\iota^* x^* x y_\iota = x^* x y_\iota^* y_\iota = e_\iota$, thus w_ι is a partial isometry with initial projection e_ι. Let $f_\iota = w_\iota w_\iota^* = x y_\iota y_\iota^* x^*$; thus $f_\iota = \mathrm{LP}(w_\iota) \leq \mathrm{LP}(x) = f$ for all ι. The f_ι are mutually orthogonal: if $\iota \neq \varkappa$ then

$$\begin{aligned} f_\iota f_\varkappa &= (x y_\iota y_\iota^* x^*)(x y_\varkappa y_\varkappa^* x^*) = x(y_\iota y_\iota^*)(x^* x)(y_\varkappa y_\varkappa^*) x^* \\ &= x e_\iota (y_\varkappa y_\varkappa^*) x^* = x e_\iota (e_\varkappa y_\varkappa y_\varkappa^*) x^* = 0 . \end{aligned}$$

Quoting Theorem 1, we have $e = \sup e_\iota \sim \sup f_\iota \leq f$. {Thus $e \precsim f$, that is, $\mathrm{RP}(x) \precsim \mathrm{LP}(x)$. Applying this result to x^* yields $f \precsim e$, therefore $e \sim f$ by the Schröder-Bernstein theorem [§ 1, Th. 1]. But the following argument is tidier, and its sharper conclusion is needed in the next section.}

Finally, we assert that $\sup f_\iota = f$. Write $k = 1 - \sup f_\iota$. Since k is orthogonal to every f_ι, we have

$$0 = k f_\iota x = k(x y_\iota y_\iota^* x^*) x = k x (y_\iota^* y_\iota)(x^* x) = k x e_\iota$$

for all ι; since $\sup e_\iota = e = \mathrm{RP}(x)$, it follows that $0 = kxe = kx$, hence $kf = 0$, $f \le 1 - k = \sup f_\iota$. ∎

An application (see also Exercise 6):

Corollary. *If A is a Baer *-ring satisfying the* (EP)-*axiom and the* (UPSR)-*axiom, then* $\mathrm{LP}(x) \sim \mathrm{RP}(x)$ *for all x in A.*

Proof. A has GC [§ 14, Th. 1]. ∎

This is a convenient place to record the following terminology:

Definition 1. We say that a weakly Rickart *-ring *satisfies* $\mathrm{LP} \sim \mathrm{RP}$ if $\mathrm{LP}(x) \sim \mathrm{RP}(x)$ for every element x.

Exercises

1A. If A is a Baer *-ring such that the *-ring A_2 of all 2×2 matrices over A is also a Baer *-ring, then partial isometries in A are addable.

2A. If A is a Baer *-ring satisfying the (WEP)-axiom, then the following conditions on A are equivalent: (a) A has GC; (b) $\mathrm{LP}(x) \sim \mathrm{RP}(x)$ for all x in A; (c) A satisfies the parallelogram law (P).

3A. If A is a properly infinite Baer *-ring with PC, then partial isometries in A are addable.

4A. Let A be the *-ring of bounded complex sequences that are real at infinity [§ 4, Exer. 14]; thus A is a commutative Baer *-ring satisfying the (EP)-axiom and the (UPSR)-axiom. For $n = 1, 2, 3, \dots$ let w_n be the element of A whose nth coordinate is $i = \sqrt{-1}$, all other coordinates 0; thus $w_n^* w_n = w_n w_n^* = e_n$ has 1 in the nth coordinate, 0's elsewhere. The w_n are not addable.

5D. (i) *Problem:* Is equivalence additive in every Baer *-ring? (Cf. [§ 11, Exer. 3].) {If the answer is yes, then PC and GC are equivalent conditions (see Theorem 2).}

(ii) *Problem:* Is equivalence additive in every Baer *-ring with PC? {Same remark as in (i).} In view of Exercise 3, the question is open only for finite rings.

(iii) *Problem:* Is equivalence additive in every Baer *-ring satisfying the parallelogram law (P)?

6C. The corollaries of Theorems 1 and 3 hold with (UPSR) weakened to (SR).

7C. If A is a Baer *-ring such that (i) A satisfies the (EP)-axiom, and (ii) partial isometries in A are addable, then A also satisfies the (SR)-axiom.

8A. Let A be a Baer *-ring with GC, and let e, f be projections in A such that $C(e) \le C(f)$, eAe is orthoseparable, and fAf is properly infinite. Then $e \precsim f$.

9A. Let A be an orthoseparable Baer *-ring of Type III, with orthogonal GC.

(i) If e is any nonzero projection, there exists a nonzero central projection h such that $he \sim h$.

(ii) If e is any projection with $C(e) = 1$, then $e \sim 1$.

(iii) $e \sim C(e)$ for every projection e.

(iv) $e \sim f$ iff $C(e) = C(f)$.

(Of course (iv) implies (i)–(iii), but it is convenient to prove them in the indicated order.) The conclusion (namely, (iv)) remains true with orthoseparability replaced by the following condition: there exists an orthogonal family (h_α) of central projections with $\sup h_\alpha = 1$, such that $h_\alpha A$ is orthoseparable for each α.

10A. If A is an abelian Rickart $*$-ring, then $\mathrm{LP}(x) = \mathrm{RP}(x)$ for all $x \in A$.

11D. Let A be a Rickart C^*-algebra. *Problems:*
(i) Does $\mathrm{LP}(x) \sim \mathrm{RP}(x)$ hold for all $x \in A$?
(ii) Are partial isometries $\aleph_0$-addable in A [§ 11, Def. 1]?
(iii) Is equivalence $\aleph_0$-additive in A?
(iv) Does it help to assume GC? (Recall that (P) holds in every Rickart C^*-algebra [§ 13, Th. 1].)

12C, D. Equivalence in a Baer $*$-ring A is said to be *centrally additive* if, in the notation at the beginning of the section, $e \sim f$ whenever there exists an orthogonal family of central projections $(h_\iota)_{\iota \in I}$ such that e_ι, f_ι are $\le h_\iota$ for all $\iota \in I$.
(i) The following conditions on a Baer $*$-ring A are equivalent: (a) A has GC; (b) A has PC and equivalence is additive; (c) A has PC and equivalence is centrally additive.
(ii) *Problem:* Is equivalence centrally additive in every Baer $*$-ring? In every Baer $*$-ring with PC?

§ 21. Polar Decomposition

The concept of 'polar decomposition' (or 'canonical factorization') has several axiomatic mutants. Before pinning down a specific definition, we exhibit three results of this genre.

Proposition 1. *Let A be a Baer $*$-ring satisfying the* (WEP)*-axiom and the* (WSR)*-axiom, and assume that partial isometries in A are addable. Let $x \in A$ and choose $r \in \{x^*x\}''$ with $x^*x = r^*r$. Then:*
(i) *There exists a partial isometry w such that $x = wr$, $w^*w = \mathrm{RP}(x)$, $ww^* = \mathrm{LP}(x)$.*
(ii) *This factorization of x is uniquely determined by r, in the sense that if also $x = vr$ with $v^*v = \mathrm{RP}(x)$, then $v = w$.*
(iii) $w^*x = r$.

Proof. {Note that A has PC [§ 14, Prop. 3]. If, in addition, A satisfies the parallelogram law (P), then it has GC [§ 14, Prop. 7]; in this case the addability hypothesis is redundant when A has no abelian summand [§ 20, Th. 1].}

(i) For brevity, write $C = \{x^*x\}''$. We adopt the notation in the proof of [§ 20, Th. 3]. In particular, $x^*xy_\iota^*y_\iota = e_\iota$, where y_ι, e_ι are in C and $y_\iota e_\iota = y_\iota$.

First, we revise the y_ι slightly. Since $r^*ry_\iota^*y_\iota = e_\iota$, with all elements lying in the commutative ring C, clearly re_ι is invertible in $e_\iota C$, with inverse $z_\iota = r^*y_\iota^*y_\iota$. Then $z_\iota \in e_\iota C \subset C$ and $e_\iota = (re_\iota)z_\iota = rz_\iota$, therefore

$r^* r z_\iota^* z_\iota = (r z_\iota)^*(r z_\iota) = e_\iota$; thus z_ι has the properties of y_ι. Changing notation, we can suppose, in addition, that $r y_\iota = e_\iota$.

Write $e = \mathrm{RP}(x)$, $f = \mathrm{LP}(x)$, and set $w_\iota = x y_\iota$. As noted in the proof of [§ 20, Th. 3], $e = \sup e_\iota$ and the projections $f_\iota = w_\iota w_\iota^*$ are also orthogonal, with $\sup f_\iota = f$. By the addability hypothesis, there exists a partial isometry w such that $w^* w = e$, $w w^* = f$ and $w e_\iota = w_\iota = f_\iota w$ for all ι.

Let us show that $x = wr$. For all ι, we have

$$(x - wr)e_\iota = x e_\iota - w r e_\iota = x e_\iota - w e_\iota r$$
$$= x e_\iota - w_\iota r = x e_\iota - x y_\iota r = x e_\iota - x e_\iota = 0,$$

therefore $(x - wr)e = 0$; since $e = \mathrm{RP}(x) = \mathrm{RP}(r)$, we conclude that $x - wr = 0$.

(ii) If also $x = vr$ with $v^* v = e$, then $0 = wr - vr = (w - v)r$, therefore $0 = (w - v)e = we - ve = w - v$.

(iii) In the notation of (i), $w^* x = w^* w r = e r = r$. ∎

When unique positive square roots are available, the result is sharper (see also Exercise 11):

Proposition 2. *Let A be a Baer $*$-ring satisfying the* (EP)*-axiom and the* (UPSR)*-axiom, and assume that partial isometries in A are addable. Let $x \in A$ and let r be the unique positive square root of $x^* x$. Then:*

(i) *There exists a unique partial isometry w such that $x = wr$ and $w^* w = \mathrm{RP}(x)$.*

(ii) *In addition, $w w^* = \mathrm{LP}(x)$ and $w^* x = r$.*

(iii) *If also $x = vs$ with $s \geq 0$ and $v^* v = \mathrm{RP}(s)$, then $s = r$ and $v = w$.*

Proof. (i), (ii) Let $e = \mathrm{RP}(x)$, $f = \mathrm{LP}(x)$. By Proposition 1, we may write $x = wr$ with $w^* w = e$, $w w^* = f$, and $w^* x = r$. The uniqueness assertion follows from (ii) of Proposition 1.

(iii) Suppose also $x = vs$ with the indicated properties. Then $x^* x = s v^* v s = s \mathrm{RP}(s) s = s^2$, therefore $s = r$ by the uniqueness of positive square roots. Then $v^* v = \mathrm{RP}(s) = \mathrm{RP}(r) = \mathrm{RP}(x)$, therefore $v = w$ by the uniqueness assertion of (i). ∎

The next result is elementary, but instructive as to the algebra of such factorizations:

Proposition 3. *Let A be a weakly Rickart $*$-ring, let $x \in A$, and suppose there exists an element $r \in A$ such that*

$$x \in Ar, \qquad x^* x = r^* r \quad \textit{and} \quad \mathrm{LP}(r) = \mathrm{RP}(r).$$

Then one can write $x = wr$ with $w^* w = \mathrm{RP}(x)$ and $w w^* = \mathrm{LP}(x)$; the partial isometry w is uniquely determined by r, and one has $w^* x = r$.

Proof. Let $e=\mathrm{RP}(x)$, $f=\mathrm{LP}(x)$. Thus $e=\mathrm{RP}(x)=\mathrm{RP}(x^*x)=\mathrm{RP}(r^*r)$ $=\mathrm{RP}(r)=\mathrm{LP}(r)$.

Say $x=ar$. Then $x=a(er)=(ae)r$; setting $w=ae$, we have $x=wr$ with $we=w$.

We assert that $w^*x=r$. Indeed,

$$r^*(w^*x-r)=(wr)^*x-r^*r=x^*x-r^*r=0,$$

therefore $0=e(w^*x-r)=(we)^*x-er=w^*x-r$.

Next, $w^*w=e$. For, $0=w^*x-r=w^*(wr)-er=(w^*w-e)r$, therefore $0=(w^*w-e)e=w^*w-e$.

Since the involution of A is proper [§ 5, Prop. 3], it follows that w is a partial isometry [§ 2, Prop. 2], thus $g=ww^*$ is a projection; we now show that $g=f$. Since $f=\mathrm{LP}(x)=\mathrm{LP}(wr)\le\mathrm{LP}(w)=g$, $g-f$ is a projection; moreover, $(g-f)wr=gx-fx=x-x=0$, therefore $0=(g-f)we$ $=(g-f)w$; thus $0=(g-f)g=g-f$.

If also $x=vr$ with $v^*v=e$, then $v=w$ as in the proof of Proposition 1. ∎

Motivated by Propositions 1–3, and the fact that $\mathrm{LP}(r)=\mathrm{RP}(r)$ for every normal element r in a weakly Rickart $*$-ring, we define 'polar decomposition' as follows:

Definition 1. Let A be a $*$-ring. We say that A has *polar decomposition* (PD) if, for each $x\in A$, there exists an element r such that

$$r\in\{x^*x\}'',\qquad r^*=r,\qquad x^*x=r^2\quad\text{and}\quad x\in Ar$$

(in particular, A satisfies the (SR)-axiom). We say that A has *weak polar decomposition* (WPD) if, for each $x\in A$, there exists an element r such that

$$r\in\{x^*x\}'',\qquad x^*x=r^*r\quad\text{and}\quad x\in Ar$$

(in particular, A satisfies the (WSR)-axiom). We may also speak of individual elements x having PD or WPD, even when not all elements of A do.

Even weak polar decomposition is a strong hypothesis:

Proposition 4. *If A is a weakly Rickart $*$-ring with* WPD, *then* (i) *A satisfies* $\mathrm{LP}\sim\mathrm{RP}$, *and therefore* (ii) *$A$ satisfies the parallelogram law* (P). (iii) *If, in addition, A is a Baer $*$-ring, then A has* GC.

Proof. (i) The r of Definition 1 is normal ($r^*r=rr^*$), therefore $\mathrm{LP}(r)=\mathrm{RP}(r)$; quote Proposition 3.

(ii) See [§ 13, Prop. 2].

(iii) See [§ 14, Cor. 2 of Prop. 7]. ∎

Exercises

1C. If A is a Baer $*$-ring satisfying the (EP)-axiom and the (SR)-axiom, and if A has no abelian summand, then A has PD.

2A. Every commutative $*$-ring with unity trivially has WPD.

3A. The $*$-ring of bounded, complex sequences that are real at infinity [cf. § 20, Exer. 4] has WPD but not PD.

4A. Let A be the C^*-algebra of all compact operators on a separable, infinite-dimensional Hilbert space, and let $A_1 = \{a + \lambda 1 : a \in A,\ \lambda \text{ complex}\}$ as in [§ 3, Example 1]. Then A_1 satisfies the (UPSR)-axiom, but there exist elements $a \in A$ such that $a \notin A_1(a^*a)^{\frac{1}{2}}$. In fact, A_1 does not have WPD.

5A. In a $*$-ring with PD, every ideal is self-adjoint (i.e., is a $*$-ideal).

6A. If A is a $*$-ring with unity, satisfying the (WSR)-axiom, then every invertible element of A has a WPD.

7A. In a finite Rickart $*$-ring with WPD, $yx = 1$ implies $xy = 1$.

8A. Suppose A is a Baer $*$-ring with WPD. Let $x \in A$ and choose $r \in \{x^*x\}''$ with $x \in Ar$ and $x^*x = r^*r$. One can write $x = ur$ with u an extremal partial isometry [cf. § 14, Exer. 19].

9A. Let A be an AW^*-algebra, $x \in A$, $\|x\| \le 1$. There exist extremal partial isometries u', u'' such that $x = \frac{1}{2}u' + \frac{1}{2}u''$.

10D. *Problem:* Does every Rickart C^*-algebra have PD? (Cf. [§ 20, Exer. 11].)

11C. If A is a Baer $*$-ring such that (i) A satisfies the (EP)-axiom, and (ii) partial isometries in A are addable, then A has PD.

Chapter 5

Ideals and Projections

§ 22. Ideals and p-Ideals

The dominant theme of this book is the interplay, in a $*$-ring A, between the ring structure and the set $\tilde{A}$ of projections of A. For the ring structure, the ideals of A are subsets of central importance; how may the corresponding subsets of $\tilde{A}$ be characterized? This is the question treated in the present section.

Throughout this section, A denotes a weakly Rickart $*$-ring, with additional hypotheses as needed; thus $\tilde{A}$ is a lattice [§ 5, Prop. 7].

A frequent hypothesis is the parallelogram law [§ 13]:

(P) $\qquad e-e\cap f\sim e\cup f-f$ for all projections e, f;

this holds, for example, in every weakly Rickart C^*-algebra [§ 13, Th. 1].

Usually, the following stronger condition is needed:

(LP $\sim$ RP) $\qquad \mathrm{LP}(x)\sim\mathrm{RP}(x)$ for all x in A.

This condition implies (P) [§ 13, Prop. 2]; it holds if A has WPD [§ 21, Prop. 4], or if A is a Baer $*$-ring satisfying the (EP)-axiom and the (SR)-axiom [§ 20, Exer. 6]. {Whether the condition holds in a Rickart C^*-algebra is not known; its validity would simplify some of the results in this section.} A Baer $*$-ring satisfying LP $\sim$ RP has GC [§ 14, Cor. 2 of Prop. 7].

We remark that if A has PD, then every ideal of A is a $*$-ideal; this is the case, for example, when A is an AW^*-algebra ([§ 20, Th. 1], [§ 21, Prop. 2]), or when A is a Baer $*$-ring satisfying the (EP)-axiom and the (SR)-axiom and having no abelian summand [§ 21, Exer. 1].

Proposition 1. *Assume A satisfies the parallelogram law* (P). *If I is any ideal in A, the set $\tilde{I}$ of projections in I has the following properties:*

(i) *if $e\in\tilde{I}$ and $g\le e$, then $g\in\tilde{I}$;*

(ii) *if $e\in\tilde{I}$ and $g\sim e$, then $g\in\tilde{I}$;*

(iii) *if $e, f\in\tilde{I}$ then $e\cup f\in\tilde{I}$.*

Proof. (i) $g=ge$.

(ii) If w is a partial isometry implementing $g \sim e$, then $w = ew \in IA \subset I$, therefore $g = w^* w \in AI \subset I$.

(iii) Suppose $e, f \in \tilde{I}$. By (i), $e - e \cap f \in I$; then $e \cup f - f \sim e - e \cap f$ yields $e \cup f - f \in I$ by (ii), thus $e \cup f = (e \cup f - f) + f \in I$. ∎

Definition 1. A nonempty subset $\mathfrak{p}$ of $\tilde{A}$ is called a *p-ideal* of A if it satisfies conditions (i)–(iii) of Proposition 1. A p-ideal is *proper* if it is a proper subset of $\tilde{A}$. {Note that conditions (i) and (ii) can be combined into one: if $e \in \mathfrak{p}$ and $g \lesssim e$, then $g \in \mathfrak{p}$. In fact, all three conditions can be combined into one: if $e, f \in \mathfrak{p}$ and $g \lesssim e \cup f$, then $g \in \mathfrak{p}$.}

Proposition 1 asserts that the projections in an ideal form a p-ideal. Proposition 2 is a converse. The following elementary lemma, already used implicitly on several occasions, is worth setting down explicitly in the present context:

Lemma. *For all* x, y *in* A, $\mathrm{RP}(x+y) \le \mathrm{RP}(x) \cup \mathrm{RP}(y)$ *and* $\mathrm{RP}(xy) \le \mathrm{RP}(y)$.

Proposition 2. *Let* $\mathfrak{p}$ *be a p-ideal of* A. *Write* $A\mathfrak{p}$ *for the set of all finite sums* $\sum a_i e_i$ *with* $a_i \in A$ *and* $e_i \in \mathfrak{p}$, *that is,* $A\mathfrak{p}$ *is the left ideal generated by* $\mathfrak{p}$. *Then*

$$A\mathfrak{p} = \{x \in A : \mathrm{RP}(x) \in \mathfrak{p}\},$$

therefore $(A\mathfrak{p})^{\sim} = \mathfrak{p}$.

If A *satisfies* $\mathrm{LP} \sim \mathrm{RP}$, *then* $A\mathfrak{p} = \mathfrak{p}A$ *is the ideal generated by* $\mathfrak{p}$, *and is a* $*$*-ideal.*

Proof. If $\mathrm{RP}(x) \in \mathfrak{p}$, then $x = x\mathrm{RP}(x) \in A\mathfrak{p}$. Conversely, if $x \in A\mathfrak{p}$, say $x = \sum_1^n a_i e_i$, then, citing the lemma, we have

$$\mathrm{RP}(x) \le \mathrm{RP}(a_1 e_1) \cup \cdots \cup \mathrm{RP}(a_n e_n) \le e_1 \cup \cdots \cup e_n \in \mathfrak{p},$$

therefore $\mathrm{RP}(x) \in \mathfrak{p}$. This establishes the formula for $A\mathfrak{p}$, from which it is immediate that $A\mathfrak{p}$ and $\mathfrak{p}$ contain the same projections.

Dually, $\{x \in A : \mathrm{LP}(x) \in \mathfrak{p}\} = \mathfrak{p}A$, the right ideal generated by $\mathfrak{p}$. When A satisfies $\mathrm{LP} \sim \mathrm{RP}$, it follows that $\mathrm{RP}(x) \in \mathfrak{p}$ iff $\mathrm{LP}(x) \in \mathfrak{p}$, thus $A\mathfrak{p} = \mathfrak{p}A$ and is obviously a $*$-ideal. ∎

Suppose $\mathfrak{p}$ is a nonempty subset of A, and J is the ideal generated by $\mathfrak{p}$. Then $J = A\mathfrak{p}A$, the set of all finite sums $\sum a_i x_i b_i$ with $x_i \in \mathfrak{p}$ and $a_i, b_i \in A$ (the essential point is that $A\mathfrak{p}A \supset \mathfrak{p}$ because $x = \mathrm{LP}(x)x\mathrm{RP}(x)$). If $\mathfrak{p}$ is a $*$-subset, then J is a $*$-ideal. If $\mathfrak{p}$ is a set of projections, it is obvious that J coincides with the ideal generated by its projections; such ideals have a name:

Definition 2. An ideal J of A is said to be *restricted* if the ideal generated by $\tilde{J}$ coincides with J, that is, $J = A\tilde{J}A$ (= the set of all finite sums $\sum a_i e_i b_i$ with $e_i \in \tilde{J}$ and $a_i, b_i \in A$). {As noted above, J is necessarily a $*$-ideal.}

When A satisfies $\mathrm{LP} \sim \mathrm{RP}$, its restricted ideals may be characterized as follows:

Proposition 3. *Assume A satisfies $\mathrm{LP} \sim \mathrm{RP}$. The following conditions on an ideal J of A are equivalent:*

(a) *J is restricted;*

(b) *$x \in J$ implies $\mathrm{RP}(x) \in J$;*

(b′) *$x \in J$ implies $\mathrm{LP}(x) \in J$;*

(c) *J is the left ideal generated by $\tilde{J}$;*

(c′) *J is the right ideal generated by $\tilde{J}$.*

Proof. Write $\mathfrak{p} = \tilde{J}$. By Proposition 1, $\mathfrak{p}$ is a p-ideal; moreover,

$$A\mathfrak{p}A = A\mathfrak{p} = \mathfrak{p}A$$

by Proposition 2, thus (a), (c) and (c′) are obviously equivalent.

(a) implies (b): Assuming (a), we have $J = A\mathfrak{p}$ by the foregoing, therefore

$$J = \{x \in A : \mathrm{RP}(x) \in J\}$$

by Proposition 2.

(b) implies (a): If $x \in J$ then, by hypothesis, $\mathrm{RP}(x) \in \mathfrak{p}$, therefore $x = x\mathrm{RP}(x)$ is in the ideal generated by $\mathfrak{p}$.

Thus (a) and (b) are equivalent; dually, (a) and (b′) are equivalent. ∎

Combining the foregoing results:

Theorem 1. *Let A be a weakly Rickart $*$-ring satisfying $\mathrm{LP} \sim \mathrm{RP}$. For each p-ideal $\mathfrak{p}$, write $J(\mathfrak{p}) = \{x \in A : \mathrm{RP}(x) \in \mathfrak{p}\}$. Then the correspondences*

$$\mathfrak{p} \mapsto J(\mathfrak{p}),$$
$$J \mapsto \tilde{J}$$

define mutually inverse bijections between the set of all p-ideals $\mathfrak{p}$ and the set of all restricted ideals J (necessarily $*$-ideals). The correspondences preserve inclusion and, in particular, maximality.

Proof. If $\mathfrak{p}$ is a p-ideal, then, by Proposition 2, $J(\mathfrak{p})$ is a $*$-ideal with $(J(\mathfrak{p}))^{\sim} = \mathfrak{p}$, and is generated by its projections; thus $J(\mathfrak{p})$ is a restricted ideal with $(J(\mathfrak{p}))^{\sim} = \mathfrak{p}$.

Conversely, if J is a restricted ideal, write $\mathfrak{p} = \tilde{J}$. By Proposition 1, $\mathfrak{p}$ is a p-ideal. By Proposition 2, the ideal generated by $\mathfrak{p}$ is $J(\mathfrak{p})$; since J is restricted, this means $J = J(\mathfrak{p})$. ∎

Let us reconsider Proposition 2 in the context of weakly Rickart C^*-algebras; whether LP $\sim$ RP in such an algebra is not known, but the closure operation provides a remedy if one is willing to limit attention to closed ideals:

Proposition 4. *Let A be a weakly Rickart C^*-algebra, let $\mathfrak{p}$ be a p-ideal in A, and let I be the closed ideal generated by $\mathfrak{p}$. Then*

$$(*) \qquad I=\overline{A\mathfrak{p}}=\overline{\mathfrak{p}A}$$

and $\tilde{I}=\mathfrak{p}$.

Proof. By definition, $I=(A\mathfrak{p}A)^-$ (the bar denotes closure); to verify (*), it will suffice to prove the second equality, i.e., that the closed left ideal generated by $\mathfrak{p}$ coincides with the closed right ideal generated by $\mathfrak{p}$. Given $x\in A\mathfrak{p}$, it is sufficient to show that $x\in(\mathfrak{p}A)^-$, that is, $x^*\in(A\mathfrak{p})^-$ (recall that the involution is continuous). Given any $\varepsilon>0$, we seek $z\in A\mathfrak{p}$ such that $\|x^*-z\|<\varepsilon$. Write $x^*x=r^2$ with $r\geq 0$, $r\in\{x^*x\}''$. Choose $y\in\{x^*x\}''$, $y\geq 0$, such that (i) $(x^*x)y^2=e$, e a projection, and (ii) $\|x^*x-(x^*x)e\|<\varepsilon^2$ [§ 9, Cor. of Prop. 4]. From (i) we have $r^2y^2=e$, therefore $ry=e$ (by commutativity and the uniqueness of positive square roots). Since $e=y^2x^*x$ and $x\in A\mathfrak{p}$, we have $e\in A\mathfrak{p}$, thus $e\in\mathfrak{p}$ by Proposition 2. Set $w=xy$. Then $w^*w=yx^*xy$ $=x^*xy^2=e$; writing $f=ww^*$, we have $f\sim e\in\mathfrak{p}$, therefore $f\in\mathfrak{p}$. Then $w^*=w^*f\in A\mathfrak{p}$. Since, by a straightforward calculation,

$$(x-wr)^*(x-wr)=x^*x-(x^*x)e,$$

we have $\|x-wr\|^2=\|x^*x-(x^*x)e\|<\varepsilon^2$, therefore

$$\|x^*-rw^*\|<\varepsilon;$$

writing $z=rw^*$, we have $z\in r(A\mathfrak{p})\subset A\mathfrak{p}$, thus the proof of (*) is complete.

{If A is an AW^*-algebra, the result may be sharpened: A satisfies LP $\sim$ RP [§20, Cor. of Th. 3], therefore $A\mathfrak{p}=\mathfrak{p}A$ by Proposition 2.}

It remains to show that I contains no new projections; that is, assuming $g\in\tilde{I}$, it is to be shown that $g\in\mathfrak{p}$. By (*), there exists a sequence $x_n\in A\mathfrak{p}$ such that $x_n\to g$. Then $x_n^*x_n\in A\mathfrak{p}$ and $gx_n^*x_ng\to g$. Let n be an index such that $\|gx_n^*x_ng-g\|<1$; setting $y=x_n^*x_n$, we have

$$\|gyg-g\|<1, \qquad y\in A\mathfrak{p}, \qquad y\geq 0.$$

It follows that gyg is an invertible positive element of gAg; since its inverse is also positive, there exists $s\in gAg$, $s\geq 0$, such that $(gyg)s^2=g$. Calculating commutants in gAg, we have $s^2\in\{gyg\}''$ [§ 3, Prop. 9, (6)] and $s\in\{s^2\}''$, thus

$$s\in\{s^2\}''\subset\{gyg\}'';$$

in particular, s commutes with gyg, thus

$$g = s(gyg)s = sys.$$

Setting $w = y^{\frac{1}{2}}s$, we have $w^*w = g$. Writing $f = ww^*$, we have

$$f = \mathrm{LP}(w) = \mathrm{LP}(y^{\frac{1}{2}}s) \le \mathrm{LP}(y^{\frac{1}{2}}) = \mathrm{RP}(y^{\frac{1}{2}}) = \mathrm{RP}(y) \in \mathfrak{p},$$

therefore $f \in \mathfrak{p}$; then $g \sim f \in \mathfrak{p}$ yields $g \in \mathfrak{p}$. ∎

In a sense, closed ideals are the topological analogue of restricted ideals:

Proposition 5. *If I is a closed ideal in a weakly Rickart C^*-algebra A, then I is the closed ideal generated by its projections (in particular, I is a $*$-ideal).*

Proof. Since A satisfies the parallelogram law (P) [§13, Th. 1], it follows from Proposition 1 that $\tilde{I}$ is a p-ideal.

Let $x \in I$. Given any $\varepsilon > 0$, it will suffice to produce an element x' in the ideal $A\tilde{I}A$ generated by $\tilde{I}$, such that $\|x - x'\| < \varepsilon$. Choose $y \in \{x^*x\}''$ with $y \ge 0$, $x^*xy^2 = e$ a projection, $\|x^*x - (x^*x)e\| < \varepsilon^2$ [§9, Cor. of Prop. 4]. Write $x^*x = r^2$ with $r \in \{x^*x\}''$, $r \ge 0$; setting $w = xy$, we have $w^*w = e$. Let $f = ww^* = \mathrm{LP}(w)$. Since $e = y^2x^*x \in AI \subset I$, we have $f \sim e \in \tilde{I}$, therefore $f \in \tilde{I}$; then $fw = w$ shows that $wr = fwr$ is in the ideal generated by $\tilde{I}$, and, setting $x' = wr$, we have $\|x - x'\| < \varepsilon$ as in the proof of Proposition 4.

{It is true, more generally, that every closed ideal in a C^*-algebra is a $*$-ideal [cf. **24**, §1, Prop. 1.8.2]. In an AW^*-algebra, it follows immediately from polar decomposition that every ideal is a $*$-ideal: in the notation of [§21, Prop. 2], $x^* = w^*xw^*$. The above argument uses a sort of 'approximate polar decomposition'. We remark that I is in fact the closed linear span of its projections (Exercise 7).} ∎

Combining Propositions 4 and 5, we get a topological analogue of Theorem 1:

Theorem 2. *If A is a weakly Rickart C^*-algebra, then the correspondences*

$$\mathfrak{p} \mapsto \overline{A\mathfrak{p}} = \overline{\mathfrak{p}A},$$

$$I \mapsto \tilde{I}$$

define mutually inverse bijections between the set of all p-ideals $\mathfrak{p}$ and the set of all closed ideals I (in particular, the latter are necessarily $*$-ideals). The correspondences preserve inclusion and, in particular, maximality.

Proof. If $\mathfrak{p}$ is a p-ideal, then $(A\mathfrak{p})^- = (\mathfrak{p}A)^-$ is a closed $*$-ideal I with $\tilde{I} = \mathfrak{p}$ (Proposition 4).

Conversely, suppose I is a closed ideal. Set $\mathfrak{p}=\tilde{I}$; as noted above, $\mathfrak{p}$ is a p-ideal. By Proposition 4, $(A\mathfrak{p})^{-}=(\mathfrak{p}A)^{-}$ is the closed ideal generated by $\mathfrak{p}$, thus it coincides with I by Proposition 5. ∎

Exercises

1A. The condition $\mathrm{LP}\sim\mathrm{RP}$ in Theorem 1 can be dispensed with by limiting the classes of ideals and p-ideals that are paired, as follows. Let A be any weakly Rickart $*$-ring.

(a) *A strict ideal* of A is an ideal I such that $x\in I$ implies $\mathrm{RP}(x)\in I$. If I is a strict ideal of A, then (i) $g\le e\in\tilde{I}$ implies $g\in\tilde{I}$, (ii) $\mathrm{RP}(x)\in\tilde{I}$ implies $\mathrm{LP}(x)\in\tilde{I}$, (iii) $e,f\in\tilde{I}$ implies $e\cup f\in\tilde{I}$, and (iv) I is a $*$-ideal. One has $I=\{x\in A:\mathrm{RP}(x)\in\tilde{I}\}$. A strict ideal is a restricted ideal.

(b) Let $\mathfrak{p}$ be a nonempty set of projections in A satisfying the conditions (i) $g\le e\in\mathfrak{p}$ implies $g\in\mathfrak{p}$, (ii) $\mathrm{RP}(x)\in\mathfrak{p}$ implies $\mathrm{LP}(x)\in\mathfrak{p}$, and (iii) $e,f\in\mathfrak{p}$ implies $e\cup f\in\mathfrak{p}$. Since $\mathrm{LP}(w)\sim\mathrm{RP}(w)$ for any partial isometry w, clearly $\mathfrak{p}$ is a p-ideal. Call such a set a *strict p-ideal*. If $\mathfrak{p}$ is a strict p-ideal, then

$$A\mathfrak{p}=\mathfrak{p}A=\{x\in A:\mathrm{RP}(x)\in\mathfrak{p}\}$$

is a strict ideal I such that $\tilde{I}=\mathfrak{p}$.

(c) The correspondences $I\mapsto\tilde{I}$ and $\mathfrak{p}\mapsto I$ described in (a), (b) are mutually inverse bijections between the set of all strict ideals and the set of all strict p-ideals.

(d) If A satisfies $\mathrm{LP}\sim\mathrm{RP}$, then all p-ideals and restricted ideals are strict, and (c) concides with Theorem 1.

2A. If A is a weakly Rickart $*$-ring and I is a strict ideal of A (Exercise 1), then I is also a weakly Rickart $*$-ring (with unambiguous RP's and LP's).

3A. Let A be a weakly Rickart C^*-algebra and let $\mathfrak{p}$ be a p-ideal in A that is closed under countable suprema (that is, if e_n is a sequence in $\mathfrak{p}$, then $\sup e_n\in\mathfrak{p}$). Define $I=\{x\in A:\mathrm{RP}(x)\in\mathfrak{p}\}$. Then I is a closed, restricted ideal with $\tilde{I}=\mathfrak{p}$; I is itself a weakly Rickart C^*-algebra (with unambiguous RP's and LP's). (Cf. [§3, Example 2].)

4A, C. (i) Let A be a Baer $*$-ring in which every nonzero left ideal contains a nonzero projection (a condition weaker than the (VWEP)-axiom). If L is a left ideal that contains the supremum of every orthogonal family of projections in it, then $L=Ae$ for a suitable projection e.

(ii) If A is a von Neumann algebra and L is a left ideal that is closed in the ultrastrong (or ultraweak, strong, weak) topology, then $L=Ae$ for a suitable projection e.

5A. Let A be a Rickart $*$-ring, A° its reduced ring [§3, Exer. 18]. If I is an ideal in A, write $I^\circ=I\cap A^\circ$.

(i) Although A° is generated by its projections (as a ring), it does not follow that every ideal of A° is restricted.

(ii) If I is an ideal of A, then I° is an ideal of A°.

(iii) If A satisfies $e\sim f$ iff $e\overset{*}{\sim}f$ [cf. §17, Exer. 20], then A and A° have the same p-ideals; if, in addition, A satisfies $\mathrm{LP}\sim\mathrm{RP}$, then the correspondence $I\mapsto I^\circ$ pairs bijectively the restricted ideals of A and A°. (Application: A any AW^*-algebra.)

6A. If A is a Banach algebra and I is an ideal in A, then $\bar{I}$ has no new idempotents (that is, every idempotent in $\bar{I}$ is already in I).

7A. Let A be a weakly Rickart C^*-algebra. (i) If $\mathfrak{p}$ is any p-ideal in A, then $(A\mathfrak{p})^- = (\mathfrak{p}A)^- = (\mathbb{C}\mathfrak{p})^-$ (the closed linear span of $\mathfrak{p}$). In particular, (ii) every closed ideal I is the closed linear span of its projections, that is, $I = (\mathbb{C}\tilde{I})^-$. (iii) If J is any ideal, then $\bar{J} = (\mathbb{C}\tilde{J})^-$ (and $(\bar{J})^\sim = \tilde{J}$).

8A. Let A be a Rickart $*$-ring in which any two projections e, f are comparable ($e \precsim f$ or $f \precsim e$). In particular, A is factorial.

(i) The p-ideals of A form a chain under inclusion, that is, if $\mathfrak{p}_1, \mathfrak{p}_2$ is any pair of p-ideals in A, then either $\mathfrak{p}_1 \subset \mathfrak{p}_2$ or $\mathfrak{p}_2 \subset \mathfrak{p}_1$.

(ii) If A satisfies $\mathrm{LP} \sim \mathrm{RP}$ then the restricted ideals of A form a chain under inclusion.

(iii) If A is a C^*-algebra then the closed ideals of A form a chain under inclusion, and the center of A is one-dimensional.

9A. Let A be a finite Baer $*$-ring with GC, and let $\mathfrak{p}_1, \mathfrak{p}_2$ be p-ideals in A.

(i) The set

$$\{e \in \tilde{A} : e \leq e_1 \cup e_2 \text{ for some } e_1 \in \mathfrak{p}_1, e_2 \in \mathfrak{p}_2\}$$

is the smallest p-ideal containing both $\mathfrak{p}_1$ and $\mathfrak{p}_2$.

(ii) If $\mathfrak{p}_1$ is maximal and $\mathfrak{p}_2 \not\subset \mathfrak{p}_1$, then $1 = e_1 \cup e_2$ for suitable $e_i \in \mathfrak{p}_i$.

10D. *Problem:* Does every Rickart C^*-algebra satisfy $\mathrm{LP} \sim \mathrm{RP}$?

§ 23. The Quotient Ring Modulo a Restricted Ideal

Throughout this section, A denotes a Rickart $*$-ring satisfying $\mathrm{LP} \sim \mathrm{RP}$, and I denotes a proper, restricted ideal of A [cf. §22, Prop. 3]. We study A/I, equipped with the natural quotient $*$-ring structure; the canonical mapping $A \to A/I$ is denoted $x \mapsto \tilde{x}$.

Lemma. *If* $x, y \in A$ *and* $e = \mathrm{RP}(x)$, *then* $\mathrm{RP}(xy) = \mathrm{RP}(ey)$.

Proof. It suffices to observe that xy and ey have the same right-annihilators. ∎

Proposition 1. (i) *A/I is a Rickart $*$-ring.*

(ii) $\mathrm{RP}(\tilde{x}) = (\mathrm{RP}(x))^\sim$ *and* $\mathrm{LP}(\tilde{x}) = (\mathrm{LP}(x))^\sim$ *for all* $x \in A$; *in particular, every projection in A/I has the form $\tilde{e}$, with e a projection in A.*

(iii) $(e \cup f)^\sim = \tilde{e} \cup \tilde{f}$ *and* $(e \cap f)^\sim = \tilde{e} \cap \tilde{f}$ *for all projections e, f in A.*

(iv) $e \sim f$ *implies* $\tilde{e} \sim \tilde{f}$, *and* $e \precsim f$ *implies* $\tilde{e} \precsim \tilde{f}$.

(v) *A/I satisfies* $\mathrm{LP} \sim \mathrm{RP}$.

(vi) *If e, f are projections in A such that $\tilde{e} \sim \tilde{f}$, then there exist subprojections $e_0 \leq e$, $f_0 \leq f$ such that $\tilde{e}_0 = \tilde{e}$, $\tilde{f}_0 = \tilde{f}$ and $e_0 \sim f_0$.*

Proof. Note that, since I is restricted, $\tilde{x} = 0$ iff $\mathrm{RP}(x) \in I$ [§22, Prop. 3].

(i), (ii) If $x \in A$ and $e = \mathrm{RP}(x)$, then, citing the lemma, $\tilde{x}\tilde{y} = 0$ iff $xy \in I$ iff $\mathrm{RP}(xy) \in I$ iff $\mathrm{RP}(ey) \in I$ iff $ey \in I$ iff $\tilde{e}\tilde{y} = 0$, thus the right-annihilator of $\tilde{x}$ in A/I is $(1 - \tilde{e})A/I$. This shows that A/I is a Rickart $*$-ring and that $\mathrm{RP}(\tilde{x}) = \tilde{e} = (\mathrm{RP}(x))^\sim$. If, in particular, $\tilde{x}$ is a projection, then $\tilde{x} = \mathrm{RP}(\tilde{x}) = \tilde{e}$.

(iii) This is immediate from (ii) and [§ 3, Prop. 7].

(iv) Obvious.

(v) Immediate from (ii), (iv) and the fact that A satisfies LP$\sim$RP.

(vi) By assumption, there exists $x \in A$ such that $\tilde{x}^* \tilde{x} = \tilde{e}$, $\tilde{x}\tilde{x}^* = \tilde{f}$. Then $\tilde{x} = \tilde{f}\tilde{x}\tilde{e} = (fxe)^{\sim}$; replacing x by fxe, we can suppose $fx = x = xe$. Let $e_0 = \mathrm{RP}(x)$, $f_0 = \mathrm{LP}(x)$. Then $e_0 \leq e$, $f_0 \leq f$, $e_0 \sim f_0$ and, citing (ii), we have

$$\tilde{e}_0 = (\mathrm{RP}(x))^{\sim} = \mathrm{RP}(\tilde{x}) = \tilde{e}$$

and similarly $\tilde{f}_0 = \tilde{f}$. {Warning: If w is a partial isometry implementing $e_0 \sim f_0$, it does not follow that $\tilde{w}$ implements the original equivalence $\tilde{e} \sim \tilde{f}$ (that is, $\tilde{w}$ need not equal $\tilde{x}$).} ∎

Proposition 2. (i) *If u, v are projections in A/I such that $u \leq v$, and if $v = \tilde{f}$ with f a projection in A, then there exists a projection e in A such that $u = \tilde{e}$ and $e \leq f$.*

(ii) *If u_n is an orthogonal sequence of projections in A/I, then there exists an orthogonal sequence of projections e_n in A such that $u_n = \tilde{e}_n$ for all n.*

Proof. (i) Write $u = \tilde{g}$, g a projection in A. Then $u = uv$ yields $u = (gf)^{\sim}$; setting $e = \mathrm{RP}(gf)$, we have $e \leq f$ and $u = \tilde{e}$ by Proposition 1, (ii).

(ii) Let e_1 be any projection in A with $u_1 = \tilde{e}_1$. Since $u_2 \leq 1 - u_1 = (1 - e_1)^{\sim}$, by (i) there exists a projection $e_2 \leq 1 - e_1$ such that $u_2 = \tilde{e}_2$. Since $u_3 \leq 1 - (u_1 + u_2) = (1 - e_1 - e_2)^{\sim}$, there exists a projection $e_3 \leq 1 - e_1 - e_2$ such that $u_3 = \tilde{e}_3$, etc. ∎

Proposition 3. *If A has* GC *(e.g., if A is a Baer $*$-ring), then so does A/I.*

Proof. If u, v are projections in A/I, lift them to projections e, f in A, apply GC to e, f and pass to quotients (note that if h is a central projection in A, then $\tilde{h}$ is central in A/I). For example, if A is a Baer $*$-ring, then it follows from LP$\sim$RP that A has GC [§ 14, Cor. 2 of Prop. 7]. ∎

Proposition 4. *If A is finite and has* GC, *then A/I is finite.*

Proof. If u, v are projections in A/I such that $u \sim v$, it will suffice to show that $1 - u \sim 1 - v$ [§ 17, Prop. 4, (i)]. Write $u = \tilde{e}$, $v = \tilde{f}$ with $e \sim f$ (Proposition 1). Since A has GC and is finite, it follows that $1 - e \sim 1 - f$; passing to quotients, $1 - u \sim 1 - v$. ∎

Proposition 5. *If A has* GC *(e.g., if A is a Baer $*$-ring), then every central projection in A/I has the form $\tilde{h}$ with h a central projection in A.*

Proof. Let u be a central projection in A/I. Write $u=\tilde{e}$, e a projection in A, and let h be a central projection in A such that

$$(*) \qquad h(1-e) \precsim he,$$

$$(**) \qquad (1-h)e \precsim (1-h)(1-e).$$

Passing to quotients in $(*)$, we have $\tilde{h}(1-u) \precsim \tilde{h}u$; since $\tilde{h}u$ is central, it follows that $\tilde{h}(1-u) \leq \tilde{h}u$, therefore $\tilde{h}(1-u)=0$. Similarly, $(**)$ yields $(1-\tilde{h})u=0$, thus $u=\tilde{h}$. ∎

Definition 1. We call I (or the p-ideal $\tilde{I}$) *factorial* if A/I is a factor, that is, if the only central projections in A/I are 0 and 1.

Corollary. *If A has* GC, *then the following conditions on I are equivalent:* (a) *I is factorial;* (b) *if h is any central projection in A, then either $h \in I$ or $1-h \in I$.*

Proof. (b) implies (a): If u is a central projection in A/I then, by Proposition 5, there exists a central projection h in A such that $u=\tilde{h}$; by hypothesis, $h \in I$ or $1-h \in I$, thus $u=0$ or 1.

It is obvious that (a) implies (b). ∎

Exercises

1A. Let A be any weakly Rickart $*$-ring and let I be a strict ideal of A [§ 22, Exer. 1]. Equip A/I with the natural $*$-ring structure, and write $x \mapsto \tilde{x}$ for the canonical mapping $A \to A/I$.

(i) A/I is a weakly Rickart $*$-ring.

(ii) $\mathrm{RP}(\tilde{x}) = (\mathrm{RP}(x))^{\sim}$ and $\mathrm{LP}(\tilde{x}) = (\mathrm{LP}(x))^{\sim}$ for all $x \in A$; in particular, every projection in A/I has the form $\tilde{e}$ with e a projection in A.

(iii) $(e \cup f)^{\sim} = \tilde{e} \cup \tilde{f}$ and $(e \cap f)^{\sim} = \tilde{e} \cap \tilde{f}$ for all projections e, f in A.

(iv) $e \sim f$ implies $\tilde{e} \sim \tilde{f}$, and $e \precsim f$ implies $\tilde{e} \precsim \tilde{f}$.

(v) If u, v are projections in A/I such that $u \leq v$ and if $v=\tilde{f}$, f a projection in A, then there exists a projection e in A such that $u=\tilde{e}$ and $e \leq f$.

(vi) If u, v are orthogonal projections in A/I and if $v=\tilde{f}$, f a projection in A, then there exists a projection e in A such that $u=\tilde{e}$ and e is orthogonal to f.

(vii) If u_n is an orthogonal sequence of projections in A/I, then there exists an orthogonal sequence of projections e_n in A such that $u_n=\tilde{e}_n$ for all n.

(viii) If A has GC, then every central projection in A/I has the form $\tilde{h}$ with h a central projection in A.

2B. Let A be a Rickart C^*-algebra, I a closed ideal of A, and write $x \mapsto \overline{x}$ for the canonical mapping $A \to A/I$.

(i) I is a $*$-ideal of A, and A/I is a C^*-algebra.

(ii) If u is a projection in A/I, then there exists a projection e in A such that $u=\overline{e}$.

(iii) If u, v are projections in A/I such that $u \leq v$, and if $v=\overline{f}$ with f a projection in A, then there exists a projection e in A such that $u=\overline{e}$ and $e \leq f$.

(iv) If u_n is an orthogonal sequence of projections in A/I, then there exists an orthogonal sequence of projections e_n in A such that $u_n=\overline{e}_n$ for all n.

3A. With notation as in Exercise 2, suppose, in addition, that A is an AW^*-algebra.

(i) Let u, v be projections in A/I, say $u=\bar{e}$, $v=\bar{f}$, with e, f projections in A. If $u \sim v$, then there exist subprojections $e_0 \le e$, $f_0 \le f$ and a partial isometry $w \in A$, such that $u=\bar{e}_0$, $v=\bar{f}_0$, $w^* w=e_0$, $ww^*=f_0$ and $\bar{w}$ is the given partial isometry implementing $u \sim v$.

(ii) If A is finite, then equivalent projections in A/I are unitarily equivalent, therefore A/I is finite.

4A. If A is a factorial Baer $*$-ring satisfying LP $\sim$ RP, then every restricted ideal in A is factorial.

5A. Consider the category whose objects are the weakly Rickart $*$-rings, and whose morphisms are the $*$-homomorphisms φ satisfying $\mathrm{RP}(\varphi(x))=\varphi(\mathrm{RP}(x))$.

(i) The subobjects are the $*$-subrings containing RP's. {Examples: [§ 5, Props. 5, 6].}

(ii) The kernels of morphisms are the strict ideals (Exercise 1).

6A. Consider the category whose objects are the Baer $*$-rings, and whose morphisms are the $*$-homomorphisms φ satisfying $\mathrm{RP}(\varphi(x))=\varphi(\mathrm{RP}(x))$ and $\varphi(\sup e_\iota)=\sup \varphi(e_\iota)$ for families (e_ι) of projections.

(i) The subobjects are the Baer $*$-subrings [§ 4, Def. 3]. {Examples: [§ 4, Props. 6, 7].}

(ii) The kernels of morphisms are the direct summands.

In a sense, the reduction theory [§ 44, Th. 1] mixes certain objects of this exercise with certain morphisms of the preceding one.

7A. Consider the category whose objects are the Baer $*$-rings, and whose morphisms are the $*$-homomorphisms φ satisfying $\mathrm{RP}(\varphi(x))=\varphi(\mathrm{RP}(x))$ and $\varphi(\sup e_\iota)=\sup \varphi(e_\iota)$ for orthogonal families (e_ι) of projections.

The subobjects, and kernels of morphisms, are the same as for Exercise 6 [cf. § 4, Exer. 27].

8A. Consider the category whose objects are the Baer $*$-rings satisfying the (WEP)-axiom, and whose morphisms are the $*$-homomorphisms φ such that $\varphi(\sup e_\iota)=\sup \varphi(e_\iota)$ for orthogonal families (e_ι) of projections.

(i) If A is an object, then the subobjects of A are the objects B such that B is a $*$-subring of A containing orthogonal sups (as calculated in A); it follows that B is a Baer $*$-subring of A [§ 7, Exer. 8].

(ii) The kernels of morphisms are the direct summands.

(iii) Every morphism φ satisfies $\mathrm{RP}(\varphi(x))=\varphi(\mathrm{RP}(x))$.

A subcategory: the AW^*-algebras, with same morphisms [cf. § 7, Exer. 9].

9C. Consider the category whose objects are the von Neumann algebras (on all possible Hilbert spaces), and whose morphisms are the $*$-homomorphisms φ such that $\varphi(I)=I$ (I stands for the identity operator on the respective spaces) and $\varphi(\sup E_\iota)=\sup \varphi(E_\iota)$ for orthogonal families (E_ι) of projections.

(i) If $\mathscr{A}$ is an object—say $\mathscr{A}$ is a von Neumann algebra on the Hilbert space $\mathscr{H}$—then the subobjects of $\mathscr{A}$ are the von Neumann algebras $\mathscr{B}$ on $\mathscr{H}$ such that $\mathscr{B} \subset \mathscr{A}$.

(ii) The kernels of morphisms are the direct summands.

(iii) Every morphism φ satisfies $\mathrm{RP}(\varphi(T))=\varphi(\mathrm{RP}(T))$.

(iv) The morphisms are the $*$-homomorphisms φ, with $\varphi(I)=I$, that are normal (ultrastrongly continuous, ultraweakly continuous).

(v) Every morphism φ satisfies $\varphi(\sup E_\iota)=\sup \varphi(E_\iota)$ for arbitrary families (E_ι) of projections.

With 'von Neumann algebra on $\mathscr{H}$' replaced by 'weakly closed $*$-subalgebra of $\mathscr{L}(\mathscr{H})$', the condition $\varphi(I)=I$ can be dispensed with.

§ 24. Maximal-Restricted Ideals, Weak Centrality

In this section A is a Baer $*$-ring satisfying LP$\sim$RP, and Z is the center of A.

We pass freely between restricted ideals of A and p-ideals [§ 22, Th. 1]. A *maximal-restricted* ideal of A is a proper, restricted ideal that is maximal among such ideals. Since A has GC [§ 14, Cor. 2 of Prop. 7], the results of Section 23 (except Proposition 4) are applicable to A.

Proposition 1. *If I is a maximal-restricted ideal of A, then I is factorial.*

Proof. Write $\mathscr{I}=\tilde{I}$; thus $\mathscr{I}$ is a maximal p-ideal. Assuming h is a central projection of A such that $h\notin\mathscr{I}$, it suffices to show that $1-h\in\mathscr{I}$ [§ 23, Cor. of Prop. 5]. Let

$$\mathscr{I}'=\{e\in\tilde{A}: he\in\mathscr{I}\};$$

obviously $\mathscr{I}\subset\mathscr{I}'$ and $1-h\in\mathscr{I}'$, so it will suffice to show that $\mathscr{I}=\mathscr{I}'$. In turn, it is enough (by the maximality of $\mathscr{I}$) to show that $\mathscr{I}'$ is a proper p-ideal.

If $f\precsim e\in\mathscr{I}'$, then $hf\precsim he\in\mathscr{I}$, therefore $hf\in\mathscr{I}$, thus $f\in\mathscr{I}'$. If $e, f\in\mathscr{I}'$, then $h(e\cup f)=he\cup hf\in\mathscr{I}$, thus $e\cup f\in\mathscr{I}'$. Finally $1\notin\mathscr{I}'$ (because $h\notin\mathscr{I}$), thus $\mathscr{I}'$ is proper. ∎

The next lemmas lead up to the proposition that $\mathscr{I}\mapsto\mathscr{I}\cap Z$ pairs bijectively the maximal p-ideals of A with those of Z.

Lemma 1. *Let $\mathscr{I}_1, \mathscr{I}_2$ be proper p-ideals such that $\mathscr{I}_1\cap Z\subset\mathscr{I}_2\cap Z$, and suppose that $\mathscr{I}_1$ is factorial. Then $\mathscr{I}_1\subset\mathscr{I}_2$ or $\mathscr{I}_2\subset\mathscr{I}_1$, and in either case $\mathscr{I}_1\cap Z=\mathscr{I}_2\cap Z$.*

Proof. If $\mathscr{I}_2\subset\mathscr{I}_1$ then $\mathscr{I}_2\cap Z\subset\mathscr{I}_1\cap Z$, thus $\mathscr{I}_1\cap Z=\mathscr{I}_2\cap Z$.

Suppose $\mathscr{I}_2\not\subset\mathscr{I}_1$. Fix $f\in\mathscr{I}_2$ with $f\notin\mathscr{I}_1$. We are to show that $\mathscr{I}_1\subset\mathscr{I}_2$. Suppose $e\in\mathscr{I}_1$. Let h be a central projection such that

$$(*)\qquad he\precsim hf,$$

$$(**)\qquad (1-h)f\precsim(1-h)e.$$

Since $e\in\mathscr{I}_1$, it results from $(**)$ that $(1-h)f\in\mathscr{I}_1$; since $f\notin\mathscr{I}_1$, it follows from $f=hf+(1-h)f$ that $hf\notin\mathscr{I}_1$. All the more, $h\notin\mathscr{I}_1$; since $\mathscr{I}_1$ is factorial, $1-h\in\mathscr{I}_1$ [§ 23, Cor. of Prop. 5], thus

$$1-h\in\mathscr{I}_1\cap Z\subset\mathscr{I}_2\cap Z\subset\mathscr{I}_2.$$

Then also $(1-h)e\in\mathscr{I}_2$; also, citing $(*)$, we have $he\precsim hf\leq f\in\mathscr{I}_2$, therefore $he\in\mathscr{I}_2$; thus $\mathscr{I}_2$ contains both he and $(1-h)e$, therefore their sum e. Thus $\mathscr{I}_1\subset\mathscr{I}_2$. It remains to show that $\mathscr{I}_2\cap Z\subset\mathscr{I}_1\cap Z$. Suppose

$h \in \mathscr{I}_2 \cap Z$. If, on the contrary, $h \notin \mathscr{I}_1 \cap Z$, then $1-h \in \mathscr{I}_1$ by factoriality, therefore $1-h \in \mathscr{I}_1 \cap Z \subset \mathscr{I}_2 \cap Z$; thus $\mathscr{I}_2$ contains both h and $1-h$, therefore $1 \in \mathscr{I}_2$, contrary to the hypothesis that $\mathscr{I}_2$ is proper. ∎

Lemma 2. *Let $\mathscr{I}_1, \mathscr{I}_2$ be proper p-ideals with $\mathscr{I}_1 \cap Z \subset \mathscr{I}_2 \cap Z$, and suppose that $\mathscr{I}_1$ is factorial and $\mathscr{I}_2$ is maximal. Then $\mathscr{I}_1 \subset \mathscr{I}_2$ and $\mathscr{I}_1 \cap Z = \mathscr{I}_2 \cap Z$.*

Proof. By Lemma 1, $\mathscr{I}_1 \cap Z = \mathscr{I}_2 \cap Z$ and either $\mathscr{I}_1 \subset \mathscr{I}_2$ or $\mathscr{I}_2 \subset \mathscr{I}_1$; in the latter case, $\mathscr{I}_2 = \mathscr{I}_1$ by maximality, thus $\mathscr{I}_1 \subset \mathscr{I}_2$ in both cases. ∎

The following result is known as *weak centrality* (see also Exercises 2 and 4):

Proposition 2. *If $\mathscr{I}_1, \mathscr{I}_2$ are maximal p-ideals with $\mathscr{I}_1 \cap Z \subset \mathscr{I}_2 \cap Z$, then $\mathscr{I}_1 = \mathscr{I}_2$.*

Proof. Since $\mathscr{I}_1$ is factorial (Proposition 1) and $\mathscr{I}_2$ is maximal, Lemma 2 yields $\mathscr{I}_1 \subset \mathscr{I}_2$, therefore $\mathscr{I}_1 = \mathscr{I}_2$ by maximality of $\mathscr{I}_1$. ∎

For ideals, Proposition 2 means the following:

Corollary 1. *If I_1, I_2 are maximal-restricted ideals of A with $I_1 \cap \tilde{Z} \subset I_2 \cap \tilde{Z}$, then $I_1 = I_2$.*

Proof. Writing $\mathscr{I}_\nu = \tilde{I}_\nu (\nu = 1, 2)$, the hypothesis reads $\mathscr{I}_1 \cap Z \subset \mathscr{I}_2 \cap Z$, therefore $\mathscr{I}_1 = \mathscr{I}_2$ by Proposition 2; since a restricted ideal is generated by its projections, $I_1 = I_2$. ∎

Corollary 2. *If $\mathscr{I}_1$ is a factorial p-ideal, then there exists a unique maximal p-ideal $\mathscr{I}_2$ such that $\mathscr{I}_1 \subset \mathscr{I}_2$. Moreover, $\mathscr{I}_1 \cap Z = \mathscr{I}_2 \cap Z$.*

Proof. By a straightforward application of Zorn's lemma, there exists a maximal p-ideal $\mathscr{I}_2$ such that $\mathscr{I}_1 \subset \mathscr{I}_2$. Necessarily $\mathscr{I}_2 \cap Z = \mathscr{I}_1 \cap Z$ by Lemma 2. If also $\mathscr{I}_1 \subset \mathscr{I}_2'$, $\mathscr{I}_2'$ a maximal p-ideal, then $\mathscr{I}_2' \cap Z = \mathscr{I}_1 \cap Z = \mathscr{I}_2 \cap Z$, therefore $\mathscr{I}_2' = \mathscr{I}_2$ by Proposition 2. ∎

We now relate the p-ideals of A to those of Z. In a commutative ring (or even in an abelian ring), equivalent projections are equal; in such a ring, a nonempty set of projections is a p-ideal iff (i) along with e, it contains every $g \le e$, and (ii) along with e, f, it contains $e \cup f = e + f - ef$. If $\mathscr{I}$ is a p-ideal of A, it is obvious that $\mathscr{I} \cap Z$ is a p-ideal of Z and that $\mathscr{I}$ is proper iff $\mathscr{I} \cap Z$ is proper. In the next proposition, we show that the correspondence $\mathscr{I} \mapsto \mathscr{I} \cap Z$ pairs bijectively the maximal p-ideals of A with those of Z.

Lemma. *If $\mathscr{J}$ is a proper p-ideal of Z and if $\mathscr{I} = \{e \in \tilde{A} : C(e) \in \mathscr{J}\}$, then $\mathscr{I}$ is a proper p-ideal of A such that $\mathscr{I} \cap Z = \mathscr{J}$.*

Proof. Since $C(e \cup f) = C(e) \cup C(f)$ and since $g \lesssim e$ implies $C(g) \leq C(e)$, $\mathscr{I}$ is a p-ideal. Obviously $\mathscr{I} \cap Z = \mathscr{J}$; in particular, $1 \notin \mathscr{I}$. ∎

Proposition 3. *The correspondence* $\mathscr{I} \mapsto \mathscr{I} \cap Z$ *maps the set of maximal p-ideals of A bijectively onto the set of all maximal p-ideals of Z.*

Proof. (i) Suppose $\mathscr{I}$ is a maximal p-ideal of A. Then $\mathscr{I} \cap Z$ is a proper p-ideal of Z; let us show that it is maximal. Assuming $\mathscr{J}$ is a proper p-ideal of Z with $\mathscr{I} \cap Z \subset \mathscr{J}$, it is to be shown that $\mathscr{I} \cap Z = \mathscr{J}$. By the lemma, there exists a proper p-ideal $\mathscr{I}'$ of A such that $\mathscr{I}' \cap Z = \mathscr{J}$. Let $\mathscr{I}''$ be a maximal p-ideal of A with $\mathscr{I}' \subset \mathscr{I}''$. Then $\mathscr{I}'' \cap Z \supset \mathscr{I}' \cap Z = \mathscr{J} \supset \mathscr{I} \cap Z$, therefore $\mathscr{I}'' = \mathscr{I}$ by Proposition 2, thus $\mathscr{J} = \mathscr{I} \cap Z$.

(ii) The mapping $\mathscr{I} \mapsto \mathscr{I} \cap Z$ is injective by Proposition 2.

(iii) To show that it is surjective, suppose $\mathscr{J}$ is any maximal p-ideal of Z. As argued in (i), there exists a maximal p-ideal $\mathscr{I}$ of A with $\mathscr{I} \cap Z \supset \mathscr{J}$; since $\mathscr{I} \cap Z$ is a proper p-ideal of Z, $\mathscr{I} \cap Z = \mathscr{J}$ by the maximality of $\mathscr{J}$. ∎

Proposition 3 has topological overtones; the details are as follows. The projections of Z form a complete Boolean algebra $\tilde{Z}$; let $\mathscr{X}$ be the Stone representation space of $\tilde{Z}$. Thus $\mathscr{X}$ is a Stonian space (see Section 7) whose Boolean algebra of clopen sets is isomorphic to $\tilde{Z}$. Let us identify a projection $h \in \tilde{Z}$ with the characteristic function of the clopen set in $\mathscr{X}$ to which it corresponds. If $\sigma \in \mathscr{X}$, set

$$\mathscr{J}_\sigma = \{h \in \tilde{Z} : h(\sigma) = 0\};$$

then $\sigma \mapsto \mathscr{J}_\sigma$ maps $\mathscr{X}$ bijectively onto the set of all maximal p-ideals of Z. It follows from Proposition 3 that to each $\sigma \in \mathscr{X}$ there corresponds a unique maximal p-ideal $\mathscr{I}_\sigma$ of A such that

$$\mathscr{I}_\sigma \cap Z = \mathscr{J}_\sigma,$$

and $\sigma \leftrightarrow \mathscr{I}_\sigma$ pairs $\mathscr{X}$ bijectively with the set of all maximal p-ideals of A.

This is a good place to preview Chapter 7. Suppose A is a finite Baer $*$-ring satisfying LP$\sim$RP. It is shown in Chapter 6 that there exists a 'dimension function' $e \mapsto D(e)$ on $\tilde{A}$, with values in $C(\mathscr{X})$, having various pleasant properties (see Section 25). The technical core of Chapter 7 is the proof that

$$\mathscr{I}_\sigma = \{e \in \tilde{A} : D(e)(\sigma) = 0\}$$

for each $\sigma \in \mathscr{X}$ (Section 39); writing

$$I_\sigma = \{x \in A : \mathrm{RP}(x) \in \mathscr{I}_\sigma\}$$

(this is the typical maximal-restricted ideal of A), it results that A/I_σ is a Baer $*$-ring (finite and factorial). This, together with the fact that

$$\bigcap_{\sigma \in \mathscr{X}} I_\sigma = \{0\}$$

(proved in Section 36), constitutes the 'reduction theory' of such a ring A [§ 44, Th. 1].

Exercises

1A. A factorial p-ideal need not be maximal.

2A. Let A be an AW^*-algebra with center Z. If M_1, M_2 are maximal ideals of A such that $M_1 \cap Z \subset M_2 \cap Z$, then $M_1 = M_2$. (See also Exercise 4.)

3C. Let A be a Banach algebra with unity.

(i) If M is a maximal ideal of A, then M is closed and A/M is a simple Banach algebra.

(ii) If P is a primitive ideal of A, then P is closed and A/P is a (primitive) Banach algebra; the center of A/P is one-dimensional.

(iii) If I is an ideal of A such that A/I is a semisimple ring, then I is the intersection of a family of primitive ideals, thus I is closed and A/I is a semisimple Banach algebra.

(iv) If A is a C^*-algebra, then every primitive ideal of A is a closed $*$-ideal, and every closed ideal is the intersection of a family of primitive ideals.

4C. Let A be a Rickart C^*-algebra with GC (for example, let A be any AW^*-algebra), and let Z be the center of A.

(i) If I_1, I_2 are closed, proper ideals of A such that $I_1 \cap Z \subset I_2 \cap Z$, and if 0, 1 are the only central projections of A/I_1 (as is the case when I_1 is maximal—or, more generally, when I_1 is a primitive ideal), then either $I_1 \subset I_2$ or $I_2 \subset I_1$; in either case, $I_1 \cap Z = I_2 \cap Z$.

(ii) If P is a primitive ideal and M is a maximal ideal such that $P \cap Z \subset M \cap Z$, then $P \subset M$ and $P \cap Z = M \cap Z$.

(iii) If M_1, M_2 are maximal ideals such that $M_1 \cap Z \subset M_2 \cap Z$, then $M_1 = M_2$.

(iv) If P is a primitive ideal, there exists a unique maximal ideal M such that $P \subset M$ (in other words, A/P has a unique maximal ideal). Necessarily $P \cap Z = M \cap Z$.

(v) If N is a closed, proper ideal of Z, then there exists a closed, proper ideal I of A such that $I \cap Z = N$ (in particular, $I \supset N$).

(vi) The correspondence $M \mapsto M \cap Z$ maps the set of all maximal ideals M of A bijectively onto the set of all maximal ideals of Z.

5A. Let A be a Rickart $*$-ring, with center Z, and let $\mathscr{X}$ be the Stone representation space of the Boolean algebra $\tilde{Z}$. (Thus $\mathscr{X}$ is a totally disconnected, compact space whose Boolean algebra of clopen sets is isomorphic to $\tilde{Z}$.) Identify each $h \in \tilde{Z}$ with the characteristic function of the clopen set in $\mathscr{X}$ to which it corresponds. Fix $\sigma \in \mathscr{X}$. We say that $x \in A$ *vanishes in a neighborhood of* σ if there exists $h \in \tilde{Z}$ with $h(\sigma) = 1$ and $xh = 0$. Let J be the set of all such x.

(i) J is a strict ideal of A (cf. [§ 22, Exer. 1], [§ 23, Exer. 1]).

(ii) $J \cap \tilde{Z} = \{h \in \tilde{Z} : h(\sigma) = 0\}$.

(iii) If A has GC, then J is factorial.

(iv) If A is a Baer $*$-ring, then $e \in \tilde{J}$ iff $C(e) \in \tilde{J}$.

6A. Let A be a Rickart $*$-ring, $\mathscr{I}$ a proper, strict p-ideal of A [§ 22, Exer. 1].

(i) If h is a central projection such that $h \notin \mathscr{I}$, and if $\mathscr{I}' = \{e \in \tilde{A} : he \in \mathscr{I}\}$, then $\mathscr{I}'$ is a proper, strict p-ideal such that $\mathscr{I} \subset \mathscr{I}'$ and $1 - h \in \mathscr{I}'$.

(ii) If $\mathscr{I}$ is maximal-strict (that is, maximal among proper, strict p-ideals), and if A has GC, then $\mathscr{I}$ is factorial. This generalizes Proposition 1.

Part 3: Finite Rings

Chapter 6

Dimension in Finite Baer $*$-Rings

Throughout this chapter (except in some of the exercises), A denotes a finite Baer $*$-ring with GC. For the most part, the Type I and Type II summands of A require different techniques and are treated separately. A salient feature of the exposition is that virtually all results are obtained without assuming the parallelogram law (P); it is only in the final section on modularity (Section 34) that (P) is invoked.

We write Z for the center of A. The projection lattice $\tilde{Z}$ is a complete Boolean algebra; we write $\mathscr{X}$ for its Stone representation space, and $C(\mathscr{X})$ for the algebra of continuous complex-valued functions on $\mathscr{X}$ (but only real-valued functions are needed in this chapter); thus $\mathscr{X}$ is a Stonian space, $\tilde{Z}$ may be identified with the complete Boolean algebra of clopen sets in $\mathscr{X}$ (by identifying a central projection h of A with the characteristic function of the corresponding clopen subset of $\mathscr{X}$), and $C(\mathscr{X})$ is a commutative AW^*-algebra [§ 7, Prop. 1]. In general, the rings Z and $C(\mathscr{X})$ have in common only their projection lattices (also, certain rational-linear combinations of projections may be regarded as common to the rings, as described in Section 26 below). {If Z is an AW^*-algebra—as it is when A is an AW^*-algebra—then it is the closed linear span of its projections [§ 8, Prop. 3] and we may identify Z with $C(\mathscr{X})$. Nevertheless, the considerations of the present chapter would not be materially simplified by assuming that A is a finite AW^*-algebra.}

§ 25. Statement of the Results

The central preoccupation of the chapter is with $C(\mathscr{X})$-valued functions of the following kind:

Definition 1. A *(finite) dimension function* for A is a function $e \mapsto D(e)$ defined on the set $\tilde{A}$ of projections of A, with values in $C(\mathscr{X})$, such that
(D1) $e \sim f$ implies $D(e) = D(f)$,
(D2) $D(e) \geq 0$,
(D3) $D(h) = h$ when h is central,
(D4) $ef = 0$ implies $D(e+f) = D(e) + D(f)$.

(In connection with (D3), recall that we are identifying h with the characteristic function in $C(\mathscr{X})$ corresponding to it.)

Our principal aim is to show that such a function exists, and is uniquely determined by the properties (D1)–(D4). Along the way, we develop a number of properties that are implicit in these four, of which the most striking is complete additivity (a generalization of (D4) to orthogonal families of arbitrary cardinality). Assuming, in addition, that A satisfies the parallelogram law (P), it is shown in Section 34 that the projection lattice of A is modular, and it follows from complete additivity of dimension that the lattice operations are 'infinitely distributive' in a certain sense (the official term is that they are 'continuous'); thus the projection lattice of a finite Baer $*$-ring, satisfying GC and (P), is a 'continuous geometry' in the sense of von Neumann.

The actual order of events is as follows.

§ 26. Rudimentary dimension theory for the 'simple' projections, i.e., the projections that neatly divide their central covers; proof that there exist sufficiently many simple projections to serve as building blocks for the general dimension theory.

§ 27. First consequences of (D1)–(D4); proof that the general case may be reduced to the Type I and Type II cases.

§ 28. Type I case: complete additivity and uniqueness of dimension.

§ 29. Type I case: existence of dimension.

§ 30. Type II case: dimension theory for the 'fundamental' projections.

§ 31. Type II case: existence of a completely additive dimension function.

§ 32. Type II case: uniqueness of dimension.

§ 33. Dimension in an arbitrary finite Baer $*$-ring with GC.

§ 34. Modularity, continuity of the lattice operations (assuming the parallelogram law).

§ 26. Simple Projections

For the first part of the section, no restrictions are needed on the type of A (a finite Baer $*$-ring with GC); the deeper results at the end of the section require separate discussions for Types I and II.

Proposition 1. *If e is a nonzero projection and if*

$$e=e_1+\cdots+e_m=f_1+\cdots+f_n$$

are orthogonal decompositions with $e_i \sim f_j$ for all i and j, then $m=n$.

Proof. Say $m \leq n$. By finite additivity of equivalence [§ 1, Prop. 8],

$$e=e_1+\cdots+e_m \sim f_1+\cdots+f_m \leq e\,,$$

therefore $f_1+\cdots+f_m=e$ by finiteness [§ 17, Prop. 4]; it follows that $m=n$ ($m<n$ would imply $ef_n=0$). ∎

Definition 1. Let e and f be nonzero projections. We say that f *divides* e (n times) if there exists an orthogonal decomposition

$$e=f_1+\cdots+f_n$$

with $f\sim f_i$ for $i=1,\ldots,n$. The integer n is unique by Proposition 1; it is denoted $(e:f)$. To indicate briefly that f divides e, we say that '$(e:f)$ exists'. Whenever we write $(e:f)$, it is understood that the condition 'f divides e' is either being assumed or is on the verge of being verified.

In Definition 1 it is not required (nor in general possible) that $f=f_i$ for some i, or even that $f\leq e$; divisibility is really a property of equivalence classes of projections, as the next propositions show.

Proposition 2. *If* $(e:f)$ *exists and* $f\sim g$, *then* $(e:g)$ *exists and* $(e:f)=(e:g)$.

Proof. Immediate from Definition 1. ∎

Proposition 3. *If* $(e:g)=(f:g)$ *then* $e\sim f$.

Proof. Immediate from finite additivity of equivalence. ∎

Proposition 4. *If* $(e:g)$ *exists and* $e\sim f$, *then* $(f:g)$ *exists and* $(e:g)=(f:g)$.

Proof. Let $n=(e:g)$ and suppose $e=e_1+\cdots+e_n$ with $g\sim e_i$ for all i. The equivalence $e\sim f$ induces an orthogonal decomposition $f=f_1+\cdots+f_n$ with $f_i\sim e_i\sim g$ [§ 1, Prop. 9]. ∎

Generalized comparability makes its entrance here:

Proposition 5. *If* $(e:f)=(e:g)$ *then* $f\sim g$.

Proof. Say $e=f_1+\cdots+f_n=g_1+\cdots+g_n$ with $f\sim f_i$ and $g\sim g_i$ for all i. In view of GC, there exists a central projection h such that

$$hf_1\precsim hg_1,\qquad (1-h)g_1\precsim(1-h)f_1\,.$$

For $i=1,\ldots,n$ we have

$$hf_i\sim hf_1\precsim hg_1\sim hg_i\,,$$

say $hf_i\sim g_i'\leq hg_i$; adding the latter equivalences, we have

$$he=\sum_1^n hf_i\sim\sum_1^n g_i'\leq\sum_1^n hg_i=he\,,$$

therefore $\sum_1^n g_i'=\sum_1^n hg_i$ by finiteness. It follows that $g_i'=hg_i$ for all i, thus $hf_i \sim hg_i$. Similarly $(1-h)f_i \sim (1-h)g_i$, therefore $f_i \sim g_i$. ∎

Proposition 6. *Suppose* $(e:g)$ *and* $(f:g)$ *exist.*
(i) *If* $ef=0$, *then* $(e+f:g)$ *exists and*

$$(e+f:g)=(e:g)+(f:g).$$

(ii) *If* $e \leq f, e \neq f$, *then* $(f-e:g)$ *exists and*

$$(f-e:g)=(f:g)-(e:g).$$

Proof. (i) Obvious.
(ii) Write $m=(e:g)$, $n=(f:g)$. Say

$$e=e_1+\cdots+e_m, \quad f=f_1+\cdots+f_n$$

with $e_i \sim f_j \sim g$. Necessarily $m<n$; for, $n \leq m$ would imply

$$f=f_1+\cdots+f_n \sim e_1+\cdots+e_n \leq e \leq f,$$

therefore $e=f$ by finiteness, a contradiction. Set $e'=f_1+\cdots+f_m$; then $e \sim e'$, therefore

$$f-e \sim f-e'=f_{m+1}+\cdots+f_n$$

[§17, Prop. 4, (ii)]. Citing Proposition 4, we have $(f-e:g)=(f-e':g)$ $=n-m$. ∎

Propositions 3 and 4 show that if $(e:g)$ and $(f:g)$ exist, then $(e:g)$ $=(f:g)$ iff $e \sim f$. More generally,

Proposition 7. *If* $(e:g)$ *and* $(f:g)$ *exist, then* $(e:g) \leq (f:g)$ *iff* $e \precsim f$.

Proof. If $e \precsim f$, say $e \sim e' \leq f$, then $(e:g)=(e':g) \leq (f:g)$ by Propositions 4 and 6. The converse is obvious from finite additivity of equivalence. ∎

Proposition 8. *If* $(e:f)$ *and* $(f:g)$ *exist, then* $(e:g)$ *exists and* $(e:g)$ $=(e:f)(f:g)$.

Proof. Let $m=(e:f)$, $n=(f:g)$. Say

$$e=f_1+\cdots+f_m, \quad f=g_1+\cdots+g_n$$

with $f \sim f_i$ and $g \sim g_j$ for all i and j. For each i, the equivalence $f \sim f_i$ induces an orthogonal decomposition of f_i into n projections equivalent to g, thus e is the sum of mn projections equivalent to g. ∎

Conversely:

Proposition 9. *If* $(e:f)$ *and* $(e:g)$ *exist, and if* $(e:f)$ *divides* $(e:g)$, *then* $(f:g)$ *exists and* $(e:g)=(e:f)(f:g)$.

Proof. Say $(e{:}f)=m$, $(e{:}g)=mn$. Write

$$e=\sum_{i=1}^{m}\sum_{j=1}^{n} g_{ij}$$

with $g\sim g_{ij}$. Define

$$f_i=\sum_{j=1}^{n} g_{ij} \quad (i=1,\ldots,m).$$

Since $(f_i{:}g)=n$ for all i, we have $f_1\sim\cdots\sim f_m$ by Proposition 3; then $e=f_1+\cdots+f_m$ shows that $(e{:}f_1)=m=(e{:}f)$, therefore $f_1\sim f$ by Proposition 5. Since $(f_1{:}g)=n$ and $f_1\sim f$, we infer from Proposition 4 that $(f{:}g)$ exists and is equal to n. ∎

The most important case of Definition 1 is when the 'numerator' is central:

Proposition 10. *If h is a central projection and $(h{:}e)$ exists, then $h=C(e)$ and there exists an orthogonal decomposition $h=e_1+\cdots+e_n$, $n=(h{:}e)$, with $e\sim e_i$ for all i and $e=e_1$.*

Proof. Say $h=f_1+\cdots+f_n$ with $e\sim f_i$ for all i. Then $C(e)=C(f_i)$, thus

$$h=C(h)=C(f_1\cup\cdots\cup f_n)=C(f_1)\cup\cdots\cup C(f_n)=C(e)$$

[§6, Prop. 1]. In particular, $e\le h$, therefore $e\sim f_1$ implies $h-e\sim h-f_1$ $=f_2+\cdots+f_n$; the latter equivalence induces an orthogonal decomposition of $h-e$ into $n-1$ projections equivalent to e. ∎

Definition 2. A nonzero projection e is called *simple* if it divides its central cover, that is, $(C(e){:}e)$ exists. The integer $(C(e){:}e)$ is called the *order* of e.

Suppose e is a simple projection, $h=C(e)$, $n=(h{:}e)$. Write $h=e_1+\cdots+e_n$ with $e=e_1$ and $e\sim e_i$ for all i (Proposition 10). Since hA is $*$-isomorphic to the $n\times n$ matrix ring $(eAe)_n$ [§16, Prop. 1], one can show that nh is invertible in hA [§16, Exer. 1], thus $(1/n)h$ may be defined (as the inverse of nh in hA). Actually, it is more convenient to identify h with the characteristic function in $C(\mathscr{X})$ corresponding to it in the Stone representation; then $(1/n)h$ is just a scalar multiple of the function h.

Definition 3. If e is a simple projection, say $h=C(e)$, $n=(h{:}e)$, we define

$$T(e)=\left(\frac{1}{n}\right)h$$

(the notation T alludes to the 'normalized trace' of a matrix); by the preceding remarks, $T(e)$ may be regarded as an element of Z or of

$C(\mathscr{X})$. We also define $T(0)=0$. Thus $T(h)=h$ for every central projection h.

Proposition 11. *If e is simple and $e\sim f$, then f is simple and $T(e)=T(f)$.*

Proof. Let $h=C(e)$, $n=(h:e)$. Then $C(f)=C(e)=h$ and $(h:f)=(h:e)=n$ by Proposition 2, thus $T(f)=(1/n)h=T(e)$. ∎

Conversely:

Proposition 12. *If e, f are simple and $T(e)=T(f)$, then $e\sim f$.*

Proof. Writing $h=C(e)$, $m=(h:e)$ and $k=C(f)$, $n=(k:f)$, the hypothesis is that $(1/m)h=(1/n)k$. It follows easily (for example, from the functional representation) that $h=k$ and $m=n$; thus $(h:e)=m=n=(h:f)$, therefore $e\sim f$ by Proposition 5. ∎

Proposition 13. *If e is simple, then $T(he)=hT(e)$ for every central projection h.*

Proof. If $he=0$, then $hC(e)=0$ and the desired relation reduces to $T(0)=0$. Assuming $he\neq 0$, let $k=C(e), n=(k:e)$ and write $k=e_1+\cdots+e_n$ with $e\sim e_i$ for all i; then $hk=he_1+\cdots+he_n$ with $he\sim he_i$ for all i, thus he is simple and $T(he)=(1/n)hk=hT(e)$. ∎

We conclude the section with the key results on the *existence* of simple projections. The Type I and Type II cases are treated separately.

Proposition 14. *If A is a finite Baer $*$-ring of Type* I, *with* GC, *and if e is any nonzero projection in A, then there exists an orthogonal family $(e_\iota)_{\iota\in I}$ of simple abelian projections such that $e=\sup e_\iota$.*

Proof. By an obvious exhaustion argument, it suffices to show that e has a simple abelian subprojection. We can suppose A is homogeneous [§18, Th. 2], say of order n. Write $1=e_1+\cdots+e_n$ with $e_1\sim\cdots\sim e_n$ and the e_i abelian. From $e\neq 0$ and $C(e_1)=1$ it follows that $eAe_1\neq 0$ [§6, Cor. 1 of Prop. 3]; in view of GC, there exist nonzero subprojections $f\leq e$, $f_1\leq e_1$ such that $f\sim f_1$ [§14, Prop. 2]. Let $h=C(f_1)=C(f)$. Since e_1 is abelian, $f_1=he_1$ [§ 15, Prop. 6]. Define $f_i=he_i$ for $i=2,\ldots,n$. Then

$$h=h(e_1+\cdots+e_n)=f_1+\cdots+f_n,$$

and $f\sim f_1=he_1\sim he_i=f_i$ for $i=2,\ldots,n$, thus f is simple. Moreover, $f\precsim e_1$ shows that f is abelian [§15, Prop. 7]. ∎

Definition 4. A simple projection is called *fundamental* if its order is a power of 2. (In particular, a nonzero central projection is fundamental of order 2^0.)

The concept is germane only to the continuous case:

Proposition 15. *Let A be a finite Baer $*$-ring of Type* II, *with* GC. *If e is any nonzero projection and r is any nonnegative integer, there exists a subprojection f of e such that $(e:f)=2^r$. In particular, if e is central then f is fundamental and $T(f)=2^{-r}e$.*

Proof. Immediate from [§ 19, Th. 1]. ∎

Lemma. *Let A be as in Proposition 15. Suppose*

$$1=g+\sum_1^n g_i,$$

where $g_1\sim\cdots\sim g_n$ and $g\precsim g_1$. Then g_1 has a fundamental subprojection.

Proof. Let r be a positive integer such that $2^r>n+1$, and write $m=2^r$. By Proposition 15, there exists an orthogonal decomposition

$$1=\sum_1^m d_j$$

with $d_1\sim\cdots\sim d_m$. Let h be a central projection such that

$$hd_1\precsim hg_1,\quad (1-h)g_1\precsim(1-h)d_1\,. \tag{1}$$

Say $hd_1\sim g_0\le hg_1$; it will suffice to show that hd_1 (and therefore g_0) is fundamental. Since $h=\sum_1^m hd_j$ with $hd_1\sim\cdots\sim hd_m$, it is enough to show that $h\ne 0$. Assume to the contrary that $h=0$. Then $g_1\precsim d_1$ by (1), therefore

$$g_i\precsim d_i\quad\text{for } i=1,\dots,n\,. \tag{2}$$

Also,

$$g\precsim g_1\sim g_{n+1}\precsim d_{n+1}\,. \tag{3}$$

Adding (2) and (3), we have

$$1=g+\sum_1^n g_i\precsim\sum_1^{n+1} d_i\ne 1\,,$$

contrary to finiteness. ∎

Proposition 16. *Let A be a finite Baer $*$-ring of Type* II, *with* GC. *If e is any nonzero projection, there exists an orthogonal family $(e_\iota)_{\iota\in I}$ of fundamental projections such that $e=\sup e_\iota$.*

Proof. It will suffice to show that e has a fundamental subprojection. Expand $\{e\}$ to a maximal homogeneous partition—necessarily with finitely many terms [§ 17, Prop. 1, (ii)]—say $e_1,\dots,e_n$, where $e=e_1$. Set

$f=1-\sum_1^n e_i$ and let h be a central projection such that

$$hf \precsim he, \quad (1-h)e \precsim (1-h)f.$$

Necessarily $h \neq 0$ ($h=0$ would imply $e \precsim f$, contradicting maximality). Then

$$h=hf+h(1-f)=hf+\sum_1^n he_i,$$

where $he_1 \sim \cdots \sim he_n$ and $hf \precsim he = he_1$. Applying the lemma in hA (which satisfies the same hypotheses as A), we see that he contains a projection g that is fundamental relative to hA—hence relative to A. ∎

Exercises

1A. Let A be a finite Baer $*$-ring with GC, of Type I_n. If f is any nonzero abelian projection in A, then f is simple of order n.

2A. Proposition 5 fails in the ring of 2×2 matrices over the field of three elements [cf. § 1, Exer. 18].

3A, D. Proposition 14 holds with GC weakened to PC.
Problem: Does Proposition 16?

§ 27. First Properties of a Dimension Function

In this section we develop some of the direct consequences of the defining properties (D1)–(D4) of a dimension function [§25, Def. 1] (see also Exercise 1); the deeper consequences (complete additivity and uniqueness) come later.

Proposition 1. *Let A be a finite Baer $*$-ring with* GC, *and suppose D is a dimension function for A; thus*
(D1) $e \sim f$ *implies* $D(e)=D(f)$,
(D2) $D(e) \geq 0$,
(D3) $D(h)$ *when h is central,*
(D4) $ef=0$ *implies* $D(e+f)=D(e)+D(f)$.
Then D also has the following properties:
(D5) $0 \leq D(e) \leq 1$,
(D6) $D(he)=hD(e)$ *when h is central,*
(D7) $D(f)=T(f)$ *when f is simple,*
(D8) $D(e)=0$ *iff* $e=0$,
(D9) $e \sim f$ *iff* $D(e)=D(f)$,
(D10) $e \precsim f$ *iff* $D(e) \leq D(f)$.

Proof. (D5) $0 \leq D(e) \leq D(e)+D(1-e)=D(e+(1-e))=1$. More generally, $e \leq f$ implies $D(e) \leq D(f)$; indeed, $f=e+(f-e)$ yields $D(f) = D(e)+D(f-e)$, thus $D(f)-D(e)=D(f-e) \geq 0$.

(D6) Since h commutes with e, we have $h \cup e - e = h - he$ [§1, Prop. 3]; by the preceding remark,

$$D(h \cup e) - D(e) = D(h) - D(he) = h - D(he),$$

and multiplication by h yields

$$(*) \qquad hD(h \cup e) - hD(e) = h - hD(he).$$

Since $he \le h$, we have $D(he) \le D(h) = h$, therefore $hD(he) = D(he)$ by the functional representation. Also, $h \le h \cup e$ implies $h = D(h) \le D(h \cup e) \le 1$, and multiplication by h yields $h = hD(h \cup e)$. Thus (*) simplifies to

$$h - hD(e) = h - D(he),$$

which proves (D6).

(D7) Let $h = C(f)$, $n = (h:f)$ and write $h = f_1 + \cdots + f_n$ with $f \sim f_i$ for all i. Then

$$h = D(h) = \sum_1^n D(f_i) = nD(f),$$

thus $D(f) = (1/n)h = T(f)$.

(D8) If $e \ne 0$ there exists a simple projection f such that $f \le e$ [§26, Props. 14 and 16], therefore $D(e) \ge D(f) = T(f) \ne 0$. On the other hand, $D(0) = 0$ by either (D3) or (D4).

(D9) Suppose $D(e) = D(f)$. Let h be a central projection such that

$$he \precsim hf, \qquad (1-h)f \precsim (1-h)e.$$

Say $he \sim f' \le hf$. Then

$$D(he) = D(f') \le D(hf) = hD(f) = hD(e) = D(he),$$

thus $D(f') = D(hf)$, $D(hf - f') = 0$, hence $hf - f' = 0$ by (D8). Thus $he \sim f' = hf$. Similarly $(1-h)e \sim (1-h)f$, therefore $e \sim f$.

(D10) If $e \precsim f$, say $e \sim e_1 \le f$, then $D(e) = D(e_1) \le D(f)$.

Conversely, suppose $D(e) \le D(f)$. Let h be a central projection such that

$$(**) \qquad (1-h)e \precsim (1-h)f, \qquad hf \precsim he.$$

Then $D(hf) \le D(he) = hD(e) \le hD(f) = D(hf)$, thus $D(he) = D(hf)$, therefore $he \sim hf$ by (D9); adding this to the first relation in (**), we have $e \precsim f$. ∎

The deeper properties of dimension depend on the fact that the set of all real-valued functions in $C(\mathscr{X})$—that is, the real algebra $C_{\mathbb{R}}(\mathscr{X})$—is a boundedly complete lattice with respect to the usual pointwise ordering. To put it another way, the positive unit ball of $C(\mathscr{X})$—that is, the set of all continuous functions c such that $0 \leq c \leq 1$—is a complete lattice. These completeness assertions concerning the real function lattice are equivalent to the extremal disconnectedness of $\mathscr{X}$ by Stone's theory [**85**]. (Caution: The lattice supremum of an infinite set of functions is $\geq$, but in general $\neq$, the pointwise supremum.)

If c_j is an increasingly directed family in $C_{\mathbb{R}}(\mathscr{X})$, bounded above by some element of $C_{\mathbb{R}}(\mathscr{X})$ (equivalently, by some real constant), and if $c = \sup c_j$ (in the lattice sense just described), we write $c_j \uparrow c$. The following is a sample of the kind of elementary facts about such suprema that we shall need:

Lemma 1. *For positive functions in* $C(\mathscr{X})$:
(i) *If* $c_j \uparrow c$ *then* $ac_j \uparrow ac$.
(ii) *If* $c_j \uparrow c$ *and* $d_j \uparrow d$, *then* $c_j + d_j \uparrow c + d$.
(iii) *If* $c_j \uparrow c$ *and* $d_k \uparrow d$, *then* $c_j + d_k \uparrow c + d$.

Proof. (i) For all j, $ac_j \leq ac$, therefore $s = \sup ac_j$ exists and $s \leq ac$. The assertion is that $s = ac$. Assume to the contrary; then there exists an $\varepsilon > 0$ and a nonzero projection h in $C(\mathscr{X})$ such that $ac - s \geq \varepsilon h$, therefore ha is invertible in $hC(\mathscr{X})$; let b be the element of $hC(\mathscr{X})$ such that $b(ha) = h$, that is, $ba = h$. For each index j,

$$ac \geq s + \varepsilon h \geq ac_j + \varepsilon h;$$

multiplying by b, we have $(ba)c \geq (ba)c_j + \varepsilon bh$, thus

$$(*) \qquad hc \geq hc_j + \varepsilon b;$$

on the other hand,

$$(**) \qquad (1-h)c \geq (1-h)c_j;$$

adding $(*)$ and $(**)$, we have $c \geq c_j + \varepsilon b$. Thus $c_j \leq c - \varepsilon b$ for all j, therefore $c \leq c - \varepsilon b$; this implies $b = 0$, a contradiction.

(ii) Let $a = \sup(c_j + d_j)$; obviously $a \leq c + d$ and $c_j + d_j \uparrow a$. The assertion is that $a = c + d$; it is enough to show that $d \leq a - c$. Fix an index j; it will suffice to show that $d_j \leq a - c$. For all $k \geq j$ we have $c_k \leq a - d_k \leq a - d_j$, therefore $c \leq a - d_j$.

(iii) Write $a_{(j,k)} = c_j + d_k$ and order the index pairs (j,k) by writing $(j,k) \leq (j',k')$ iff $j \leq j'$ and $k \leq k'$. Clearly $a_{(j,k)} \uparrow a$ with $a \leq c + d$. The proof that $a = c + d$ is similar to that in (ii). ∎

Let $(a_\iota)_{\iota \in I}$ be a family in $C_{\mathbb{R}}(\mathscr{X})$ such that $a_\iota \geq 0$ for all $\iota \in I$. For each finite subset $J \subset I$, write

$$a_J = \sum_{\iota \in J} a_\iota;$$

with the indices J ordered by inclusion, the family (a_J) is increasingly directed. If the family (a_J) is bounded above by some element of $C_{\mathbb{R}}(\mathscr{X})$, we write $\sum_{\iota\in I} a_\iota$ for the supremum of the a_J; thus

$$\sum_{\iota\in I} a_\iota = \sup\left\{\sum_{\iota\in J} a_\iota : J\subset I,\ J \text{ finite}\right\}.$$

Such sums enter dimension theory via the following lemma:

Lemma 2. *If D is a dimension function for A, $(e_\iota)_{\iota\in I}$ is an orthogonal family of projections in A, and $e=\sup e_\iota$, then $\sum_{\iota\in I} D(e_\iota)$ exists and is $\leq D(e)$.*

Proof. If J is any finite subset of I, write $e_J=\sum_{\iota\in J} e_\iota$; then e_J is a projection $\leq e$, thus

$$\sum_{\iota\in J} D(e_\iota) = D\Big(\sum_{\iota\in J} e_\iota\Big) = D(e_J) \leq D(e)\,. \qquad \blacksquare$$

Definition 2. A dimension function D for A is said to be *completely additive* if, in the notation of Lemma 2, one always has $\sum_{\iota\in I} D(e_\iota)=D(e)$. {It is shown later in the chapter that *every* dimension function is completely additive [§ 33, Th. 1].}

Our study of general finite Baer $*$-rings with GC will be reduced to the Type I and Type II cases by writing $A=hA+(1-h)A$, where h is the unique central projection such that hA is Type I and $(1-h)A$ is continuous [§ 15, Th. 1, (4)]. The pieces are put back together by means of the following proposition; although the result is needed only for the case of two summands, that is, for the case $\Lambda=\{1,2\}$, it is more instructive—and not essentially harder—to consider arbitrarily many summands:

Proposition 2. *Let $(h_\alpha)_{\alpha\in\Lambda}$ be an orthogonal family of nonzero central projections in A such that $\sup h_\alpha=1$.*

(i) *If, for each $\alpha\in\Lambda$, D_α is a dimension function for $h_\alpha A$, then the D_α may be assembled in a natural way to form a dimension function D for A. If every D_α is completely additive, then so is D.*

(ii) *If one knows that each $h_\alpha A$ has a unique dimension function, it follows that the dimension function of A is also unique.*

Proof. {The main task of the chapter is to remove the "if"'s; thus, eventually, the proposition has content for every finite Baer $*$-ring with GC.}

For each α, let P_α be the clopen subset of $\mathscr{X}$ corresponding to h_α in the Stone representation of the complete Boolean algebra $\tilde{Z}$ (thus h_α is the characteristic function of P_α). The central projections of $h_\alpha A$ are the projections in $h_\alpha Z$, that is, the projections h in Z such that $h\leq h_\alpha$;

these correspond to the clopen sets in $\mathscr{X}$ that are contained in P_α, in other words, to the clopen subsets of the Stonian space P_α. Thus we may identify P_α with the Stone representation space of the complete Boolean algebra of central projections of $h_\alpha A$. At the same time, $C(P_\alpha)$ may be identified with the ideal $h_\alpha C(\mathscr{X})$ of $C(\mathscr{X})$: a function in $C(P_\alpha)$ is regarded as a function in $h_\alpha C(\mathscr{X})$ by defining it to be zero on $\mathscr{X}-P_\alpha$; and a function in $h_\alpha C(\mathscr{X})$ is regarded as a function in $C(P_\alpha)$ by restricting it to P_α.

Since $\sup h_\alpha=1$ (in other words, the union of the P_α is dense in $\mathscr{X}$), the commutative AW^*-algebra $C(\mathscr{X})$ is the C^*-sum of the $h_\alpha C(\mathscr{X})$ [§ 10, Prop. 2], which in turn may be identified with the C^*-sum of the $C(P_\alpha)$; this is the technical core of the proposition.

(i) Suppose that for each α, D_α is a dimension function for $h_\alpha A$. If e is any projection in A, then $h_\alpha e$ is a projection in $h_\alpha A$ for all $\alpha\in\Lambda$, and we may define

$$D(e)=(D_\alpha(h_\alpha e))_{\alpha\in\Lambda},$$

an element in the C^*-sum of the $C(P_\alpha)$. In view of the identifications made above, $D(e)$ may be regarded as the unique element of $C(\mathscr{X})$ such that $h_\alpha D(e)=D_\alpha(h_\alpha e)$ for all $\alpha\in\Lambda$. It follows routinely that D is a dimension function for A.

Suppose, moreover, that every D_α is completely additive. Let $(e_\iota)_{\iota\in I}$ be an orthogonal family of projections in A, $e=\sup e_\iota$. Write

$$c=\sum_{\iota\in I} D(e_\iota), \qquad c_\alpha=\sum_{\iota\in I} D_\alpha(h_\alpha e_\iota)$$

(these sums exist by Lemma 2). By hypothesis, $c_\alpha=D_\alpha(h_\alpha e)$ for each $\alpha\in\Lambda$; it is to be shown that $c=D(e)$. Citing Lemma 1, (i), we have

$$\begin{aligned} h_\alpha c &= h_\alpha \sum_{\iota\in I} D(e_\iota)=\sum_{\iota\in I} h_\alpha D(e_\iota)\\ &=\sum_{\iota\in I} D_\alpha(h_\alpha e_\iota)=c_\alpha=D_\alpha(h_\alpha e)=h_\alpha D(e), \end{aligned}$$

thus $h_\alpha(D(e)-c)=0$ for all α, therefore $D(e)-c=0$.

(ii) Suppose D_1 and D_2 are both dimension functions for A. For each α, the restrictions of D_1, D_2 to the set of projections of $h_\alpha A$ evidently define dimension functions for $h_\alpha A$; by the assumed uniqueness for $h_\alpha A$, we conclude that

$$D_1(h_\alpha e)=D_2(h_\alpha e)$$

for all projections e in A. Thus $h_\alpha D_1(e)=h_\alpha D_2(e)$ for all α and all e, therefore $D_1(e)=D_2(e)$ for all e. ∎

In principle, one could use Proposition 2 to reduce the Type I case to the Type I_n case $(n=1,2,3,\ldots)$ [cf. § 18, Th. 1]; it turns out to be simpler to give a unified treatment of the Type I case, thereby avoiding the necessity of putting the homogeneous pieces back together.

Exercises

1A. Let A be a Baer $*$-ring with GC, and suppose there exists a function $e \mapsto D(e)$ defined on the set of projections of A, with values in a group (notated additively), satisfying conditions (D1), (D3) and (D4) of [§ 25, Def. 1]. Then A is finite.

2A. Let A be a purely infinite (i.e., Type III) Baer $*$-ring with orthogonal GC. Assume, moreover, that A is orthoseparable [§ 18, Exer. 10]. Then $e \sim f$ if and only if $C(e)=C(f)$.

3A. Notation as in Proposition 1.
(i) For every projection e, $C(e)$ is the right projection of $D(e)$.
(ii) $D(e)$ is a projection iff e is a central projection.

4A. The ring A of all 2×2 matrices over the field of three elements is a finite Baer $*$-factor of Type I_2 that possesses a dimension function D, although A does not have GC. If e, f are the projections described in [§ 1, Exer. 17], then $D(e)=D(f)=\frac{1}{2}$ but e and f are not equivalent.

5D. *Problem:* Does every finite Baer $*$-ring have a dimension function?

§ 28. Type I_{fin}: Complete Additivity and Uniqueness of Dimension

Proposition 1. *Let A be a finite Baer $*$-ring of Type* I, *with* GC. *Suppose D is a dimension function for A [§ 25, Def. 1]. Then D is completely additive; that is, if $(e_\iota)_{\iota \in I}$ is any orthogonal family of projections and $e=\sup e_\iota$, then $D(e)=\sum_{\iota \in I} D(e_\iota)$.*

Proof. For each finite subset $J \subset I$, write $e_J=\sum_{\iota \in J} e_\iota$, and let $c=\sum_{\iota \in I} D(e_\iota)$. Thus $D(e_J)\uparrow c$ and $c \le D(e)$ [§ 27, Lemma 2]; the problem is to show that $c=D(e)$.

Let (h_α) be an orthogonal family of nonzero central projections with $\sup h_\alpha=1$, such that for each α, the set of indices

$$I_\alpha=\{\iota \in I: h_\alpha e_\iota \neq 0\}$$

is finite [§ 18, Prop. 5]; it will suffice to show that $h_\alpha c=h_\alpha D(e)$ for all α.

Fix an index α. We have

$$h_\alpha e=h_\alpha \sup_{\iota \in I} e_\iota=\sup_{\iota \in I} h_\alpha e_\iota=\sum_{\iota \in I_\alpha} h_\alpha e_\iota=h_\alpha e_{I_\alpha},$$

thus $h_\alpha D(e)=h_\alpha D(e_{I_\alpha})$ [§ 27, Prop. 1, (D6)]; the proof will be concluded by showing that

$$(*) \qquad h_\alpha c=h_\alpha D(e_{I_\alpha}).$$

At any rate,

$$(**) \qquad h_\alpha D(e_J) \uparrow h_\alpha c$$

[§ 27, Lemma 1, (i)]. If J is any finite subset of I such that $J \supset I_\alpha$, then $h_\alpha e_\iota = 0$ for $\iota \in J - I_\alpha$, thus $h_\alpha e_J = h_\alpha e_{I_\alpha}$ and so

$$h_\alpha D(e_J) = h_\alpha D(e_{I_\alpha});$$

passing to the limit along J (the right side remains constant), we infer from $(**)$ that $h_\alpha c = h_\alpha D(e_{I_\alpha})$, which is the desired relation $(*)$. ∎

Uniqueness in the Type I case follows at once:

Proposition 2. *Let A be a finite Baer $*$-ring of Type* I, *with* GC. *If D_1 and D_2 are dimension functions for A, then $D_1 = D_2$.*

Proof. Let e be any nonzero projection, and let $(e_\iota)_{\iota \in I}$ be an orthogonal family of simple projections such that $e = \sup e_\iota$ [§ 26, Prop. 14]. Since $D_1(e_\iota) = T(e_\iota) = D_2(e_\iota)$ for all ι [§ 27, Prop. 1], it follows from Proposition 1 that

$$D_1(e) = \sum_{\iota \in I} D_1(e_\iota) = \sum_{\iota \in I} D_2(e_\iota) = D_2(e). \quad ∎$$

§ 29. Type $\mathrm{I}_{\mathrm{fin}}$: Existence of a Dimension Function

Throughout this section, A is a finite Baer $*$-ring of Type I, with GC. A feature of the exposition is that we do not explicitly resort to a decomposition of A into homogeneous summands (implicitly, this was done in the proof of [§ 26, Prop. 14]).

Lemma 1. *If $e_1, \ldots, e_n$ are orthogonal, simple abelian projections, then $\sum_1^n T(e_i) \le 1$.*

Proof. There exist orthogonal central projections $h_1, \ldots, h_\tau$ with $h_1 + \cdots + h_\tau = 1$, such that, on setting

$$e_{\alpha i} = h_\alpha e_i \qquad (\alpha = 1, \ldots, \tau;\ i = 1, \ldots, n),$$

either $e_{\alpha i} = 0$ or $C(e_{\alpha i}) = h_\alpha$ [§ 6, Prop. 5]. It will suffice to show that for each α, $h_\alpha \sum_1^n T(e_i) \le h_\alpha$, that is, $\sum_1^n T(h_\alpha e_i) \le h_\alpha$ [§ 26, Prop. 13]. Dropping down to $h_\alpha A$ and changing notation, we can suppose that the e_i are faithful.

We are reduced to showing that if $e_1, \ldots, e_n$ are orthogonal simple abelian projections such that $C(e_i) = 1$ for all i, then $\sum_1^n T(e_i) \le 1$. We have $e_1 \sim \cdots \sim e_n$ [§ 18, Prop. 1], therefore $T(e_1) = \cdots = T(e_n)$ [§ 26, Prop. 11]. Let $r = (1 : e_i)$; then $T(e_i) = (1/r)1$ and

$$\sum_1^n T(e_i) = \left(\frac{n}{r}\right) 1.$$

The problem is to show that $n \leq r$. Since $(1 : e_1) = r$ we may write $1 = f_1 + \cdots + f_r$ with $e_1 \sim f_j$ $(j = 1, \ldots, r)$; if one had $r < n$, then

$$1 = f_1 + \cdots + f_r \sim e_1 + \cdots + e_r \neq 1$$

would contradict finiteness. ∎

Lemma 2. *If $(e_\iota)_{\iota \in I}$ is an orthogonal family of simple abelian projections, then $\sum_{\iota \in I} T(e_\iota)$ exists and is ≤ 1.*

Proof. This is immediate from Lemma 1 and the definition of such sums (cf. the discussion preceding [§ 27, Lemma 2]). ∎

Lemma 3. *If $e_1 + \cdots + e_m = f_1 + \cdots + f_n$, where the e_i and f_j are simple abelian projections, the e_i are orthogonal and the f_j are orthogonal, then*

$$\sum_1^m T(e_i) = \sum_1^n T(f_j).$$

Proof. We can suppose, without loss of generality, that the e_i and f_j are faithful (cf. the proof of Lemma 1). Then $e_i \sim f_j$ for all i and j [§ 18, Prop. 1], therefore $m = n$ and $T(e_i) = T(f_j)$ [§ 26, Props. 1 and 11]. ∎

Lemma 4. *If $e = \sup e_\iota = \sup f_\varkappa$, where each of $(e_\iota)_{\iota \in I}$ and $(f_\varkappa)_{\varkappa \in K}$ is an orthogonal family of simple abelian projections, then*

$$\sum_{\iota \in I} T(e_\iota) = \sum_{\varkappa \in K} T(f_\varkappa).$$

Proof. Write $c = \sum_{\iota \in I} T(e_\iota)$, $d = \sum_{\varkappa \in K} T(f_\varkappa)$ (the sums exist by Lemma 2). Let (h_α) be an orthogonal family of central projections with $\sup h_\alpha = 1$, such that for each α, the sets of indices

$$I_\alpha = \{\iota \in I : h_\alpha e_\iota \neq 0\}, \qquad K_\alpha = \{\varkappa \in K : h_\alpha f_\varkappa \neq 0\}$$

are both finite [§ 18, Prop. 5]. It will suffice to show that $h_\alpha c = h_\alpha d$ for all α.

Fix an index α. Clearly

$$h_\alpha e = \sum_{\iota \in I_\alpha} h_\alpha e_\iota = \sum_{\varkappa \in K_\alpha} h_\alpha f_\varkappa;$$

applying Lemma 3 in $h_\alpha A$, it follows that

$$(*) \qquad \sum_{\iota\in I_\alpha} T(h_\alpha e_\iota) = \sum_{\varkappa\in K_\alpha} T(h_\alpha f_\varkappa)\,.$$

If $\iota\notin I_\alpha$ then $h_\alpha e_\iota = 0$, therefore $h_\alpha T(e_\iota) = T(h_\alpha e_\iota) = 0$; it follows that

$$h_\alpha c = \sum_{\iota\in I_\alpha} h_\alpha T(e_\iota) = \sum_{\iota\in I_\alpha} T(h_\alpha e_\iota)$$

and similarly
$$h_\alpha d = \sum_{\varkappa\in K_\alpha} T(h_\alpha f_\varkappa)\,,$$

thus $h_\alpha c = h_\alpha d$ by (*). ∎

Lemma 5. *If* $\{c_{ij}\colon i\in I, j\in J\}$ *is a doubly indexed family in* $C_{\mathbb{R}}(\mathscr{X})$ *such that* $0\le c_{ij}\le 1$ *for all i and j, then*

$$\sup\{c_{ij}\colon i\in I, j\in J\} = \sup_{i\in I}\big(\sup_{j\in J} c_{ij}\big)\,.$$

Proof. A routine argument, valid in any complete lattice—in the present instance, the lattice of continuous functions c, on the Stonian space $\mathscr{X}$, such that $0\le c\le 1$. ∎

Proposition 1. *If A is a finite Baer *-ring of Type* I, *with* GC, *then there exists a unique dimension function D for A. Moreover, D is completely additive.*

Proof. Uniqueness and complete additivity are covered by Section 28; we are concerned here with existence.

Define $D(0)=0$. If e is any nonzero projection, write $e=\sup e_\iota$ with $(e_\iota)_{\iota\in I}$ an orthogonal family of simple abelian projections [§ 26, Prop. 14] and define

$$D(e) = \sum_{\iota\in I} T(e_\iota)\,;$$

the sum exists and is ≤ 1 (Lemma 2), and it is independent of the particular such partition of e (Lemma 4). It remains to verify properties (D1)–(D4) of [§ 25, Def. 1].

(D2) By construction, $D(e)\in C(\mathscr{X})$, $0\le D(e)\le 1$.

(D1) If e is partitioned as above and if $e\sim f$, then the equivalence induces a partition $(f_\iota)_{\iota\in I}$ of f such that $e_\iota\sim f_\iota$ for all ι; since $T(e_\iota)=T(f_\iota)$ [§ 26, Prop. 11], it follows that

$$D(e) = \sum_{\iota\in I} T(e_\iota) = \sum_{\iota\in I} T(f_\iota) = D(f)\,.$$

(D3) If h is a nonzero central projection, it is to be shown that $D(h)=h$. Since the definition of D for hA is consistent with that for A, we can suppose without loss of generality that $h=1$. Thus, assuming $(e_\iota)_{\iota\in I}$ is an orthogonal family of simple abelian projections such that $\sup e_\iota=1$, and setting $c=\sum_{\iota\in I} T(e_\iota)$ (that is, $c=D(1)$), it is to be shown that $c=1$.

Let (h_α) be an orthogonal family of nonzero central projections with $\sup h_\alpha=1$, such that for each α, the set of indices

$$I_\alpha=\{\iota\in I: h_\alpha e_\iota\neq 0\}$$

is finite [§ 18, Prop. 5]; it will suffice to show that $h_\alpha c=h_\alpha$ for all α. For any fixed index α,

$$h_\alpha=\sup h_\alpha e_\iota=\sum_{\iota\in I_\alpha} h_\alpha e_\iota$$

and

$$h_\alpha c=\sum_{\iota\in I_\alpha} T(h_\alpha e_\iota)$$

(see the proof of Lemma 4); thus it suffices to consider the case that the family (e_ι) is finite.

Changing notation, we are in the following situation: we are given orthogonal, simple abelian projections $e_1, \ldots, e_n$ such that

$$(*) \qquad e_1+\cdots+e_n=1\,,$$

and we are to show that $\sum_1^n T(e_i)=1$. Dropping down to each of a finite number of direct summands (as in the proof of Lemma 1), we can suppose that the e_i are faithful. Then $e_1\sim\cdots\sim e_n$ [§ 18, Prop. 1] and it follows from $(*)$ that $(1:e_i)=n$; thus $T(e_i)=(1/n)1$, $\sum_1^n T(e_i)=1$.

(D4) Assuming e, f are nonzero projections such that $ef=0$, it is to be shown that $D(e+f)=D(e)+D(f)$. Write

$$e=\sup e_i, \qquad f=\sup f_j\,,$$

where $(e_i)_{i\in I}$ and $(f_j)_{j\in J}$ are orthogonal families of simple abelian projections. Let K be the disjoint union of I and J; for $k\in K$, define $g_k=e_k$ or f_k according as $k\in I$ or $k\in J$. By the associativity of suprema, $(g_k)_{k\in K}$ is a partition of $e+f$ into simple abelian projections.

Write E, F, G for generic finite subsets of I, J, K, and write

$$a_E=\sum_{i\in E} T(e_i)\,,$$
$$b_F=\sum_{j\in F} T(f_j)\,,$$
$$c_G=\sum_{k\in G} T(g_k)\,;$$

by the definition of D, we have

$$D(e)=\sup_E a_E, \qquad D(f)=\sup_F b_F, \qquad D(e+f)=\sup_G c_G.$$

Each G can be written uniquely in the form $G=E\cup F$, and one has

$$c_G=a_E+b_F;$$

conversely, given any E and F we may define $G=E\cup F$. Thus, the c_G's may be regarded as being doubly indexed by E's and F's:

$$c_G=c_{E,F}=a_E+b_F.$$

Citing Lemma 5 at the appropriate step, we have

$$\begin{aligned} D(e+f)&=\sup_G c_G=\sup_{E,F}(a_E+b_F)\\ &=\sup_E\big(\sup_F(a_E+b_F)\big)\\ &=\sup_E\big(a_E+\sup_F b_F\big)\\ &=\sup_E(a_E+D(f))\\ &=\big(\sup_E a_E\big)+D(f)=D(e)+D(f). \qquad \blacksquare \end{aligned}$$

Exercises

1A. If A is a finite Baer $*$-ring of Type I_n, with GC, and if e is any projection in A, then $D(e)$ is a simple function on $\mathscr{X}$ whose values are contained in the set $\{0, 1/n, 2/n, \dots, (n-1)/n, 1\}$.

2A. Proposition 1 yields an alternative proof of the additivity of equivalence for the case of a Baer $*$-ring of Type $\mathrm{I}_{\mathrm{fin}}$ with GC [cf. § 20, Th. 1].

§ 30. Type $\mathrm{II}_{\mathrm{fin}}$: Dimension Theory of Fundamental Projections

Throughout this section, A is a finite Baer $*$-ring of Type II with GC.

The techniques developed in this section are the continuous substitutes for the special properties of abelian projections (notably [§ 18, Props. 1, 4 and 5]). The exposition is patterned after that of J. Dixmier [**18**] (see also [**2**, Appendix III]).

Lemma 1. *Suppose that*

$$T(f)+\sum_1^p T(f_j)\le\sum_1^n T(e_i),$$

where each of the projections $f, f_1, \ldots, f_p, e_1, \ldots, e_n$ *is either* 0 *or fundamental; assume that the* e_i *are orthogonal, the* f_j *are orthogonal, and that*

$$\sum_1^p f_j \leq \sum_1^n e_i \,.$$

Then $f \precsim \sum_1^n e_i - \sum_1^p f_j$.

Proof. For any central projection h,

$$T(hf) + \sum_1^p T(hf_j) \leq \sum_1^n T(he_i)$$

[§26, Prop. 13]; therefore, dropping down to each of a finite number of direct summands, we can suppose that each of the given projections is either 0 or faithful [§6, Prop. 5].

If $f=0$ there is nothing to prove; assume $f \neq 0$. Those e_i, f_j that are 0 contribute nothing to the hypothesis or the conclusion; discarding them, we can suppose that the e_i, f_j are nonzero (but we accept the possibility that no f_j's remain—in which case, both $\sum_1^p T(f_j)$ and $\sum_1^p f_j$ are interpreted as 0). {We write out the proof assuming that the f_j are present; if they are not, the same proof works provided all reference to the f_j is suppressed.}

Say $(1:e_i)=2^{r_i}$, $(1:f_j)=2^{s_j}$, $(1:f)=2^s$. The hypothesis reads

$$2^{-s}1 + \sum_1^p 2^{-s_j}1 \leq \sum_1^n 2^{-r_i}1 \,,$$

thus

(i) $$2^{-s} + \sum_1^p 2^{-s_j} \leq \sum_1^n 2^{-r_i} \,.$$

Let r be a positive integer greater than all of r_i, s_j, s, and let g be a projection such that $(1:g)=2^r$ [§26, Prop. 15]. Since $(1:g)$ is divisible by each of $(1:e_i)$, $(1:f_j)$ and $(1:f)$, it follows from [§26, Prop. 9] that g divides each of e_i, f_j, f and that

$$(e_i:g) = (1:g)(1:e_i)^{-1} = 2^{r-r_i} \,,$$

$$(f_j:g) = 2^{r-s_j}, \quad (f:g) = 2^{r-s} \,.$$

Multiplying through (i) by 2^r, we have

(ii) $$(f:g) + \sum_1^p (f_j:g) \leq \sum_1^n (e_i:g) \,.$$

Setting $\bar{f}=\sum_1^p f_j$ and $\bar{e}=\sum_1^n e_i$, the inequality (ii) may be written $(f:g)+(\bar{f}:g)\le(\bar{e}:g)$ [§26, Prop. 6], thus

$$\text{(iii)}\qquad (f:g)\le(\bar{e}:g)-(\bar{f}:g);$$

moreover, since $\bar{f}\le\bar{e}$ it follows that g divides $\bar{e}-\bar{f}$, and (iii) may be written

$$(f:g)\le(\bar{e}-\bar{f}:g),$$

therefore $f\precsim\bar{e}-\bar{f}$ [§26, Prop. 7]. ∎

Lemma 1 is the first step of an induction:

Proposition 1. *Suppose that*

$$\sum_1^q T(g_k)+\sum_1^p T(f_j)\le\sum_1^n T(e_i),$$

where each of g_k, f_j, e_i is either 0 or fundamental; assume that the e_i are orthogonal, the f_j are orthogonal, and that

$$\sum_1^p f_j\le\sum_1^n e_i\,.$$

Then there exist projections $d_1,\dots,d_q$, orthogonal to each other and to the f_j, such that $d_k\sim g_k$ $(k=1,\dots,q)$ and

$$\sum_1^q d_k\le\sum_1^n e_i-\sum_1^p f_j\,.$$

Proof. For $q=1$ this is the lemma. Assume inductively that all is well with $q-1$, and consider q. The given inequality implies that

$$\text{(i)}\qquad \sum_1^{q-1} T(g_k)+\sum_1^p T(f_j)\le\sum_1^n T(e_i)\,.$$

By the induction hypothesis, there exist orthogonal projections $d_1,\dots,d_{q-1}$ such that $d_k\sim g_k$ $(k=1,\dots,q-1)$ and

$$\text{(ii)}\qquad \sum_1^{q-1} d_k+\sum_1^p f_j\le\sum_1^n e_i\,.$$

Since $T(d_k)=T(g_k)$ $(k=1,\dots,q-1)$, the given inequality may be written

$$\text{(iii)}\qquad T(g_q)+\left(\sum_1^{q-1} T(d_k)+\sum_1^p T(f_j)\right)\le\sum_1^n T(e_i);$$

in view of the orthogonality of the projections on the left side of (ii), the lemma applied to (iii) yields

$$g_q \precsim \sum_1^n e_i - \left(\sum_1^{q-1} d_k + \sum_1^p f_j\right),$$

thus g_q is equivalent to a projection d_q meeting the requirements. ∎

Lemma 2. *If* $c \in C(\mathscr{X})$, $c \geq 0$, $c \neq 0$, *then there exists a fundamental projection* f *such that* $T(f) \leq c$.

Proof. Since $\mathscr{X}$ is extremally disconnected, there exists an $\varepsilon > 0$ and a nonzero projection h in $C(\mathscr{X})$ such that $c \geq \varepsilon h$. Let r be a positive integer such that $2^{-r} < \varepsilon$ and let f be a projection such that $(h{:}f) = 2^r$ [§ 26, Prop. 15]. Then $T(f) = 2^{-r} h \leq \varepsilon h \leq c$. ∎

Proposition 2. *Suppose*

$$e_1 + \cdots + e_n = f_1 + \cdots + f_p,$$

where the e_i *are orthogonal, the* f_j *are orthogonal, and each of the* e_i, f_j *is either* 0 *or fundamental. Then*

$$\sum_1^n T(e_i) = \sum_1^p T(f_j).$$

Proof. It suffices to show, e.g., that

$$\sum_1^n T(e_i) \leq \sum_1^p T(f_j).$$

Assume to the contrary; then there exists an $\varepsilon > 0$ and a nonzero central projection h such that

$$\left[\sum_1^n T(e_i) - \sum_1^p T(f_j)\right] h \geq \varepsilon h,$$

thus

$$\varepsilon h + \sum_1^p T(h f_j) \leq \sum_1^n T(h e_i).$$

Dropping down to hA and changing notation, we can suppose that

$$(*) \qquad \varepsilon 1 + \sum_1^p T(f_j) \leq \sum_1^n T(e_i).$$

Let f be a fundamental projection such that $T(f) \leq \varepsilon 1$ (Lemma 2); from $(*)$,

$$T(f) + \sum_1^p T(f_j) \leq \sum_1^n T(e_i).$$

Citing Lemma 1, we have

$$f \precsim \sum_1^n e_i - \sum_1^p f_j = 0,$$

thus $f=0$, a contradiction. ∎

Lemma 3. *If $e_1, \ldots, e_n$ are orthogonal fundamental projections, then* $\sum_1^n T(e_i) \le 1$.

Proof. We can suppose, without loss of generality, that the e_i are faithful (cf. the proof of [§29, Lemma 1]). Say $(1:e_i)=2^{r_i}$; thus $T(e_i) = 2^{-r_i}1$. Assume to the contrary that $\sum_1^n 2^{-r_i} > 1$. Let r be an integer such that $r > r_i$ for all i and let g be a projection such that $(1:g)=2^r$. Then

$$\sum_1^n 2^{r-r_i} > 2^r,$$

and, arguing as in the proof of Lemma 1, this may be written

$$\sum_1^n (e_i:g) > (1:g);$$

setting $e = \sum_1^n e_i$, we have

$$(*) \qquad (e:g) > (1:g)$$

[§26, Prop. 6]. From $(*)$ it follows that $1 \precsim e$ [§26, Prop. 7], therefore $e=1$ by finiteness; but this contradicts $(*)$. ∎

Proposition 3. *If $(e_\iota)_{\iota \in I}$ is any orthogonal family of fundamental projections, then $\sum_{\iota \in I} T(e_\iota)$ exists and is ≤ 1.*

Proof. For every finite subset J of I, we have $\sum_{\iota \in J} T(e_\iota) \le 1$ by Lemma 3. (Cf. the discussion preceding [§ 27, Lemma 2].) ∎

Lemma 4. *If $(e_\iota)_{\iota \in I}$ is a family of projections such that $h = \inf C(e_\iota) \ne 0$ and if e is any projection such that $he \ne 0$, then there exists a nonzero central projection k such that ke and the ke_ι are faithful in kA.*

Proof. Let $k = hC(e) = C(he)$. Since $k \le h \le C(e_\iota)$ for all ι, we have $C(ke_\iota) = kC(e_\iota) = k$; also $C(ke) = kC(e) = k$. ∎

The following proposition is the Type II substitute for [§18, Prop. 5]:

Proposition 4. *Let* $(e_\iota)_{\iota\in I}$ *be an orthogonal family of fundamental projections such that the orders* $(C(e_\iota):e_\iota)$ *are bounded. There exists an orthogonal family* (h_α) *of nonzero central projections with* $\sup h_\alpha=1$, *such that for each* α, *the set* $\{\iota\in I\colon h_\alpha e_\iota\neq 0\}$ *is finite.*

Proof. If h is a central projection, then either $he_\iota=0$ or he_ι is fundamental with $(C(he_\iota):he_\iota)=(C(e_\iota):e_\iota)$ (see the proof of [§26, Prop. 13]). By an obvious exhaustion argument, it will suffice to find a nonzero central projection h such that $\{\iota\in I\colon he_\iota\neq 0\}$ is finite.

Assume to the contrary that no such h exists. Let r be an integer such that $2^r>(C(e_\iota):e_\iota)$ for all $\iota\in I$, and write $m=2^r$. We construct m indices $\iota_1,\dots,\iota_m$ as follows. Choose any $\iota_1\in I$ and set $h_1=C(e_{\iota_1})$. By supposition, there are infinitely many indices ι with $h_1e_\iota\neq 0$; let ι_2 be such an index, $\iota_2\neq\iota_1$; by Lemma 4 there exists a nonzero central projection h_2 such that $C(h_2e_{\iota_1})=C(h_2e_{\iota_2})=h_2$. Continuing inductively, we arrive at indices $\iota_1,\iota_2,\dots,\iota_m$ and a nonzero central projection h_m such that

$$C(h_m e_{\iota_1})=C(h_m e_{\iota_2})=\cdots=C(h_m e_{\iota_m})=h_m\,.$$

Dropping down to h_mA and changing notation, we have the following situation: $e_1,\dots,e_m$ are orthogonal, faithful, fundamental projections, and

$$m=2^r>(1:e_i)\quad\text{for } i=1,\dots,m.$$

Say $(1:e_i)=2^{r_i}$; then $T(e_i)=2^{-r_i}1$, and Lemma 3 yields the absurdity

$$1\geq\sum_1^m 2^{-r_i}>\sum_1^m 2^{-r}=m2^{-r}=1\,.\qquad\blacksquare$$

The most unpleasant (and the last) computation in the chapter is as follows:

Lemma 5. *Suppose that*

$$\sum_1^n T(f_j)\geq\sum_{\iota\in I}T(e_\iota)\,,$$

where $(e_\iota)_{\iota\in I}$ *is an orthogonal family of fundamental projections, and* $f_1,\dots,f_n$ *are orthogonal fundamental projections such that*

$$f_1+\cdots+f_n\leq\sup e_\iota\,.$$

Then $\sum_1^n T(f_j)=\sum_{\iota\in I}T(e_\iota)$.

Proof. Assuming to the contrary, Lemma 2 yields a fundamental projection g such that

$$\text{(i)}\qquad \sum_1^n T(f_j)\geq T(g)+\sum_{\iota\in I}T(e_\iota)\,.$$

In particular, $T(g) \leq \sum_1^n T(f_j)$, thus $g \precsim f_1 + \cdots + f_n$ by Lemma 1; replacing g by an equivalent projection, we can suppose that

$$g \leq f_1 + \cdots + f_n \leq \sup e_\iota .$$

The plan of the proof is to construct an orthogonal family $(g_\iota)_{\iota \in I}$, with $g_\iota \sim e_\iota$, such that $g g_\iota = 0$ and

$$g_\iota \leq f_1 + \cdots + f_n$$

for all $\iota \in I$; this will imply $g \sup g_\iota = 0$, and it will then follow from additivity of equivalence [§20, Prop. 4] that

$$\sup e_\iota \sim \sup g_\iota \leq g + \sup g_\iota \leq f_1 + \cdots + f_n \leq \sup e_\iota ,$$

whence $\sup g_\iota = g + \sup g_\iota$ by finiteness, $g = 0$, a contradiction.

The construction of the g_ι is by induction; at the mth stage $(m = 0, 1, 2, \ldots)$ one constructs the g_ι corresponding to those e_ι whose order is 2^m. For $m = 0, 1, 2, \ldots$ write

$$I_m = \{\iota \in I : (C(e_\iota) : e_\iota) = 2^m\} ;$$

thus I is the disjoint union of $I_0, I_1, I_2, \ldots$. Note that $\iota \in I_0$ iff e_ι is central.

Suppose $\varkappa \in I_0$. Citing (i) and Lemma 3, we have

$$T(g) + e_\varkappa = T(g) + T(e_\varkappa) \leq T(g) + \sum_{\iota \in I} T(e_\iota) \leq \sum_1^n T(f_j) \leq 1 ,$$

thus $T(g) + e_\varkappa \leq 1$; it is then clear from the functional representation in $C(\mathscr{X})$ that $e_\varkappa T(g) = 0$, thus $e_\varkappa C(g) = 0$, $g e_\varkappa = 0$. Moreover, it follows from (i) that

$$T(e_\varkappa) \leq \sum_1^n T(f_j) ,$$

therefore $e_\varkappa \precsim \sum_1^n f_j$ by Lemma 1; since $e_\varkappa$ is central, it results from finiteness that $e_\varkappa \leq \sum_1^n f_j$ [§17, Exer. 2]. Defining $g_\varkappa = e_\varkappa$ $(\varkappa \in I_0)$, we have an orthogonal family of subprojections of $\sum_1^n f_j$ that are orthogonal to g; this meets the requirements for the indices in I_0.

Assume inductively that suitable g_ι have been constructed for all ι in $I^* = I_0 \cup \cdots \cup I_m$; thus, for $\iota \in I^*$, the g_ι are orthogonal subprojections of $\sum_1^n f_j$, $g_\iota g = 0$ and $g_\iota \sim e_\iota$. Since $T(g_\iota) = T(e_\iota)$ $(\iota \in I^*)$, the inequality (i) may be written

$$\text{(ii)} \qquad \sum_1^n T(f_j) \geq \Big\{ T(g) + \sum_{\iota \in I^*} T(g_\iota) \Big\} + \sum_{\iota \in I - I^*} T(e_\iota)$$

(the juggling with infinite sums is justified as in the proof of (D4) in [§ 29, Prop. 1]). If I_{m+1} is empty, there is nothing to be done, and the induction is complete. Otherwise, since $I_{m+1} \subset I - I^*$, it follows from (ii) that

$$\text{(iii)} \qquad \sum_1^n T(f_j) \geq \left\{ T(g) + \sum_{\iota \in I^*} T(g_\iota) \right\} + \sum_{\iota \in I_{m+1}} T(e_\iota) .$$

The projections $(e_\iota)_{\iota \in I^* \cup I_{m+1}}$ have bounded orders (bounded by 2^{m+1}). By Proposition 4, there exists an orthogonal family $(h_\alpha)_{\alpha \in \Lambda}$ of nonzero central projections with $\sup h_\alpha = 1$, such that for each α, the set

$$\{\iota \in I^* \cup I_{m+1} : h_\alpha e_\iota \neq 0\}$$

is finite; since $e_\iota \sim g_\iota$ for $\iota \in I^*$, this means that for each α, the sets

$$\{\iota \in I^* : h_\alpha g_\iota \neq 0\}, \qquad \{\iota \in I_{m+1} : h_\alpha e_\iota \neq 0\}$$

are finite.

Fix an index $\alpha \in \Lambda$. Multiplying through (iii) by h_α, we have

$$\text{(iv)} \qquad \sum_1^n T(h_\alpha f_j) \geq \left\{ T(h_\alpha g) + \sum_{\iota \in I^*} T(h_\alpha g_\iota) \right\} + \sum_{\iota \in I_{m+1}} T(h_\alpha e_\iota) ,$$

and all but finitely many terms in (iv) are 0. Applying Proposition 1 in $h_\alpha A$ to (iv), there exist orthogonal projections g_ι^α $(\iota \in I_{m+1})$—all but finitely many of them 0—such that, for each $\iota \in I_{m+1}$, $g_\iota^\alpha \sim h_\alpha e_\iota$, $g_\iota^\alpha \leq \sum_1^n h_\alpha f_j$, and g_ι^α is orthogonal to $h_\alpha g$ and to the $h_\alpha g_\varkappa$ $(\varkappa \in I^*)$.

Do this for each $\alpha \in \Lambda$. Then, for each $\iota \in I_{m+1}$, define

$$g_\iota = \sup\{g_\iota^\alpha : \alpha \in \Lambda\} ;$$

by additivity of equivalence,

$$g_\iota \sim \sup\{h_\alpha e_\iota : \alpha \in \Lambda\} = (\sup h_\alpha) e_\iota = e_\iota .$$

Since, for each $\alpha \in \Lambda$, the g_ι^α $(\iota \in I_{m+1})$ are orthogonal, it follows that the g_ι $(\iota \in I_{m+1})$ are also orthogonal.

Fix an index $\iota \in I_{m+1}$. Since $g_\iota^\alpha \leq \sum_1^n h_\alpha f_j \leq \sum_1^n f_j$ for all α, we have

$$g_\iota \leq \sum_1^n f_j .$$

Moreover, $g_\iota g = 0$, because

$$h_\alpha(g_\iota g) = (h_\alpha g_\iota)(h_\alpha g) = g_\iota^\alpha (h_\alpha g) = 0$$

for all $\alpha \in \Lambda$. Similarly, $g_\iota g_\varkappa = 0$ for $\varkappa \in I^*$.

Thus the g_ι $(\iota \in I_{m+1})$ have the required properties. This completes the induction, thereby achieving the desired contradiction. ∎

Proposition 5. *Let e be a nonzero projection, and suppose*

$$e = \sup e_\iota = \sup f_\varkappa ,$$

where $(e_\iota)_{\iota \in I}$ *and* $(f_\varkappa)_{\varkappa \in K}$ *are orthogonal families of fundamental projections. Then*

$$\sum_{\iota \in I} T(e_\iota) = \sum_{\varkappa \in K} T(f_\varkappa) .$$

Proof. The sums exist by Proposition 3. By symmetry, it is enough to show that

$$\sum_{\iota \in I} T(e_\iota) \geq \sum_{\varkappa \in K} T(f_\varkappa) ;$$

thus, if J is any finite subset of K, it will suffice to show that

$$\sum_{\iota \in I} T(e_\iota) \geq \sum_{\varkappa \in J} T(f_\varkappa) .$$

Assume to the contrary. Then (as in the proof of Proposition 2) there exists a nonzero central projection h such that

$$h \sum_{\iota \in I} T(e_\iota) \leq h \sum_{\varkappa \in J} T(f_\varkappa)$$

without having equality. Thus

$$(*) \qquad \sum_{\varkappa \in J} T(h f_\varkappa) \geq \sum_{\iota \in I} T(h e_\iota) ,$$

equality does not hold in $(*)$, and

$$\sum_{\varkappa \in J} h f_\varkappa = h \sum_{\varkappa \in J} f_\varkappa \leq h e = \sup h e_\iota ;$$

this contradicts Lemma 5 (applied in hA). ∎

§ 31. Type $\mathrm{II}_{\mathrm{fin}}$: Existence of a Completely Additive Dimension Function

As in the preceding section, A is a finite Baer $*$-ring of Type II with GC.

Definition 1. If e is a nonzero projection in A, let $(e_\iota)_{\iota \in I}$ be an orthogonal family of fundamental projections with $\sup e_\iota = e$ [§ 26, Prop. 16] and define

$$D(e) = \sum_{\iota \in I} T(e_\iota) ;$$

the sum exists [§ 30, Prop. 3] and is independent of the particular decomposition [§ 30, Prop. 5], thus $D(e)$ is well-defined. Define $D(0) = 0$.

Proposition 1. *If A is a finite Baer $*$-ring of Type* II, *with* GC, *then the function D defined above is a dimension function for A. Moreover, D is completely additive.*

Proof. We verify the conditions (D1)–(D4) of [§ 25, Def. 1].

(D2) Obvious from Definition 1.

(D3) If e is fundamental, it is clear from Definition 1 that $D(e)=T(e)$. In particular, if h is a central projection then $D(h)=T(h)=h$ [§ 26, Def. 3].

(D1), (D4) The proofs follow the same format as in the Type I case [§ 29, Prop. 1], with 'simple abelian projection' replaced by 'fundamental projection'.

Finally, suppose $e=\sup e_\iota$, where $(e_\iota)_{\iota\in I}$ is an orthogonal family of nonzero projections (not necessarily fundamental). We know that $\sum_{\iota\in I} D(e_\iota)$ exists and that

$$(*)\qquad \sum_{\iota\in I} D(e_\iota)\leq D(e)$$

[§ 27, Lemma 2]; it is to be shown that equality holds in $(*)$. For each $\iota\in I$ write

$$e_\iota=\sup\{e_{\iota\varkappa}\colon \varkappa\in K_\iota\},$$

where $(e_{\iota\varkappa})_{\varkappa\in K_\iota}$ is an orthogonal family of fundamental projections. Then the $e_{\iota\varkappa}$ are a partition of e into fundamental projections, therefore $D(e)$ is the supremum of all finite sums of the form

$$(**)\qquad T(e_{\iota_1\varkappa_1})+\cdots+T(e_{\iota_n\varkappa_n}),$$

where it is understood that $\varkappa_\nu\in K_{\iota_\nu}$ and the ordered pairs $(\iota_\nu,\varkappa_\nu)$ are distinct. Given such a sum, let

$$J=\{\iota_\nu\colon \nu=1,\dots,n\};$$

thus J is a finite subset of I with

$$e_{\iota_\nu\varkappa_\nu}\leq\sum_{\iota\in J} e_\iota$$

for $\nu=1,\dots,n$, therefore

$$\sum_{\nu=1}^{n} e_{\iota_\nu\varkappa_\nu}\leq\sum_{\iota\in J} e_\iota$$

(note that the terms on the left are orthogonal); then

$$\begin{aligned}\sum_{\nu=1}^{n} T(e_{\iota_\nu\varkappa_\nu})=\sum_{\nu=1}^{n} D(e_{\iota_\nu\varkappa_\nu})=D\Big(\sum_{\nu=1}^{n} e_{\iota_\nu\varkappa_\nu}\Big)\leq D\Big(\sum_{\iota\in J} e_\iota\Big)\\ =\sum_{\iota\in J} D(e_\iota)\leq\sum_{\iota\in I} D(e_\iota).\end{aligned}$$

Thus $\sum_{\iota\in I} D(e_\iota)$ is $\geq$ each expression of the form $(**)$, therefore it is $\geq$ their supremum $D(e)$. ∎

§ 32. Type $\mathrm{II}_{\mathrm{fin}}$: Uniqueness of Dimension

Proposition 1. *Let A be a finite Baer $*$-ring of Type* II, *with* GC. *If D_1 and D_2 are dimension functions for A, then $D_1 = D_2$ and is completely additive.*

Proof. Let D be the completely additive dimension function constructed in the preceding section; it suffices to show that $D_1 = D$. Since every nonzero projection is the supremum of an orthogonal family of fundamental projections [§ 26, Prop. 16], and since D_1 and D agree on fundamental projections [§ 27, Prop. 1, (D7)], it will suffice to show that D_1 is completely additive.

Suppose $e = \sup e_\iota$, where $(e_\iota)_{\iota \in I}$ is an orthogonal family. By [§ 27, Lemma 2],

$$\sum_{\iota \in I} D_1(e_\iota) \leq D_1(e).$$

We assert that equality holds. Set

$$c = D_1(e) - \sum_{\iota \in I} D_1(e_\iota) \geq 0$$

and assume to the contrary that $c \neq 0$. Let f be a fundamental projection such that $T(f) \leq c$ [§ 30, Lemma 2]. Since $D_1(f) = T(f)$, we have

$$D_1(f) \leq c = D_1(e) - \sum_{\iota \in I} D_1(e_\iota),$$

thus

$$D_1(f) + \sum_{\iota \in I} D_1(e_\iota) \leq D_1(e).$$

All the more, if J is any finite subset of I, then

$$D_1(f) + \sum_{\iota \in J} D_1(e_\iota) \leq D_1(e),$$

thus

$$D_1(f) \leq D_1(e) - \sum_{\iota \in J} D_1(e_\iota) = D_1\Big(e - \sum_{\iota \in J} e_\iota\Big);$$

it follows that

$$f \lesssim e - \sum_{\iota \in J} e_\iota$$

[§ 27, Prop. 1, (D10)], therefore

$$D(f) \leq D\Big(e - \sum_{\iota \in J} e_\iota\Big) = D(e) - \sum_{\iota \in J} D(e_\iota),$$

thus

$$\sum_{\iota \in J} D(e_\iota) \leq D(e) - D(f).$$

Since J is an arbitrary finite subset of I,

$$\sum_{\iota\in I} D(e_\iota) \leq D(e) - D(f),$$

and since D is completely additive this may be written

$$D(e) \leq D(e) - D(f);$$

then $D(f)=0$, $f=0$, a contradiction. ∎

§ 33. Dimension in an Arbitrary Finite Baer *-Ring with GC

Theorem 1. *If A is any finite Baer *-ring with* GC, *there exists a unique dimension function D for A. Moreover, D is completely additive.*

Proof. Since A is the direct sum of a Type I ring and a Type II ring [§ 15, Th. 2], it is enough to consider these cases separately [§ 27, Prop. 2]. For the Type I case, see [§ 29, Prop. 1]. For the Type II case, existence is proved in Section 31, uniqueness and complete additivity in Section 32. ∎

An important application of complete additivity (used in the proofs of [§ 34, Prop. 2] and [§ 47, Lemma 1]):

Theorem 2. *Let A and D be as in Theorem 1. If $e_\rho \uparrow e$ then $D(e_\rho) \uparrow D(e)$ (the notation is explained in the proof). Dually, $e_\rho \downarrow e$ implies $D(e_\rho) \downarrow D(e)$.*

Proof. We assume that (e_ρ) is a family of projections indexed by the ordinals $\rho < \lambda$, λ a limit ordinal; the notation $e_\rho \uparrow e$ means that $\sigma < \rho$ implies $e_\sigma \leq e_\rho$ and that $\sup e_\rho = e$. The notation $e_\rho \downarrow e$ is defined dually. In either case, we say that (e_ρ) is a *well-directed* family.

To exploit complete additivity, we replace (e_ρ) by an orthogonal family (f_ρ), also with supremum e, defined inductively as follows: $f_1 = e_1$, and, for $\rho > 1$,

$$f_\rho = e_\rho - \sup\{e_\sigma : \sigma < \rho\}$$

(in particular, $f_\rho = e_\rho - e_{\rho-1}$ when ρ is not a limit ordinal). Observe that $f_\rho \leq e_\rho$ for all ρ, and that $f_\rho e_\sigma = 0$ whenever $\sigma < \rho$; it follows at once that $\sup f_\rho \leq e$ and that the f_ρ are orthogonal. On the other hand, $e_\sigma \leq \sup f_\rho$ for all σ (by a routine transfinite induction), thus $e \leq \sup f_\rho$ and therefore $e = \sup f_\rho$. Then

$$D(e) = \sum D(f_\rho)$$

by complete additivity. Let $c = \sup D(e_\rho)$; obviously $D(e_\rho) \uparrow c \leq D(e)$ (see Section 27 for the notation). The proof is concluded by showing that $\sum D(f_\rho) \leq c$; it is enough to show that the inequality holds for any finite subsum. Consider any finite set of indices $\rho_1 < \cdots < \rho_n$. Then $f_{\rho_i} \le e_{\rho_i} \le e_{\rho_n}$ for $i = 1, \ldots, n$, thus

$$f_{\rho_1} + \cdots + f_{\rho_n} \le e_{\rho_n}$$

and therefore $\sum_{i=1}^{n} D(f_{\rho_i}) \le D(e_{\rho_n}) \le c$. ∎

In the Type II case, the following consequence of complete additivity is very useful (see the proof of [§ 41, Th. 1]):

Theorem 3. *Let A and D be as in Theorem 1, and assume that A is of Type* II. *Given any $c \in C_{\mathbb{R}}(\mathscr{X})$, $0 \le c \le 1$, there exists a projection e in A such that $D(e) = c$.*

Proof. We can suppose $c \ne 0$. Let $(e_\iota)_{\iota \in I}$ be a maximal orthogonal family of nonzero projections such that

$$(*) \qquad \sum_{\iota \in J} D(e_\iota) \le c \text{ for every finite subset } J \text{ of } I.$$

(Such a family exists by an evident maximality argument, using [§ 30, Lemma 2] to get started.) Let $e = \sup e_\iota$. By the complete additivity of D, $D(e)$ is the supremum of the finite sums $\sum_{\iota \in J} D(e_\iota)$; in view of $(*)$, we have $D(e) \le c$. We assert that $D(e) = c$.

Assuming to the contrary, let f be a fundamental projection such that $T(f) \le c - D(e)$ [§ 30, Lemma 2]. Since $D(f) = T(f)$, we have

$$D(f) \le c - D(e) \le 1 - D(e) = D(1 - e),$$

therefore $f \lesssim 1 - e$ [§ 27, Prop. 1]. Say $f \sim f_0 \le 1 - e$. Then f_0 is orthogonal to every e_ι, and

$$D(f_0) = D(f) \le c - D(e),$$

thus $D(f_0) + D(e) \le c$, that is,

$$D(f_0) + \sum_{\iota \in I} D(e_\iota) \le c,$$

contradicting maximality.

{So to speak, D assumes all values between 0 and 1. For this reason, the term *Type* II_1 is also used to indicate finite Type II.} ∎

The next result could have been proved immediately following [§ 30, Prop. 4], but to do so would have interrupted the development of the dimension theory; this is a convenient place to record it:

Theorem 4. *Let A be a finite Baer $*$-ring with* GC, *and let Z be the center of A. Then A is orthoseparable if and only if Z is orthoseparable.*

Proof. A $*$-ring is said to be orthoseparable if every orthogonal family of nonzero projections is countable. Thus the 'only if' part is trivial.

Conversely, suppose Z is orthoseparable; assuming $(e_\iota)_{\iota\in I}$ is an orthogonal family of nonzero projections in A, it is to be shown that I is countable. It suffices to consider the cases that A is Type I or Type II [§ 15, Th. 2].

Suppose first that A is Type I. Let $(h_\alpha)_{\alpha\in\Lambda}$ be an orthogonal family of nonzero central projections with $\sup h_\alpha = 1$, such that for each α, the set

$$I_\alpha = \{\iota \in I : h_\alpha e_\iota \neq 0\}$$

is finite [§ 18, Prop. 5]. Since $I = \bigcup_{\alpha\in\Lambda} I_\alpha$ (because $\sup h_\alpha = 1$) and since, by hypothesis, Λ must be countable, I is the union of countably many finite sets.

Suppose finally that A is Type II. Replacing each e_ι by a fundamental subprojection of it [§ 26, Prop. 16], we can suppose the e_ι to be fundamental. Partitioning I according to the countably many possible orders, we can assume that the e_ι all have the same order. Let $(h_\alpha)_{\alpha\in\Lambda}$ be an orthogonal family of nonzero central projections with $\sup h_\alpha = 1$, such that for each α, the set

$$I_\alpha = \{\iota \in I : h_\alpha e_\iota \neq 0\}$$

is finite [§ 30, Prop. 4]; the proof is concluded as in the Type I case. ∎

As remarked at the beginning of the chapter, a salient feature of the exposition is an avoidance of the use of the parallelogram law (P). Instead, we make do with GC; perhaps the key technical point is the fact, used in the proof of [§ 30, Lemma 5], that additivity of equivalence is available in any continuous Baer $*$-ring with GC (indeed, in any Baer $*$-ring with GC [§ 20, Th. 1]). {To be sure, the examples of rings with GC offered in Section 14 are obtained via (P) [§ 14, Prop. 7 and Th. 1].} At any rate, without (P) this seems to be about the end of the line.

Exercises

1A. Let A be a finite Baer $*$-ring with GC. If $e_\rho \uparrow e$ (with notation as in Theorem 2) and if f is a projection such that $e_\rho \precsim f$ for all ρ, then $e \precsim f$.

2A. If A is a finite Baer $*$-ring with GC, and if A is a factor (hence, for every pair of projections e, f, either $e \precsim f$ or $f \precsim e$), then the dimension function D is real-valued and A is orthoseparable.

3A. Let A be a finite Baer $*$-ring with GC, let Z be the center of A, let $\aleph$ be an infinite cardinal number, and suppose that every orthogonal family of nonzero projections in Z has cardinality $\leq \aleph$. Then every orthogonal family of nonzero projections in A has cardinality $\leq \aleph$.

4C. Theorem 2 holds more generally with directed families in place of well-directed families. That is, $e_\iota \uparrow e$ implies $D(e_\iota)\uparrow D(e)$ for increasingly directed families (e_ι); dually, $e_\iota \downarrow e$ implies $D(e_\iota)\downarrow D(e)$.

§ 34. Modularity, Continuous Geometry

We now adjoin the parallelogram law to the foregoing hypotheses: in this section, *A is a finite Baer $*$-ring with* GC, *satisfying the parallelogram law* (P). {In connection with Baer $*$-rings satisfying GC and (P), see [§ 14, Prop. 7 and Cor. 1, 2; Th. 1; Exer. 5, 9, 21, 22] and [§ 20, Exer. 2].}

Proposition 1. *The projection lattice of A is modular; that is, if e, f, g are projections in A and if* $e \le g$, *then* $(e\cup f)\cap g = e\cup(f\cap g)$.

Proof. Let $h=(e\cup f)\cap g$, $k=e\cup(f\cap g)$. Obviously $h\ge k$, and it is straightforward to check that

$$h\cup f = e\cup f = k\cup f, \qquad h\cap f = f\cap g = k\cap f.$$

Invoking (P), it follows that

$$h-f\cap g = h-h\cap f \sim h\cup f - f = k\cup f - f \sim k-k\cap f = k-f\cap g\,,$$

thus $h-f\cap g\sim k-f\cap g$; adding $f\cap g$, we have $h\sim k$. Thus $h\sim k\le h$, therefore $k=h$ by finiteness. ∎

Proposition 2. *Let (e_ρ) be a well-directed family of projections in A. If $e_\rho\uparrow e$ then $e_\rho\cap f\uparrow e\cap f$ for all projections f. Dually, $e_\rho\downarrow e$ implies $e_\rho\cup f\downarrow e\cup f$ for all f.*

Proof. Suppose $e_\rho\uparrow e$ (for the notation, see the proof of [§ 33, Th. 2]), and let f be any projection. Set $g=\sup(e_\rho\cap f)$. Obviously $e_\rho\cap f\uparrow g \le e\cap f$; it is to be shown that $g=e\cap f$. Citing (P), we have

$$e\cap f - e_\rho\cap f = e\cap f - e_\rho\cap(e\cap f) \sim (e\cap f)\cup e_\rho - e_\rho \le e-e_\rho,$$

therefore

$$D(e\cap f - e_\rho\cap f)\le D(e-e_\rho),$$

where D is the dimension function for A; thus

$$D(e_\rho)\le D(e)+D(e_\rho\cap f)-D(e\cap f)\le D(e)+D(g)-D(e\cap f)$$

for all ρ. Since $\sup D(e_\rho)=D(e)$ [§ 33, Th. 2], it follows that

$$D(e)\le D(e)+D(g)-D(e\cap f)\,,$$

that is, $D(e\cap f-g)\le 0$, therefore $D(e\cap f-g)=0$, $e\cap f-g=0$.

{A direct proof can be given, avoiding the use of the dimension function [**47**, Th. 6.5], [**54**, Th. 69].} ∎

Definition 1. A *continuous geometry* is a complete, complemented, modular lattice, such that if (e_ρ) is a well-directed family then $e_\rho \uparrow e$ implies $e_\rho \cap f \uparrow e \cap f$ for all f, and $e_\rho \downarrow e$ implies $e_\rho \cup f \downarrow e \cup f$ for all f.

Since the projection lattice of a Baer $*$-ring is complemented (even orthocomplemented, with canonical complementation $e \mapsto e' = 1 - e$), Propositions 1 and 2 may be summarized as follows:

Theorem 1. *If A is a finite Baer $*$-ring with* GC, *satisfying the parallelogram law* (P), *then the projection lattice of A is a continuous geometry.*

Continuous geometries were invented by von Neumann ([**69**], [**71**]), to whom Theorem 1 is due for the case of a factorial finite von Neumann algebra on a separable Hilbert space. For finite AW^*-algebras, the theorem is due to Kaplansky [**47**, Th. 6.5]. In a subsequent paper, Kaplansky also showed that if a Baer $*$-ring A is *regular* (for each $a \in A$ there exists $x \in A$ with $axa = a$) then the projection lattice of A is a continuous geometry [**51**, Th. 3]; along the way, he proved that A is finite (in the strong sense that $yx = 1$ implies $xy = 1$ [cf. §17, Exer. 4, 5, 6]) and that A satisfies the variants of GC and the parallelogram law (P) with respect to 'algebraic equivalence' [§1, Exer. 6]. Incidentally, one can arrive at Theorem 1 via the following 'shortcut' through the literature: prove modularity as in Proposition 1, and quote Kaplansky's theorem that any orthocomplemented complete modular lattice is a continuous geometry [**51**]. A more general version of Theorem 1 is given in [**54**, Th. 69].

Exercises

1A. If $\mathscr{H}$ is an infinite-dimensional Hilbert space, then the conclusion of Proposition 2 does not hold in the projection lattice of $\mathscr{L}(\mathscr{H})$.

2C. The projection lattice of an AW^*-algebra A is modular iff A is finite.

3C. Let L be a complete lattice in which increasing well-directed families have the property indicated in Proposition 2. If $e_\iota \uparrow e$, where (e_ι) is any increasingly directed family in L, then $e_\iota \cap f \uparrow e \cap f$ for all f; that is, one can drop the well-ordering of the indices. It follows that if $e_\iota \uparrow e$ and $f_\varkappa \uparrow f$ then $e_\iota \cap f_\varkappa \uparrow e \cap f$.

4A. The projection lattice is modular for any finite Rickart $*$-ring satisfying the parallelogram law (P).

Chapter 7

Reduction of Finite Baer $*$-Rings

Throughout the chapter, including the exercises, A denotes a *finite* Baer $*$-ring such that $\mathrm{LP}(x) \sim \mathrm{RP}(x)$ for all x in A; in the final section of the chapter, A is a finite AW^*-algebra.

{For generalities on the condition LP $\sim$ RP, see [§14, Cor. 2 of Prop. 7] and [§20, Exer. 2, 6].}

§ 35. Introduction

The principal result of the chapter is easy to state: If I is a maximal-restricted ideal of A [cf. §24], then the quotient ring A/I is a Baer $*$-ring (hence is a finite Baer $*$-factor). In addition, the intersection of all such ideals I is $\{0\}$ [§36], thus A may be 'embedded' in the complete direct product of the A/I. These two facts constitute the 'reduction theory' alluded to in the chapter heading.

The hypothesis LP $\sim$ RP implies that A has GC and satisfies the parallelogram law (P) [§14, Cor. 2 of Prop. 7]. As at the beginning of Chapter 6, we write Z for the center of A and $\mathscr{X}$ for the Stone representation space of the complete Boolean algebra $\tilde{Z}$, and we identify a central projection h with the characteristic function of the clopen subset of $\mathscr{X}$ to which it corresponds. Since A is a finite Baer $*$-ring with GC, the results of Chapter 6 are applicable to A; in particular, A has a dimension function D [§33, Th. 1]. {In view of the parallelogram law, we know in addition that the projection lattice of A is modular—indeed, it is a continuous geometry [§34, Th. 1]—but the discussion in the present chapter avoids such matters.}

§ 36. Strong Semisimplicity

A ring with unity is said to be *strongly semisimple* if the intersection of all the maximal ideals (two-sided) is $\{0\}$. A related concept, relevant to the reduction theory, is as follows:

Proposition 1. *The intersection of all the maximal-restricted ideals of A is* $\{0\}$.

Proof. Since, for a restricted ideal I, $x \in I$ iff $\mathrm{RP}(x) \in I$, it is the same to show that the intersection of all maximal p-ideals is $\{0\}$ [§22, Th. 1].

Suppose to the contrary that some nonzero projection e belongs to every maximal p-ideal. Since e contains a simple projection [§26, Props. 14 and 16], we can suppose without loss of generality that e is simple; then $h = C(e)$ also belongs to every maximal p-ideal. Since $h \neq 0$, the p-ideal

$$(1-h)\tilde{A} = ((1-h)A)^{\sim} = \{f \in \tilde{A} : f \leq 1-h\}$$

is proper; let $\mathscr{I}$ be a maximal p-ideal such that $(1-h)\tilde{A} \subset \mathscr{I}$ (Zorn's lemma). In particular, $1-h \in \mathscr{I}$; but $h \in \mathscr{I}$ by the supposition; thus $1 \in \mathscr{I}$, contradicting properness. ∎

The following corollary is not needed for the main line of development, but it is important enough to be noted; it applies, in particular, to finite AW^*-algebras:

Corollary. *Suppose, in addition, that A satisfies the following condition: for each nonzero* $x \in A$, *the ideal* AxA *generated by x contains a nonzero projection. Then A is strongly semisimple.*

Proof. {We remark that the condition is implied by any axiom of 'EP' type [cf. §7, Def. 3].} Assume to the contrary that A is not strongly semisimple, and let x be a nonzero element that belongs to every maximal ideal M. Choose a nonzero projection $e \in AxA$ (= the set of all finite sums $\sum a_i x b_i$); then $e \in AxA \subset M$ for every M. A contradiction to Proposition 1 will be obtained by showing that if I is any maximal-restricted ideal of A, then $e \in I$. Let M be a maximal ideal such that $I \subset M$ (Zorn's lemma). Then $\tilde{I} \subset \tilde{M}$; since $\tilde{M}$ is a p-ideal [§22, Prop. 1] and $\tilde{I}$ is a maximal p-ideal [§22, Th. 1], it follows that $\tilde{I} = \tilde{M}$. Then $e \in \tilde{M} = \tilde{I}$, achieving the contradiction. ∎

Exercises

1A. A finite AW^*-algebra is strongly semisimple.

2C. Every strongly semisimple ring with unity is semisimple.

3A. If A is factorial then it is restricted-simple, in the sense that $\{0\}$ is the only restricted proper ideal of A.

4A. In any finite Baer $*$-ring with GC, the intersection of all the maximal-strict ideals is $\{0\}$. (A *maximal-strict* ideal is one that is maximal among proper, strict ideals [cf. § 24, Exer. 6].)

§ 37. Description of the Maximal p-Ideals of A: The Problem

Definition 1. If $\sigma \in \mathscr{X}$, we define

$$\mathscr{I}_\sigma = \{e \in \tilde{A}: D(e)(\sigma) = 0\},$$

and

$$\mathscr{J}_\sigma = \{h \in \tilde{Z}: h(\sigma) = 0\}.$$

Proposition 1. *For each $\sigma \in \mathscr{X}$, $\mathscr{I}_\sigma$ is a proper p-ideal of A, and $\mathscr{I}_\sigma \cap Z = \mathscr{J}_\sigma$.*

Proof. Since A satisfies the parallelogram law (P), we have

$$D(e \cup f) = D(e) + D(f) - D(e \cap f) \leq D(e) + D(f)$$

for every pair of projections e, f; in particular, it is clear that if $e, f \in \mathscr{I}_\sigma$ then $e \cup f \in \mathscr{I}_\sigma$. If $f \precsim e \in \mathscr{I}_\sigma$ then $D(f) \leq D(e)$ shows that $f \in \mathscr{I}_\sigma$. Thus $\mathscr{I}_\sigma$ is a p-ideal. Since $D(h) = h$ for central projections h, obviously $\mathscr{I}_\sigma \cap Z = \mathscr{J}_\sigma$; in particular, $1 \notin \mathscr{I}_\sigma$, thus $\mathscr{I}_\sigma$ is proper. ∎

Definition 2. If $\sigma \in \mathscr{X}$, we write I_σ for the restricted ideal of A generated by $\mathscr{I}_\sigma$, that is, $I_\sigma = \{x \in A: \mathrm{RP}(x) \in \mathscr{I}_\sigma\}$ [§ 22, Th. 1].

The problem, alluded to in the section heading, is to show that every $\mathscr{I}_\sigma$ is a maximal p-ideal, and that every maximal p-ideal has this form.

It turns out that the crux of the matter is to show the following: if $\mathscr{I}$ is any proper p-ideal, then $\mathscr{I} \subset \mathscr{I}_\sigma$ for some $\sigma \in \mathscr{X}$. Suppose, to the contrary, that $\mathscr{I}$ is a proper p-ideal for which no such σ exists; let us pursue the matter as far as possible by naive techniques. By supposition, for each $\sigma \in \mathscr{X}$ there exists an $e \in \mathscr{I}$ such that $D(e)(\sigma) > 0$, hence $D(e) > 0$ on a neighborhood of σ; the obvious compactness argument produces a finite set $e_1, \ldots, e_k$ in $\mathscr{I}$ such that $D(e_1) + \cdots + D(e_k) > 0$ on $\mathscr{X}$. Setting $e = e_1 \cup \cdots \cup e_k$, we have $e \in \mathscr{I}$ and $\mathrm{D}(e) > 0$ on $\mathscr{X}$. Since $\mathscr{X}$ is compact, there exists $\varepsilon > 0$ with $\varepsilon 1 \leq D(e)$. Suppose first that A is of Type II; choose a positive integer r with $2^{-r} < \varepsilon$, and a fundamental projection f such that $(1:f) = 2^r$ [§ 26, Prop. 15]; then $D(f) = 2^{-r}1 \leq \varepsilon 1 \leq D(e)$, thus $f \precsim e \in \mathscr{I}$ and therefore $f \in \mathscr{I}$; since f is simple, $1 = C(f) \in \mathscr{I}$, a contradiction. Next, suppose that A is of Type I_n, and let f be a simple abelian projection with $(1:f) = n$; since $C(e) = 1$ [§ 27, Exer. 3], we have $f \precsim e$ [§ 18, Cor. of Prop. 1]; one argues as above that $1 \in \mathscr{I}$, a contradiction. The discussion extends easily to cover the case that A is the direct sum of finitely many homogeneous rings. There remains the general Type I case, in which A may have homogeneous summands of arbitrarily large order [cf. § 18, Th. 2]; it is to the solution of this stubborn case that the strategem of the next section is directed.

Exercise

1A. Let $\mathscr{I}$ be a proper p-ideal of A, and let $e \in \mathscr{I}$.

(i) If A is homogeneous, then $C(e) \neq 1$ (in particular, $D(e)$ is singular).

(ii) If A is the direct sum of finitely many homogeneous rings, then $C(e) \neq 1$.

(iii) If A does not have homogeneous summands of arbitrarily large order, then $D(e)$ is singular; it follows that $\mathscr{I} \subset \mathscr{I}_\sigma$ for some $\sigma \in \mathscr{X}$.

(iv) If e is abelian then $D(e)$ is singular.

It is shown in Section 39 that $D(e)$ is singular regardless of type [§ 39, Exer. 1].

§ 38. Multiplicity Analysis of a Projection

For motivation, see the preceding section.

Definition 1. If f is a projection in A, h is a central projection, and n is a positive integer, we say that *h contains n copies of f* in case there exist orthogonal projections $f_1, \dots, f_n$ such that $f \sim f_1 \sim \dots \sim f_n$ and $h \geq f_1 + \dots + f_n$. (Since h is central, the latter condition is equivalent to $h \geq f$ [§ 1, Exer. 15].)

Remarks. Suppose h contains n copies of f.

1. If k is any central projection, then kh contains n copies of kf.

2. By the properties of the dimension function, $D(f) \leq (1/n)h$ [§ 27, Prop. 1].

Proposition 1. *If e is any projection in A and n is a positive integer, then exists a (unique) largest central projection h such that h contains n copies of he (that is, for a central projection k, $k \leq h$ if and only if k contains n copies of ke). Denoting it by h_n, we have $h_1 = 1$ and $h_n \downarrow 1 - C(e)$.*

Proof. If k is any central projection then k contains one copy of ke, thus $h_1 = 1$ has the required properties. Assume $n \geq 2$. If no nonzero central projection h contains n copies of he, set $h_n = 0$. Otherwise, let $(h_\iota)_{\iota \in I}$ be a maximal orthogonal family of nonzero central projections such that h_ι contains n copies of $h_\iota e$. Say

$$h_\iota \geq e_{\iota 1} + \dots + e_{\iota n},$$

where $h_\iota e \sim e_{\iota 1} \sim \dots \sim e_{\iota n}$. Define $h = \sup h_\iota$ and $e_\nu = \sup_{\iota \in I} e_{\iota \nu}$ $(\nu = 1, \dots, n)$. Then

$$he \sim e_1 \sim \dots \sim e_n$$

by additivity of equivalence [§ 20, Th. 1], thus h contains n copies of he. We show that h has the required properties. If k is a central projection with $k \leq h$, then k contains n copies of ke (Remark 1 above). Conversely, suppose k is a central projection such that k contains n copies of ke; it is to be shown that $k \leq h$. Indeed, since $(1-h)k$ contains n copies of $(1-h)ke$, and since $(1-h)k$ is orthogonal to every h_ι, it results from maximality that $(1-h)k = 0$, thus $k \leq h$. We define $h_n = h$.

For all n, $1-C(e)$ trivially contains n copies of $(1-C(e))e=0$, thus $1-C(e)\le h_n$; writing $h'=\inf h_n$, we thus have $1-C(e)\le h'$. On the other hand, $h'\le h_n$ implies that h' contains n copies of $h'e$, thus

$$h'D(e)=D(h'e)\le(1/n)h'$$

for all n; it follows that $h'D(e)=0$, $D(h'e)=0$, $h'e=0$, $h'C(e)=0$, $h'\le 1-C(e)$. Thus $h'=1-C(e)$. Finally, since h_{n+1} contains $n+1$—and therefore n—copies of $h_{n+1}e$, we have $h_{n+1}\le h_n$, thus $h_n\downarrow h'$. ∎

Proposition 2. *With notation as in Proposition 1, define* $k_n=h_n-h_{n+1}$ $(n=1,2,3,\ldots)$.

(1) *k_n is an orthogonal sequence of central projections with* $\sup k_n=C(e)$.

(2) *For each n, there exists an orthogonal decomposition*

$$k_n=e_1+\cdots+e_n+g_n$$

such that $k_n e\sim e_1\sim\cdots\sim e_n$ *and* $g_n\precsim k_n e$.

Proof. (1) This is immediate from $h_n\downarrow 1-C(e)$ and $h_1=1$ [§ 12, Lemma].

(2) Since $k_n\le h_n$, we know that k_n contains n copies of $k_n e$, say

$$k_n\ge e_1+\cdots+e_n$$

with $k_n e\sim e_1\sim\cdots\sim e_n$. Define $g_n=k_n-(e_1+\cdots+e_n)$; it will suffice to show that $g_n\precsim k_n e$. The proof is based on the fact that, since $k_n\le 1-h_{n+1}$, no nonzero central projection $k\le k_n$ can contain $n+1$ copies of ke.

Apply GC to the pair $k_n e$ and $k_n g_n=g_n$: let h be a central projection such that

$$(*)\qquad h k_n e\precsim h k_n g_n=h g_n,$$

$$(**)\qquad (1-h)k_n g_n\precsim(1-h)k_n e.$$

Setting $k=hk_n$, $(*)$ reads

$$ke\precsim kg_n=k-(ke_1+\cdots+ke_n),$$

thus k contains $n+1$ copies of ke; since $k\le k_n$, it follows, as noted above, that $k=0$. Thus $hk_n=0$ and $(**)$ yields $k_n g_n\precsim k_n e$, that is, $g_n\precsim k_n e$. ∎

The following term is convenient:

Definition 2. With notation as in Proposition 2, we call the sequence k_n of central projections the *multiplicity sequence* of e.

We remark that, in the proof of Proposition 1, the use of the dimension function is easily avoided [cf. § 17, Exer. 23]. The parallelogram law figures in the preceding and the following sections, but the results of the present section are valid for any finite Baer $*$-ring with GC.

§ 39. Description of the Maximal p-Ideals of A: The Solution

For motivation, see Section 37.

Lemma 1. *Let e be a projection in A and let k_n be the multiplicity sequence of e [§ 38, Def. 2]. If $\mathscr{I}$ is a p-ideal such that $k_n e \in \mathscr{I}$ for all n, then $k_n \in \mathscr{I}$ for all n.*

Proof. This is clear from [§ 38, Prop. 2, (2)]. ∎

Lemma 2. *Let e be a projection in A and let k_n be the multiplicity sequence of e. If $\sigma \in \mathscr{X}$ and if $k_n e \in \mathscr{I}_\sigma$ for all n, then $e \in \mathscr{I}_\sigma$.*

Proof. Recall that $\mathscr{I}_\sigma = \{f \in \tilde{A}: D(f)(\sigma) = 0\}$ is a proper p-ideal [§ 37, Prop. 1]. By Lemma 1 we have $k_n \in \mathscr{I}_\sigma$ for all n, that is, $k_n(\sigma) = 0$; it is to be shown that $D(e)(\sigma) = 0$. By definition, $k_n = h_n - h_{n+1}$ [§ 38, Prop. 2], thus

$$k_1 + \cdots + k_{n-1} = h_1 - h_n = 1 - h_n;$$

since $k_1, \dots, k_{n-1} \in \mathscr{I}_\sigma$, it results that $1 - h_n \in \mathscr{I}_\sigma$, thus

$$(*) \qquad h_n(\sigma) = 1 \quad \text{for } n = 1, 2, 3, \dots .$$

By the definition of h_n [§ 38, Prop. 1], h_n contains n copies of $h_n e$, therefore $h_n D(e) = D(h_n e) \leq (1/n) h_n$; evaluating this at σ, it results from $(*)$ that $D(e)(\sigma) \leq 1/n$ for all n, thus $D(e)(\sigma) = 0$. ∎

The next result is the central one:

Lemma 3. *If $\sigma \in \mathscr{X}$ then $\mathscr{I}_\sigma$ is a maximal p-ideal.*

Proof. At any rate, $\mathscr{I}_\sigma$ is a proper p-ideal of A and $\mathscr{I}_\sigma \cap Z = \mathscr{J}_\sigma = \{h \in \tilde{Z}: h(\sigma) = 0\}$ is a maximal p-ideal of Z [§ 37, Prop. 1]. Let $\mathscr{I}$ be a maximal p-ideal of A with $\mathscr{I}_\sigma \subset \mathscr{I}$ (Zorn's lemma). Then $\mathscr{I}_\sigma \cap Z \subset \mathscr{I} \cap Z$; since $\mathscr{I} \cap Z$ is a proper p-ideal of Z, it results from the maximality of $\mathscr{J}_\sigma$ that $\mathscr{I}_\sigma \cap Z = \mathscr{I} \cap Z$.

It will suffice to show that $\mathscr{I}_\sigma = \mathscr{I}$. Suppose $e \in \mathscr{I}$ and k_n is the multiplicity sequence of e. Since $k_n e \in \mathscr{I}$, we have $k_n \in \mathscr{I}$ by Lemma 1. Then $k_n \in \mathscr{I} \cap Z = \mathscr{I}_\sigma \cap Z \subset \mathscr{I}_\sigma$ for all n, therefore $e \in \mathscr{I}_\sigma$ by Lemma 2. ∎

Conversely:

Lemma 4. *If $\mathscr{I}$ is a maximal p-ideal of A, then there exists a unique $\sigma \in \mathscr{X}$ such that $\mathscr{I} = \mathscr{I}_\sigma$.*

Proof. Since $\mathscr{I} \cap Z$ is a maximal p-ideal of Z [§ 24, Prop. 3], there exists $\sigma \in \mathscr{X}$ such that $\mathscr{I} \cap Z = \mathscr{J}_\sigma = \{h \in \tilde{Z}: h(\sigma) = 0\}$. But $\mathscr{I}_\sigma$ is also a maximal p-ideal of A (Lemma 3) and $\mathscr{I}_\sigma \cap Z = \mathscr{J}_\sigma = \mathscr{I} \cap Z$ [§ 37, Prop. 1], therefore $\mathscr{I}_\sigma = \mathscr{I}$ by weak centrality [§ 24, Prop. 2]. If also $\mathscr{I} = \mathscr{I}_\tau$, then $\mathscr{J}_\tau = \mathscr{I}_\tau \cap Z = \mathscr{I}_\sigma \cap Z = \mathscr{J}_\sigma$, therefore $\tau = \sigma$. ∎

Combining Lemmas 3 and 4 with [§ 24, Prop. 3], we have the promised description:

Theorem 1. *The mapping* $\sigma \mapsto \mathscr{I}_\sigma$ *is a bijection of* $\mathscr{X}$ *onto the set of all maximal p-ideals of* A.

Alternatively, writing $I_\sigma = \{x \in A : \mathrm{RP}(x) \in \mathscr{I}_\sigma\}$ [§ 37, Def. 2], we have [§ 22, Th. 1]:

Corollary. *The mapping* $\sigma \mapsto I_\sigma$ *is a bijection of* $\mathscr{X}$ *onto the set of all maximal-restricted ideals of* A.

This is a convenient place to record the following proposition, which is a device for dropping down to direct summands:

Proposition 1. *Let* I *be a maximal-restricted ideal of* A *and let* h *be a central projection such that* $h \notin I$. *Then* hI *is a maximal-restricted ideal of* hA, *and* A/I *is* $*$*-isomorphic to* hA/hI.

Proof. Write $\pi: A \to A/I$ for the canonical mapping. Since $h \notin I$ we have $1-h \in I$ [§ 24, Prop. 1], thus $x - hx \in I$ for all $x \in A$. It follows that if π_0 is the restriction of π to hA, then $\pi_0 : hA \to A/I$ is surjective. The kernel of π_0 is $(hA) \cap I = hI$, thus A/I is $*$-isomorphic with hA/hI. It is elementary that hI is a restricted ideal of hA [cf. § 22, Def. 2]; it remains to show that there are no restricted ideals properly between hI and hA. Recall that $\mathrm{RP}(\pi(x)) = \pi(\mathrm{RP}(x))$ for all $x \in A$ [§ 23, Prop. 1].

Some general remarks: Suppose B and C are Rickart $*$-rings and $\varphi: B \to C$ is a $*$-homomorphism of B onto C such that $\mathrm{RP}(\varphi(x)) = \varphi(\mathrm{RP}(x))$ for all $x \in B$. It is routine to check that if J is a restricted ideal of B then $\varphi(J)$ is a restricted ideal of C; and if K is a restricted ideal of C, then $\varphi^{-1}(K)$ is a restricted ideal of B containing the kernel of φ (the latter is even a strict ideal of B [cf. § 22, Exer. 1]). It follows that the kernel of φ is a maximal-restricted ideal of B if and only if C is 'restricted-simple', i.e., C has no restricted ideals other that $\{0\}$ and C.

Apply the preceding paragraph to the canonical mapping $A \to A/I$: since I is maximal-restricted in A, we infer that A/I is restricted-simple. Then the $*$-isomorphic ring hA/hI is also restricted-simple; applying the preceding paragraph to the canonical mapping $hA \to hA/hI$, we conclude that hI is maximal-restricted in hA. ∎

Exercises

1A. If $\mathscr{I}$ is a proper p-ideal of A, then there exists at least one point $\sigma \in \mathscr{X}$ such that $\mathscr{I} \subset \mathscr{I}_\sigma$; in particular, for all $e \in \mathscr{I}$, $D(e)$ is a singular element of the Banach algebra $C(\mathscr{X})$.

2A. Let N be a proper ideal of $C(\mathscr{X})$ and define $\mathscr{I}_N = \{e \in \tilde{A} : D(e) \in N\}$.

(i) $\mathscr{I}_N$ is a proper p-ideal of A such that $\mathscr{I}_N \cap Z = \tilde{N}$.

(ii) If N is closed then $\mathscr{I}_N$ is the intersection of the maximal p-ideals that contain it.

(iii) Let I_N be the restricted ideal generated by $\mathscr{I}_N$. If e, f are projections in A such that $e-f\in I_N$, then $D(e)-D(f)\in N$.

3A. Suppose A is of Type II, and let $\mathscr{I}$ be a p-ideal of A. Define $P=\{\alpha D(e): \alpha>0, e\in\mathscr{I}\}_{\sim}$. Then $N=P-P+iP-iP$ is an ideal of $C(\mathscr{X})$ whose positive part is P, and $\tilde{N}=\mathscr{I}\cap Z$.

§ 40. Dimension in A/I

We fix a maximal-restricted ideal I of A, and write $I=I_\sigma$ for a suitable $\sigma\in\mathscr{X}$ [§ 39, Cor. of Th. 1].

Reviewing Section 23, we know that A/I is a finite Rickart $*$-ring with GC, satisfying LP $\sim$ RP, and the canonical mapping $x\mapsto\tilde{x}$ of A onto A/I enjoys the properties listed in [§ 23, Prop. 1]. Moreover, A/I is a factor [§ 24, Prop. 1]. {Alternatively, it is obvious from $I=I_\sigma$ that, for a central projection h, either $h\in I$ or $1-h\in I$, thus A/I is a factor by [§ 23, Cor. of Prop. 5].}

Our ultimate objective is to prove that A/I is a Baer $*$-ring. Since A/I is a Rickart $*$-ring, it will suffice to show that every orthogonal family of projections in A/I has a supremum [§ 4, Prop. 1]. In this section we show, by passing to quotients with the dimension function, that A/I is orthoseparable (that is, only countable orthogonal families occur).

Lemma. *If e and f are projections in A such that $e-f\in I$, then $D(e)(\sigma)=D(f)(\sigma)$.*

Proof. Since A satisfies the parallelogram law (P) [§ 13, Prop. 2], there exist orthogonal decompositions

$$e=e'+e'', \qquad f=f'+f''$$

such that $e'\sim f'$ and $ef''=e''f=0$ [§ 13, Prop. 5]. Since I contains $e-f$, it also contains $(e-f)e''=ee''-0=e''$ and $(f-e)f''=f''$, thus $D(e'')(\sigma)=D(f'')(\sigma)=0$. Since $D(e')=D(f')$ and D is additive, we have $D(e)-D(f)=D(e'')-D(f'')$, and evaluation at σ yields $D(e)(\sigma)-D(f)(\sigma)=0$. ∎

Definition 1. We define a real-valued function D_I on the projection lattice of A/I as follows. If u is a projection in A/I, write $u=\tilde{e}$ with e a projection in A; if also $u=\tilde{f}$, f a projection in A, then $e-f\in I$, therefore $D(e)(\sigma)=D(f)(\sigma)$ by the lemma. We define (unambiguously) $D_I(u)=D(e)(\sigma)$. Thus

$$D_I(\tilde{e})=D(e)(\sigma)$$

for all projections e in A.

Proposition 1. *The real-valued function* D_I *on* $(A/I)^{\sim}$ *has the following properties:*

(1) $0 \leq D_I(u) \leq 1$,

(2) $D_I(1) = 1$,

(3) $D_I(u) = 0$ *iff* $u = 0$.

(4) $uv = 0$ *implies* $D_I(u+v) = D_I(u) + D_I(v)$,

(5) $u \sim v$ *iff* $D_I(u) = D_I(v)$,

(6) $u \lesssim v$ *iff* $D_I(u) \leq D_I(v)$.

Proof. (1) and (2) are obvious.

(3) If $u = \tilde{e}$, $e \in \tilde{A}$, then $u = 0$ iff $e \in \tilde{I} = \mathscr{I}_\sigma$ iff $D(e)(\sigma) = 0$, that is, $D_I(u) = 0$.

(4) Suppose $uv = 0$. Write $u = \tilde{e}$, $v = \tilde{f}$ with $ef = 0$ [§ 23, Prop. 2]. Then $e+f$ is a projection, $D(e+f) = D(e) + D(f)$ and $u + v = \tilde{e} + \tilde{f} = (e+f)^{\sim}$, therefore

$$D_I(u+v) = D(e+f)(\sigma) = D(e)(\sigma) + D(f)(\sigma) = D_I(u) + D_I(v).$$

(5), (6) Suppose $u \sim v$. Write $u = \tilde{e}$, $v = \tilde{f}$ with $e \sim f$ [§ 23, Prop. 1]. Then $D(e) = D(f)$, therefore $D_I(u) = D(e)(\sigma) = D(f)(\sigma) = D_I(v)$.

Since A/I has GC and is a factor, any two projections u, v in A/I are comparable: $u \lesssim v$ or $v \lesssim u$. Moreover, A/I is finite, thus the proofs of (5), (6) may be completed by the arguments in [§ 27, Prop. 1]. ∎

When the proof that A/I is a Baer $*$-ring is completed, D_I will be its unique dimension function [§ 33, Th. 1], and in particular, D_I will be completely additive. For the present, we are content to exploit finite additivity to prove the following:

Proposition 2. *A/I is orthoseparable.*

Proof. Let $(u_\varkappa)_{\varkappa \in K}$ be any orthogonal family of nonzero projections in A/I. For $n = 1, 2, 3, \ldots$ write

$$K_n = \left\{ \varkappa \in K : D_I(u_\varkappa) > \frac{1}{n} \right\}.$$

By (3) of Proposition 1, we have

$$K = \bigcup_{n=1}^{\infty} K_n;$$

it will suffice to show that each K_n is finite. Indeed, if $\varkappa_1, \ldots, \varkappa_r \in K_n$ are distinct, then

$$1 \geq D_I(u_{\varkappa_1} + \cdots + u_{\varkappa_r}) = D_I(u_{\varkappa_1}) + \cdots + D_I(u_{\varkappa_r}) > \frac{r}{n},$$

thus $r < n$. ∎

Thus, to complete the proof that A/I is a Baer $*$-ring, it remains to show that every sequence of orthogonal projections in A/I has a supremum. For A of Type II, this is quite easy (Section 41); for A of Type I_n, it is nearly trivial (Section 42); the most complicated case, again [cf. § 37], is that where A is of Type I with homogeneous summands of arbitrarily large order (Section 43).

§ 41. A/I Theorem: Type II Case

We assume in this section that A is of Type II. Fix a maximal-restricted ideal I in A, and write $I=I_\sigma$ for a suitable $\sigma\in\mathscr{X}$ [§ 39, Cor. of Th. 1] (for a shortcut, see the discussion in Section 37). Let D_I be the dimension function for A/I introduced in the preceding section.

Theorem 1. *Suppose A is a finite Baer $*$-ring of Type* II, *satisfying* LP$\sim$RP, *and let I be a maximal-restricted ideal of A. Then A/I is a finite Baer $*$-factor of Type* II. *Moreover, A/I satisfies* LP$\sim$RP, *and any two projections in A/I are comparable.*

Proof. From the discussion in Section 40, two things remain to be shown: (1) every orthogonal sequence of projections in A/I has a supremum; (2) A/I is of Type II. Granted (1), the proof of (2) is easy: if u is any nonzero projection in A/I, say $u=\tilde{e}$ with $e\in\tilde{A}$, one can write $e=f+g$ with $f\sim g$ [§ 19, Th. 1]; then $u=\tilde{f}+\tilde{g}$ with $\tilde{f}\sim\tilde{g}$, therefore u is not abelian [§ 19, Lemma 1], thus A/I is continuous [§ 15, Def. 3, (4 b)].

Suppose $u_1, u_2, u_3, \ldots$ is an orthogonal sequence of projections in A/I. The plan is to construct a projection u such that $u_n\le u$ for all n and $\sum_1^\infty D_I(u_n)=D_I(u)$, and to infer from these properties that $u=\sup u_n$.

Let $\alpha_n=D_I(u_n)$. For all n,

$$\sum_1^n \alpha_i=\sum_1^n D_I(u_i)=D_I\left(\sum_1^n u_i\right)\le 1\,;$$

defining $\alpha=\sum_1^\infty\alpha_i$, we have $0\le\alpha\le 1$. We seek a projection u such that $u_n\le u$ for all n and $D_I(u)=\alpha$. Write $u_n=\tilde{e}_n$, with e_n an orthogonal sequence of projections in A [§ 23, Prop. 2]; in particular, $\alpha_n=D_I(u_n)$ $=D(e_n)(\sigma)$ [§ 40, Def. 1]. Since $0\le\alpha_n\le 1$ and A is of Type II, there exists, for each n, a projection $f_n\in A$ such that $D(f_n)=\alpha_n 1$ [§ 33, Th. 3]. In particular, $D(f_n)(\sigma)=\alpha_n=D(e_n)(\sigma)$, thus $D_I(\tilde{f}_n)=D_I(\tilde{e}_n)$; it follows that $\tilde{e}_n\sim\tilde{f}_n$ [§ 40, Prop. 1], hence there exist subprojections $g_n\le e_n$, $h_n\le f_n$ with

$$g_n\sim h_n,\qquad \tilde{g}_n=\tilde{e}_n=u_n,\qquad \tilde{h}_n=\tilde{f}_n$$

[§23, Prop. 1]. Then $D(g_n)=D(h_n)\le D(f_n)=\alpha_n 1$; it follows that for every finite set J of positive integers,

$$(*)\qquad \sum_{i\in J} D(g_i)\le\Big(\sum_{i\in J}\alpha_i\Big)1\le\alpha 1\,.$$

Define $g=\sup g_n$. Since the e_n are orthogonal, so are the g_n; in view of $(*)$, the complete additivity of D yields

$$(**)\qquad D(g)=\sum_1^\infty D(g_i)\le\alpha 1\,.$$

Define $u=\tilde{g}$. Since $g\ge g_n$, we have $u\ge u_n$ for all n.

We assert that $D_I(u)=\alpha$. For all n, we have $\sum_1^n u_i\le u$; then

$$\sum_1^n\alpha_i=\sum_1^n D_I(u_i)=D_I\Big(\sum_1^n u_i\Big)\le D_I(u)$$

for all n, thus $\alpha\le D_I(u)$. On the other hand, it follows from $(**)$ that $D_I(u)=D(g)(\sigma)\le\alpha$. {By the use of constant functions, we have circumnavigated the fact that the 'infinite sums' in $C(\mathscr{X})$ described in Section 27 cannot in general be evaluated pointwise.}

Finally, assuming v is a projection in A/I such that $u_n\le v$ for all n, it is to be shown that $u\le v$, that is, $u(1-v)=0$. Say $v=\tilde{f}$, f a projection in A, set $x=g(1-f)$, and assume to the contrary that $\tilde{x}\ne 0$, that is, $x\notin I$. Then $\mathrm{LP}(x)\notin I$ (because I is an ideal); writing $g_0=\mathrm{LP}(x)$, we have $g_0\le g$, $g_0\notin I$. Thus, setting $w=\tilde{g}_0$, we have $w\le u$, $w\ne 0$, and $w=(\mathrm{LP}(x))^\sim=\mathrm{LP}(\tilde{x})$. Note that w is orthogonal to every u_n; indeed,

$$u_n\tilde{x}=u_n u(1-v)=u_n(1-v)=0\,,$$

therefore $u_n\mathrm{LP}(\tilde{x})=0$, that is, $u_n w=0$. Since $w,u_1,u_2,\dots,u_n$ are orthogonal subprojections of u, we have

$$D_I(w)+\sum_1^n D_I(u_i)=D_I\Big(w+\sum_1^n u_i\Big)\le D_I(u)\,,$$

thus $D_I(w)+\sum_1^n\alpha_i\le\alpha$; since n is arbitrary, it results that $D_I(w)+\alpha\le\alpha$, thus $D_I(w)=0$, $w=0$, a contradiction. ∎

§ 42. *A/I* Theorem: Type I_n Case

We assume in this section that A is of Type I_n [§18, Def. 2] ($n=1$ is admitted, that is, A can be abelian). Fix a maximal-restricted ideal I of A, and write $I=I_\sigma$ for a suitable $\sigma\in\mathscr{X}$ [§39, Cor. of Th. 1] (for a shortcut, see the discussion in Section 37). Let D_I be the dimension function for A/I introduced in Section 40. The main result of this section:

Theorem 1. *Suppose A is a finite Baer $*$-ring of Type I_n, satisfying* LP $\sim$ RP, *and let I be a maximal-restricted ideal of A. Then A/I is a finite Baer $*$-factor of Type I_n. Moreover, A/I satisfies* LP $\sim$ RP, *and any two projections in A/I are comparable.*

We approach the proof through two lemmas.

Lemma 1. *If f is any abelian projection in A, then f is simple and $D(f)=(1/n)C(f)$.*

Proof. By hypothesis, there exists a faithful abelian projection e such that $(1:e)=n$, that is, there exists an orthogonal decomposition

$$1=e_1+\cdots+e_n$$

with $e\sim e_1\sim\cdots\sim e_n$. Then

$$C(f)=e_1C(f)+\cdots+e_nC(f)$$

with $eC(f)\sim e_1C(f)\sim\cdots\sim e_nC(f)$, and it will clearly suffice to show that $f\sim eC(f)$. Indeed, $eC(f)$ and f are abelian projections such that $C(eC(f))=C(e)C(f)=C(f)$ [§6, Prop. 1, (iii)], therefore $eC(f)\sim f$ [§18, Prop. 1]. ∎

Lemma 2. *If e is any projection in A, then the values of $D(e)$ are contained in the set $\{v/n: v=0,1,\ldots,n\}$.*

Proof. Write $e=\sup e_\iota$, with $(e_\iota)_{\iota\in J}$ an orthogonal family of abelian projections (see [§26, Prop. 14] or [§18, Exer. 2]). From Lemma 1 we know that $D(e_\iota)(\mathscr{X})\subset\{0,1/n\}$ for every $\iota\in J$.

Let (h_α) be an orthogonal family of nonzero central projections with $\sup h_\alpha=1$, such that for each α, the set $J_\alpha=\{\iota\in J: h_\alpha e_\iota\neq 0\}$ is finite [§18, Prop. 5]. Let P_α be the clopen set in $\mathscr{X}$ whose characteristic function is (identified with) h_α, thus $P_\alpha=\{\tau\in\mathscr{X}: h_\alpha(\tau)=1\}$. Since $\sup h_\alpha=1$, $\mathscr{Y}=\bigcup_\alpha P_\alpha$ is a dense open set in $\mathscr{X}$.

Write $F=\{v/n:\ v=0,1,\ldots,n\}$. Since $\mathscr{Y}$ is dense and $D(e)$ is continuous, it will suffice to show that $D(e)(\mathscr{Y})\subset F$; fixing an index α, it is enough to show that $D(e)(P_\alpha)\subset F$. We have

$$h_\alpha D(e)=D(h_\alpha e)=D\Big(\sum_{\iota\in J_\alpha}h_\alpha e_\iota\Big)=h_\alpha\sum_{\iota\in J_\alpha}D(e_\iota);$$

evaluating at any $\tau\in P_\alpha$,

$$1\geq 1\cdot D(e)(\tau)=\sum_{\iota\in J_\alpha}D(e_\iota)(\tau),$$

and since $D(e_\iota)(\tau)\in\{0, 1/n\}$, it results that $D(e)(\tau)\in F$. {Incidentally, $D(e)$ is a simple function: if k_ν is the characteristic function of the set $\{\tau: D(e)(\tau)=\nu/n\}$, then $D(e)=\sum_{\nu=0}^{n}(\nu/n)k_\nu$.} ∎

Proof of Theorem 1. If $u_1, \ldots, u_k$ are orthogonal, nonzero projections in A/I, then since $D_I(u_i)\geq 1/n$ by Lemma 2 [cf. §40, Def. 1], we have

$$1\geq D_I(u_1+\cdots+u_k)=\sum_1^k D_I(u_i)\geq \frac{k}{n},$$

thus $k\leq n$. This shows that every orthogonal family of nonzero projections in A/I is finite; since their sum serves as supremum, the discussion in Section 40 shows that A/I is a finite Baer *-factor, with comparability of projections, satisfying LP ~ RP.

It remains to show that A/I is of Type I_n. Let e_1 be an abelian projection in A such that $(1:e_1)=n$, and write

$$1=e_1+\cdots+e_n$$

with $e_1\sim\cdots\sim e_n$. Setting $u_i=\tilde{e}_i$, we have

$$1=u_1+\cdots+u_n$$

with $u_1\sim\cdots\sim u_n$; in particular, $D_I(u_1)=1/n$. The proof will be concluded by showing that u_1 is a minimal (hence trivially abelian) projection. If u is a nonzero projection with $u\leq u_1$, then $0<D_I(u)\leq D_I(u_1)=1/n$; but $D_I(u)\geq 1/n$ by Lemma 2, thus $D_I(u)=D_I(u_1)$, $D_I(u_1-u)=0$, $u_1-u=0$. ∎

Let us note a slight extension of Theorem 1. With A again a general finite Baer *-ring satisfying LP ~ RP, suppose h is a nonzero central projection in A such that hA is of Type I_n. Let P be the clopen subset of $\mathscr{X}$ corresponding to h. Fix $\sigma\in P$ and let $I=I_\sigma$; thus $h(\sigma)=1$, equivalently $1-h\in I$, equivalently, $h\notin I$. We assert that A/I has the properties listed in Theorem 1: this is immediate from [§39, Prop. 1], and Theorem 1 applied to hA.

Exercises

1A. With notation as in Theorem 1 and its proof, identify A with $(e_1Ae_1)_n$ [§16, Prop. 1].

(i) $I=\{x\in A: D(\mathrm{RP}(x))=0$ on a neighborhood of $\sigma\}$.

(ii) $I=J_n$, where $J=\{a\in e_1Ae_1: ha=0$ for some central projection h with $h(\sigma)=1\}$.

(iii) Thus $A/I=B_n$, where $B=e_1Ae_1/J$ has no divisors of zero.

2A. In order that there exist orthogonal projections $e_1,\ldots,e_n$ in A with $e_1+\cdots+e_n=1$ and $e_1\sim\cdots\sim e_n$, it is necessary and sufficient that the order of every homogeneous summand of A be a multiple of n.

§ 43. A/I Theorem: Type I Case

We assume in this section that A is of Type I. Fix a maximal-restricted ideal I of A, and write $I = I_\sigma$ for a suitable $\sigma \in \mathscr{X}$ [§39, Cor. of Th. 1]. Let D_I be the dimension function for A/I introduced in Section 40.

By the structure theory for Type I rings, there exists an orthogonal sequence (possibly finite) $h_1, h_2, h_3, \ldots$ of nonzero central projections, with $\sup h_i = 1$, such that $h_i A$ is homogeneous of Type I_{n_i} [§18, Th. 2]. We can suppose $n_1 < n_2 < n_3 < \cdots$.

Let P_i be the clopen subset of $\mathscr{X}$ corresponding to h_i and let $\mathscr{Y} = \bigcup_i P_i$; since $\sup h_i = 1$, $\mathscr{Y}$ is a dense open set in $\mathscr{X}$.

If there are only finitely many h_i—say $h_1, \ldots, h_r$—then $\mathscr{Y} = P_1 \cup \cdots \cup P_r$ is clopen, hence $\mathscr{Y} = \mathscr{X}$. Conversely, if $\mathscr{Y} = \mathscr{X}$ then since $\mathscr{X}$ is compact, the disjoint open covering (P_i) must be finite, thus there are only finitely many h_i. To put it another way, it is clear that $\mathscr{Y}$ is a proper subset of $\mathscr{X}$ iff (h_i) is an infinite sequence (iff A has homogeneous summands of arbitrarily large order), and in this case $n_i \to \infty$ as $i \to \infty$.

Theorem 1. *Suppose A is a finite Baer $*$-ring of Type* I, *satisfying* LP $\sim$ RP, *and let I be a maximal-restricted ideal of A. Then A/I is a finite Baer $*$-factor, satisfying* LP $\sim$ RP, *and any two projections in A/I are comparable. Adopt the above notations, in particular $I = I_\sigma$. If $\sigma \in \mathscr{X} - \mathscr{Y}$ then A/I is of Type* II; *if $\sigma \in \mathscr{Y}$, say $\sigma \in P_i$, then A/I is of Type* I_{n_i}.

If $\sigma \in \mathscr{Y}$ then the discussion at the end of the preceding section is applicable.

We suppose for the rest of the section that $\sigma \in \mathscr{X} - \mathscr{Y}$ (which is possible only if A has homogeneous summands of arbitrarily large order). In particular, as noted above, $n_i \to \infty$ as $i \to \infty$. We are to show that A/I is a Baer $*$-factor of Type II.

Lemma 1. *If $0 < \alpha < 1$, then there exists a projection $f \in A$ such that $D(f) \le \alpha 1$ and $D(f)(\tau) = \alpha$ for all $\tau \in \mathscr{X} - \mathscr{Y}$.*

Proof. {In the application below, we require only $D(f)(\sigma) = \alpha$, but it is no harder to get $D(f) = \alpha$ on $\mathscr{X} - \mathscr{Y}$.}

First, a topological remark: if U is any neighborhood of σ, then U intersects infinitely many of the P_i. {Suppose to the contrary that $U \cap \mathscr{Y} \subset P_1 \cup \cdots \cup P_m$. Since $\sigma \notin P_1 \cup \cdots \cup P_m$ (indeed, $\sigma \notin \mathscr{Y}$), $V = \mathscr{X} - (P_1 \cup \cdots \cup P_m)$ is a neighborhood of σ; then $U \cap V$ is a neighborhood of σ with $U \cap V \cap \mathscr{Y} = \varnothing$, contrary to the fact that $\mathscr{Y}$ is dense in $\mathscr{X}$.}

For each i, write $F_i = \{\mu/n_i : \mu = 0, 1, \ldots, n_i\}$. Since $0 < \alpha < 1$, for each i there exists $\alpha_i \in F_i$ such that

$$0 \le \alpha - \alpha_i \le \frac{1}{n_i}; \tag{1}$$

since $n_i \to \infty$ as $i \to \infty$, we have $\alpha_i \to \alpha$.

Since, for each i, h_iA is homogeneous of order n_i, there exists a projection $f_i \le h_i$ such that $D(f_i) = \alpha_i h_i$ (take f_i to be the sum of $n_i\alpha_i$ orthogonal equivalent copies of a faithful abelian projection in h_iA). Since $\alpha_i \le \alpha$ by (1), it follows that for every finite set J of positive integers,

$$\sum_{i \in J} D(f_i) = \sum_{i \in J} \alpha_i h_i \le \alpha 1 . \tag{2}$$

Define $f = \sup f_i$. Since the f_i are orthogonal and D is completely additive, it results from (2) that

$$D(f) = \sum_1^\infty D(f_i) \le \alpha 1 .$$

It remains to show that $D(f) = \alpha$ on $\mathscr{X} - \mathscr{Y}$. Fix $\tau' \in \mathscr{X} - \mathscr{Y}$; let us show that $D(f)(\tau') = \alpha$. For each i, define

$$U_i = \left\{ \tau \in \mathscr{X} : |D(f)(\tau) - D(f)(\tau')| < \frac{1}{n_i} \right\}; \tag{3}$$

by the continuity of $D(f)$, U_i is a neighborhood of τ', therefore, by the first part of the proof, U_i intersects P_j for infinitely many j.

Construct a subsequence of indices $i_1 < i_2 < i_3 < \cdots$ and a corresponding sequence of points $\tau_\nu \in U_\nu \cap P_{i_\nu}$ $(\nu = 1, 2, 3, \ldots)$ as follows. Set $i_1 = 1$ and let τ_1 be any point of $U_1 \cap P_1$. Since U_2 intersects infinitely many of the P_i, we may choose an index $i_2 > i_1$ and a point $\tau_2 \in U_2 \cap P_{i_2}$. And so on.

As $\nu \to \infty$, we have $i_\nu \to \infty$ hence $n_{i_\nu} \to \infty$; in view of (1), it follows that

$$\alpha - \alpha_{i_\nu} \to 0 \quad \text{as } \nu \to \infty . \tag{4}$$

For all ν,

$$|\alpha - D(f)(\tau')| \le |\alpha - D(f)(\tau_\nu)| + |D(f)(\tau_\nu) - D(f)(\tau')| ;$$

since $\tau_\nu \in P_{i_\nu}$ and since $h_{i_\nu} D(f) = D(h_{i_\nu} f) = D(f_{i_\nu}) = \alpha_{i_\nu} h_{i_\nu}$ is identically equal to α_{i_ν} on P_{i_ν}, we have $D(f)(\tau_\nu) = \alpha_{i_\nu}$ and thus

$$|\alpha - D(f)(\tau')| \le |\alpha - \alpha_{i_\nu}| + |D(f)(\tau_\nu) - D(f)(\tau')| ;$$

since $\tau_\nu \in U_\nu$, citing (3) we obtain

$$|\alpha - D(f)(\tau')| \le |\alpha - \alpha_{i_\nu}| + \frac{1}{n_{i_\nu}} ;$$

in view of (4), the right side tends to 0 as $\nu \to \infty$, thus $\alpha - D(f)(\tau') = 0$. ∎

Lemma 2. *Every orthogonal sequence of projections in A/I has a supremum.*

Proof. Let u_n be an orthogonal sequence of nonzero projections in A/I. Write $u_n = \tilde{e}_n$ with e_n an orthogonal sequence of projections in A [§23, Prop. 2]. Let $\alpha_n = D_I(u_n) = D(e_n)(\sigma)$; by Lemma 1 there exists a projection f_n in A such that $D(f_n) \leq \alpha_n 1$ and $D(f_n)(\sigma) = \alpha_n$. The argument proceeds exactly as in the proof of [§41, Th. 1]. ∎

Proof of Theorem 1. As noted prior to Lemma 1, we have only to dispose of the case that $\sigma \in \mathscr{X} - \mathscr{Y}$. By Lemma 2 and the discussion in Section 40, we know that A/I is a finite Baer $*$-factor, satisfying $\mathrm{LP} \sim \mathrm{RP}$, with comparability of projections. It remains only to show that A/I is of Type II, and this is immediate from the fact that if $0 < \alpha < 1$ then Lemma 1 yields a projection u in A/I with $D_I(u) = \alpha$. ∎

Exercise

1A. With notation as in Theorem 1, $\mathscr{X} - \mathscr{Y}$ is a closed subset of $\mathscr{X}$ with empty interior, therefore is rare (i. e., nowhere dense).

§ 44. Summary of Results

Gathering up the results in Sections 36 and 39–43, we have:

Theorem 1. *Let A be a finite Baer $*$-ring such that $\mathrm{LP}(x) \sim \mathrm{RP}(x)$ for all $x \in A$.*

(1) *If I is a maximal-restricted ideal of A, then A/I is a finite Baer $*$-factor, $\mathrm{LP}(u) \sim \mathrm{RP}(u)$ for all $u \in A/I$, and any two projections in A/I are comparable. In order that A/I be of Type* I, *it is necessary and sufficient that there exist a central projection h with $h \notin I$ and hA homogeneous (in which case A/I is also homogeneous, of the same order); otherwise, A/I is of Type* II.

(2) *The intersection of all the maximal-restricted ideals of A is $\{0\}$.*

(3) *Let $\mathscr{X}$ be the Stone representation space of the complete Boolean algebra of central projections of A, let D be the dimension function of A, and, for each $\sigma \in \mathscr{X}$, write*

$$I_\sigma = \{x \in A : D(\mathrm{RP}(x))(\sigma) = 0\}.$$

Then $\sigma \mapsto I_\sigma$ is a bijection of $\mathscr{X}$ onto the set of all maximal-restricted ideals of A.

Proof. (2) This is [§36, Prop. 1].

(3) This is [§39, Cor. of Th. 1].

(1) The first sentence is covered by the theorems in Sections 41–43.

If h is a central projection such that $h \notin I$ and hA is homogeneous, then, by the discussion at the end of Section 42, A/I is of Type I (and homogeneous, of the same order as hA).

Conversely, suppose A/I is of Type I. Let k be the central projection such that kA is of Type I and $(1-k)A$ is of Type II [§15, Th. 2]; in view of [§41, Th. 1], we know that A is not of Type II, thus $k \neq 0$. Let Q be the clopen subset of $\mathscr{X}$ corresponding to k. Write $I = I_\sigma$, $\sigma \in \mathscr{X}$. Necessarily $\sigma \in Q$, that is, $1-k \in I$. (If, on the contrary, $1-k \notin I$, it would follow from [§39, Prop. 1] that A/I is $*$-isomorphic to a quotient ring of the Type II ring $(1-k)A$, hence is of Type II [§41, Th. 1], contrary to hypothesis.) Then $k \notin I$, therefore A/I is $*$-isomorphic to kA/kI [§39, Prop. 1]. Let us apply the theory of Section 43 to the Type I ring kA: let h_i be an orthogonal sequence (possibly finite) of nonzero central projections with $\sup h_i = k$, such that $h_i(kA) = h_i A$ is homogeneous of Type I_{n_i}, $n_1 < n_2 < n_3 < \cdots$. Let P_i be the clopen subset of $\mathscr{X}$ corresponding to h_i, and let $\mathscr{Y} = \bigcup_i P_i$; since $\sup h_i = k$, $\mathscr{Y}$ is a dense open set in Q.

We know that $\sigma \in Q$. Regarding Q as the Stone representation space associated with the central projections of kA, observe that $kI = (kA) \cap I$ is the maximal-restricted ideal of kA corresponding to $\sigma \in Q$. (For, kI is a restricted ideal of kA, $kD(\cdot)$ is the dimension function of kA, and, for a projection $e \in A$, one has $e \in kI$ iff $e \leq k$ and $e \in I$ iff $e \in kA$ and $(kD(e))(\sigma) = D(e)(\sigma) = 0$.) It follows that $\sigma \in \mathscr{Y}$. (For, $\sigma \in Q - \mathscr{Y}$ would imply, by [§43, Th. 1] applied to kA, that kA/kI is of Type II, hence A/I is of Type II, contrary to hypothesis.) Then $\sigma \in P_i$ for some i; thus $h_i \notin I$, where $h_i A$ is homogeneous of Type I_{n_i}. ∎

Exercises

1A. Fix $\sigma \in \mathscr{X}$ and let $J = \{x \in A : xh = 0$ for some central projection h with $h(\sigma) = 1\}$ [cf. § 24, Exer. 5].

(i) $J = \{x \in A : D(\mathrm{RP}(x)) = 0$ on some neighborhood of $\sigma\}$.

(ii) A/J is a factorial Rickart $*$-ring.

(iii) A/J is a Baer $*$-ring iff $J = I_\sigma$ iff A/J is restricted-simple [cf. § 36, Exer. 3].

2D. The analogue of Part (2) of Theorem 1 holds for the maximal-strict ideals of a finite Baer $*$-ring with GC [§ 36, Exer. 4]. *Problem:* Does the analogue of Part (1) hold (explicitly, is the quotient ring a Baer $*$-ring)?

§ 45. *A/M* Theorem for a Finite *AW**-Algebra

The foregoing results provide a reduction theory for finite Baer $*$-rings, satisfying $\mathrm{LP} \sim \mathrm{RP}$, in terms of finite Baer $*$-factors. Since an AW^*-algebra satisfies $\mathrm{LP} \sim \mathrm{RP}$ [§20, Cor. of Th. 3], the results are applicable, in particular, to a finite AW^*-algebra A: regarding A as a finite Baer $*$-ring, we obtain a reduction of A into finite Baer $*$-factors A/I, where I varies over all the maximal-restricted ideals of A. However, the A/I are in general not AW^*-algebras (Exercise 9). To obtain a reduction of A in terms of finite AW^*-factors, it is necessary to replace the maximal-restricted ideals I by the maximal ideals M (the M's are just the closures of the I's [§ 22, Ths. 1, 2]); the present section is devoted to the details.

For the rest of the section, A denotes a finite AW^*-algebra. We maintain the notations established in Section 35: Z is the center of A; $\mathscr{X}$ is the Stone representation space of the complete Boolean algebra $\tilde{Z}$; D is the dimension function of A. A major new element in the picture is that Z is an AW^*-algebra [§ 4, Prop. 8, (v)], therefore Z is the closed linear span of its projections [§ 8, Prop. 3], therefore Z may be identified with $C(\mathscr{X})$; thus the dimension function D is Z-valued.

Fix a point $\sigma \in \mathscr{X}$. Let $\mathscr{I} = \{e \in \tilde{A}: D(e)(\sigma) = 0\}$, $I = \{x \in A: \mathrm{RP}(x) \in \mathscr{I}\}$, $M = \bar{I}$; thus $\mathscr{I}$, I and M are the most general maximal p-ideal, maximal-restricted ideal, and maximal ideal of A, respectively [§ 22, Ths. 1, 2] (a pertinent point is that every maximal ideal in a Banach algebra with unity is closed [cf. **75**, Cor. 2.1.4]). As in Section 40, we write $x \mapsto \tilde{x} = x + I$ for the canonical mapping $A \mapsto A/I$, and D_I for the dimension function of A/I [§ 40, Def. 1]. We write $x \mapsto \bar{x} = x + M$ for the canonical mapping $A \to A/M$. *Our objective is to prove that A/M is an AW^*-factor.* The first main step is to construct a dimension function D_M for A/M, largely imitating the techniques used in Section 40 to construct D_I.

Lemma 1. *A/M is a simple C^*-algebra.*

Proof. As noted above, the closed ideal M is maximal, therefore A/M is a simple Banach algebra with respect to the quotient norm $\|\bar{x}\| = \inf\{\|x+y\|: y \in M\}$ [cf. **75**, p. 44]. Since I is a $*$-ideal [§ 22, Prop. 2], so is its closure M (see also [§ 22, Th. 2] or [§ 21, Exer. 5]), therefore A/M admits the natural involution $(\bar{x})^* = (x^*)^-$. The proof is concluded by noting that $\|\bar{x}^* \bar{x}\| = \|\bar{x}\|^2$ for all $x \in A$; indeed, it is a standard result that any closed ideal in any C^*-algebra is a $*$-ideal, and the quotient algebra is a C^*-algebra with respect to the quotient norm [cf. **24**, Prop. 1.8.2]. ∎

Lemma 2. *If u is any projection in A/M, then $u = \bar{e}$ for a suitable projection e in A.*

Proof. Say $u = \bar{x}$. Then $u = u^* u = (x^* x)^-$; replacing x by $x^* x$, we can suppose $x \geq 0$.

Since $u^2 = u$, we have $x^2 - x = c$ for a suitable $c \in M$. If $c = 0$ we are through. Assume $c \neq 0$. Then also $x \neq 0$. Set $B = \{x, c\}''$; since $c^* = c$ and c commutes with x, B is a commutative AW^*-algebra [§ 4, Prop. 8]. Write $B = C(T)$, T a compact space. Let $N = M \cap B$. Then N is a closed ideal of B; since $1 \notin M$, N is proper; since $c \in N$, N is nonzero. It follows that there exists a nonempty closed subset S of T such that

$$N = \{b \in B: b(s) = 0 \text{ for all } s \in S\}.$$

Since $x^2-x=c\in N$, we have $x^2-x=0$ on S, thus

(1) $$s\in S \text{ implies } x(s)=0 \text{ or } x(s)=1.$$

Fix a real number ε, $0<\varepsilon<1$, and apply [§7, Prop. 3] to B and x: there exist $y\in B$ and a nonzero projection $e\in B$ such that

(2) $$xy=e,$$

(3) $$\|x-xe\|<\varepsilon.$$

Citing (2), we have $xe=x^2y=(x+c)y=e+cy$, where $cy\in NB\subset N$, thus $xe-e\in N$; then

(4) $$x(s)e(s)-e(s)=0 \quad \text{for all } s\in S.$$

We assert that $x-e\in N$ (hence $x-e\in M$, ending the proof); given any $s\in S$, it suffices to show that $x(s)-e(s)=0$. If $e(s)=0$, then $|x(s)|<\varepsilon<1$ by (3), therefore $x(s)=0$ by (1); if $e(s)=1$ then $x(s)=1$ by (4). ∎

Lemma 3. *If e, f are projections in A such that $e-f\in M$, then $D(e)(\sigma)=D(f)(\sigma)$.*

Proof. Write $e=e'+e''$, $f=f'+f''$ with $e'\sim f'$ and $ef''=e''f=0$ [§13, Th. 1, Prop. 5]. Since M contains $e-f$, it also contains $(e-f)e''=e''$ and $(f-e)f''=f''$. Then e'', $f''\in\tilde{M}=\tilde{I}=\mathscr{I}=\mathscr{I}_\sigma$ (recall that $M=\bar{I}$ contains no projections not already in I [§22, Prop. 4]), thus $D(e'')(\sigma)=D(f'')(\sigma)=0$. Then $D(e)(\sigma)=D(f)(\sigma)$ as in the proof of [§40, Lemma]. ∎

The way is prepared for defining a dimension function D_M for A/M:

Definition 1. If u is a projection in A/M, write $u=\bar{e}$ with e a projection in A (Lemma 2) and define $D_M(u)=D(e)(\sigma)$ (the definition is unambiguous by Lemma 3). Thus $D_M(\bar{e})=D_I(\tilde{e})$ [cf. §40, Def. 1]. Note that $D_M(u)=0$ iff $D_I(\tilde{e})=0$ iff $e\in I$ iff $e\in M$ iff $u=0$ (recall that $\tilde{M}=\tilde{I}=\mathscr{I}_\sigma$).

Lemma 4. *If u, v are projections in A/M such that $u\le v$, and if $v=\bar{f}$ with f a projection in A, then there exists a projection e in A such that $u=\bar{e}$ and $e\le f$.*

Proof. If x is any element of A with $u=\bar{x}$, then $u=vu^*uv=(fx^*xf)^-$; replacing x by fx^*xf, we can suppose $x\ge 0$ and $fx=x$. By the proof of Lemma 2, there exists a projection e in A such that $u=\bar{e}$ and $xy=e$ for a suitable element y. Since $fx=x$, we have $fe=e$, thus $e\le f$. ∎

Lemma 5. *If u_n is an orthogonal sequence of projections in A/M, then there exists an orthogonal sequence of projections e_n in A such that $u_n=\bar{e}_n$ for all n.*

Proof. This follows from Lemma 4, by the argument used in the proof of [§23, Prop. 2, (ii)]. ∎

Lemma 6. *If u, v are orthogonal projections in A/M, then* $D_M(u+v) = D_M(u)+D_M(v)$.

Proof. By Lemma 5, we may write $u=\bar{e}$, $v=\bar{f}$ with e, f orthogonal projections in A. Then $e+f$ is a projection with $(e+f)^- = u+v$, thus

$$D_M(u+v)=D(e+f)(\sigma)=D(e)(\sigma)+D(f)(\sigma)=D_M(u)+D_M(v). \quad ∎$$

We can now use D_M to establish orthoseparability:

Lemma 7. *A/M is orthoseparable.*

Proof. Formally the same as [§40, Prop. 2]. ∎

Since $I \subset M$, there is a natural mapping $A/I \to A/M$, namely $\tilde{x} \mapsto \bar{x}$ (an epimorphism with kernel M/I). If $I \subset M$ properly (that is, if I is not closed) then the mapping is not injective, but we are able to maneuver around this circumstance. In view of Lemma 2, and the like result for A/I [§23, Prop. 1, (ii)], the correspondence $\tilde{e} \mapsto \bar{e}\,(e \in \tilde{A})$ maps the set of all projections of A/I onto the set of all projections of A/M; it is useful to formalize this:

Definition 2. Define $\varphi:(A/I)^\sim \to (A/M)^\sim$ as follows. If $v \in (A/I)^\sim$, write $v=\tilde{e}$ with $e \in \tilde{A}$ and define $\varphi(v)=\bar{e}$. Since I and M contain the same projections, $\varphi(v)=0$ iff $v=0$.

We know that $(A/I)^\sim$ is a complete lattice [§44, Th. 1]. The general plan is to infer properties of $(A/M)^\sim$ via the mapping φ. The pertinent property of φ is the following:

Lemma 8. *If v_1, v_2 are projections in A/I such that $v_1 \le v_2$, then* $\varphi(v_1) \le \varphi(v_2)$.

Proof. Write $v_1=\tilde{e}_1$, $v_2=\tilde{e}_2$ with $e_1 \le e_2$ [§23, Prop. 2, (i)]. Then $\bar{e}_1 \le \bar{e}_2$, that is, $\varphi(v_1) \le \varphi(v_2)$. ∎

Lemma 9. *If u_n is an orthogonal sequence of projections in A/M, then there exists a projection $u \in A/M$ such that $u_n \le u$ for all n and* $D_M(u)=\sum D_M(u_n)$.

Proof. Write $u_n=\bar{e}_n$ with e_n an orthogonal sequence of projections in A (Lemma 5). Set $v_n=\tilde{e}_n$; thus v_n is an orthogonal sequence of projections in A/I with $\varphi(v_n)=u_n$. Define $v=\sup v_n$ [§44, Th. 1]. Say $v=\tilde{e}$, e a projection in A. Setting $u=\bar{e}$, we have $\varphi(v)=u$. Since $v_n \le v$ for all n, we have $\varphi(v_n) \le \varphi(v)$ by Lemma 8, that is, $u_n \le u$. Since $D_M(u)=D_M(\bar{e})$

$=D_I(\tilde{e})=D_I(v)$ and similarly $D_M(u_n)=D_I(v_n)$, it results from the countable additivity of D_I [§33, Th. 1] that

$$D_M(u)=D_I(v)=\sum D_I(v_n)=\sum D_M(u_n). \quad \blacksquare$$

The next lemma is of 'EP' type:

Lemma 10. *If* $s\in A/M$, $s\neq 0$, *then there exist* $t\in A/M$ *and a nonzero projection* $u\in A/M$ *such that* $ss^*t=u$.

Proof. Say $s=\bar{x}$, $x\in A$. Since $ss^*\neq 0$, we have $xx^*\notin M$. Write $a=xx^*$. By [§7, Prop. 3], there exists a sequence $b_n\in\{a\}''$ such that (1) $ab_n=e_n$, e_n a nonzero projection, and (2) $\|a-ae_n\|<1/n$. In particular, $ae_n\to a$; since $a\notin M$ and M is closed, there exists an index m such that $ae_m\notin M$, therefore $e_m\notin M$. Set $t=\bar{b}_m$, $u=\bar{e}_m$. $\blacksquare$

Lemma 11. *Every orthogonal sequence of projections in* A/M *has a supremum.*

Proof. Adopt the notation in the proof of Lemma 9. We assert that $u=\sup u_n$. It remains only to show that if w is a projection in A/M such that $u_n\le w$ for all n, then $u\le w$. Let $s=u(1-w)$ and assume to the contrary that $s\neq 0$. By Lemma 10, there exists $t\in A/M$ such that $st=u'$, u' a nonzero projection. Since $us=s$ we have $uu'=u'$, thus $u'\le u$. Moreover, u' is orthogonal to every u_n: $u_nu'=u_nst=u_nu(1-w)t$ $=u_n(1-w)t=0$ (because $u_n\le w$). Thus, for every n, $u'+\sum_1^n u_i$ is a subprojection of u; it follows from the additivity of D_M (Lemma 6) that

$$D_M(u)\ge D_M\left(u'+\sum_1^n u_i\right)=D_M(u')+\sum_1^n D(u_i)$$

for all n. Letting $n\to\infty$ and citing the formula in Lemma 9, we have

$$D_M(u)\ge D_M(u')+D_M(u),$$

therefore $D_M(u')=0$, $u'=0$, a contradiction. $\blacksquare$

The principal result, stated in full:

Theorem 1. *If A is a finite AW^*-algebra and M is a maximal ideal of A, then A/M is a finite AW^*-factor.*

Proof. Let us first show that A/M is a Baer $*$-ring. Every orthogonal family of nonzero projections in A/M is countable (Lemma 7) hence has a supremum (Lemma 11); thus it will suffice to show that A/M is a Rickart $*$-ring [§4, Prop. 1].

Let $s \in A/M$. We seek a projection $u \in A/M$ such that $R(\{s\}) = u(A/M)$. Consider a maximal orthogonal family of nonzero projections in $R(\{s\})$ (if no such projections exist, then a slight rearrangement of the following argument shows that the choice $u=0$ works). We know that the family must be countable, say (u_n), and we can form $u = \sup u_n$. We assert that $su=0$. If, on the contrary, $su \neq 0$, then by Lemma 10 there exists $t \in A/M$ such that $t(su)^*(su) = u'$, u' a nonzero projection. Clearly $u'u = u'$. For all n, $(su)u_n = su_n = 0$, therefore $u'u_n = 0$, $u_n \leq 1 - u'$; then $u \leq 1 - u'$, $u'u = 0$, contrary to $u'u = u' \neq 0$.

Thus $u \in R(\{s\})$, hence $u(A/M) \subset R(\{s\})$. To show the reverse inclusion, suppose $sr=0$; it is to be shown that $ur=r$. Assuming to the contrary that $(1-u)r \neq 0$, Lemma 10 provides an element t such that $(1-u)rr^*(1-u)t = w$, w a nonzero projection. Obviously $uw=0$, therefore $u_n w = 0$ for all n. But $s(1-u)r = sr - sur = 0 - 0r = 0$, hence $sw=0$, that is, $w \in R(\{s\})$, contradicting the maximality of the family (u_n).

Summarizing, A/M is a Baer $*$-ring; in view of Lemma 1, it is an AW^*-algebra. Moreover, since A/M is simple, it is obvious that 0, 1 are its only central projections, thus A/M is a factor.

It remains to show that A/M is finite. Suppose u is a projection in A/M such that $u \sim 1$. If we could show that $D_M(u) = D_M(1)$, then $D_M(1-u) = 0$ would yield $1-u=0$. Thus the proof is completed by the next lemma. ∎

Lemma 12. *If u, v are projections in A/M such that $u \sim v$, then $D_M(u) = D_M(v)$.*

Proof. Write $u = \overline{e}$, $v = \overline{f}$ with e, f projections in A. By hypothesis, there exists $x \in A$ such that $\overline{x}^*\overline{x} = \overline{e}$, $\overline{x}\overline{x}^* = \overline{f}$. Then $\overline{x} = \overline{f}\overline{x}\overline{e} = (fxe)^-$; replacing x by fxe, we can suppose $fx = x = xe$. Thus if $f_0 = \mathrm{LP}(x)$, $e_0 = \mathrm{RP}(x)$, we have $f_0 \leq f$, $e_0 \leq e$ and $e_0 \sim f_0$ (but the latter equivalence has nothing to do with the hypothesized one).

Since $xe_0 = x$ and $ee_0 = e_0$, right-multiplication of $\overline{x}^*\overline{x} = \overline{e}$ by $\overline{e}_0$ yields $\overline{e} = \overline{e}_0$. Similarly $\overline{f} = \overline{f}_0$. Thus $u = \overline{e}_0$, $v = \overline{f}_0$; since $e_0 \sim f_0$ we have $D(e_0) = D(f_0)$, therefore

$$D_M(u) = D(e_0)(\sigma) = D(f_0)(\sigma) = D_M(v). \quad ∎$$

There remains the question of type for A/M, but this is easily referred back to the discussion for A/I:

Proposition 1. *A/M and A/I have the same type; i.e., they are both Type* II *or both Type* I *(with the same order).*

Proof. We know that A/M is a finite AW^*-factor, therefore it possesses a unique real-valued dimension function [§ 33, Th. 1]; but Lemma 12 completes the proof that D_M has the properties required of a dimension function [§25, Def. 1], thus D_M is the unique dimension function for A/M. In view of the relation

$$D_M(\bar{e}) = D_I(\tilde{e}) \quad \text{for all } e \in \tilde{A}$$

(see Definition 1), the functions D_M and D_I have the same range of values, hence A/M and A/I have the same type. ∎

Exercises

1C. There exists a 'reduction theory'—more analytic and less algebraic—valid for any von Neumann algebra on a separable Hilbert space.

2C. If A is a finite von Neumann algebra and M is a maximal ideal of A, then A/M is also a von Neumann algebra (on a suitable Hilbert space).

3A. A finite AW^*-algebra A is strongly semisimple (that is, the intersection of all the maximal ideals M is $\{0\}$); thus A is $*$-isomorphic to a subalgebra of the C^*-sum of the finite factors A/M.

4A. Let A be a finite AW^*-algebra. The following assertions hold (in increasing generality):

(i) If M is a maximal ideal of A, then $e \in \tilde{M}$ implies $D(e) \in M$.

(ii) If P is a primitive ideal of A, then $e \in \tilde{P}$ implies $D(e) \in P$.

(iii) If I is a closed ideal of A, then $e \in \tilde{I}$ implies $D(e) \in I$.

(iv) If I is a closed ideal of A and if e, f are projections in A such that $e - f \in I$, then $D(e) - D(f) \in I$.

5A. Let A be a finite AW^*-algebra and let I be a closed, restricted ideal of A.

(i) $e \in \tilde{I}$ implies $C(e) \in \tilde{I}$.

(ii) If e_n is a sequence of projections in I, then $\sup e_n \in I$.

(iii) If A is orthoseparable (equivalently, its center Z is orthoseparable [§ 33, Th. 4]), then $I = hA$ for a suitable central projection h.

6A. Let A be an orthoseparable finite AW^*-algebra and let I be a maximal-restricted ideal of A. Write $I = I_\sigma$ for a suitable $\sigma \in \mathscr{X}$. Then I is closed iff σ is an isolated point of $\mathscr{X}$.

7A. If A is an orthoseparable finite AW^*-algebra of Type I possessing homogeneous summands of arbitrarily large order, and if $I = I_\sigma$ with $\sigma \in \mathscr{X} - \mathscr{Y}$ (notation as in [§ 43, Th. 1]), then I is not closed.

8A. Let A be a finite AW^*-algebra of Type II, whose center Z is infinite-dimensional (cf. Exercise 11).

(i) Z contains an element c such that $0 \le c \le 1$, c is singular (i.e., not invertible in Z) and $\mathrm{RP}(c) = 1$.

(ii) With c as in (i), choose $\sigma \in \mathscr{X}$ with $c(\sigma) = 0$. Then I_σ is not closed.

9A. Let A be a finite AW^*-algebra and let I be a nonclosed maximal-restricted ideal of A (cf. Exercises 6–8). Then (i) A/I is not semisimple; therefore (ii) A/I is not strongly semisimple; therefore (iii) A/I does not satisfy the (VWEP)-axiom. In particular, (iv) A/I is a Baer $*$-factor that cannot be normed to be a C^*-algebra.

10B. Let Z be a commutative AW^*-algebra. Write $Z=C(\mathscr{X})$, $\mathscr{X}$ a Stonian space. Fix $\sigma\in\mathscr{X}$ and define

$$N=\{c\in Z: c(\sigma)=0\},$$

$$J=\{c\in Z: c=0 \text{ on some neighborhood of } \sigma\}.$$

(i) J is a maximal-restricted ideal of Z, and $\bar{J}=N$.

(ii) Z/J is an integral domain, with unique maximal ideal N/J.

(iii) The following conditions are equivalent: (a) $J=N$ (that is, $J=\bar{J}$); (b) Z/J is semisimple; (c) Z/J is simple; (d) Z/J is a field; (e) Z/J is the field $\mathbb{C}$ of complex numbers.

Let n be a positive integer and let $A=Z_n$.

(iv) A is an AW^*-algebra of Type I_n, whose center may be identified with Z.

Let $I=I_\sigma$ be the maximal-restricted ideal of A corresponding to σ, that is, $I=\{x\in A: D(\mathrm{RP}(x))(\sigma)=0\}$.

(v) $I=\{x\in A: D(\mathrm{RP}(x))\in J\}$.

(vi) $I=J_n$.

(vii) A/I is $*$-isomorphic to $(Z/J)_n$.

(viii) Z/J is a Prüfer ring.

(ix) $\bar{I}$ is the unique maximal ideal of A containing I.

(x) The following conditions are equivalent: (a′) $I=\bar{I}$; (b′) A/I is semisimple; (c′) A/I is simple. In this case, $A/I=\mathbb{C}_n$.

In particular: For each positive integer n, $(Z/J)_n$ is a finite Baer $*$-factor of Type I_n, satisfying $\mathrm{LP}\sim\mathrm{RP}$.

11C. The following conditions on a finite AW^*-algebra A are equivalent: (a) every maximal-restricted ideal of A is closed; (a′) every maximal-restricted ideal of A is maximal; (a″) every maximal ideal of A is restricted; (b) Z is finite-dimensional; (b′) $\mathscr{X}$ is finite; (c) A is the direct sum of finitely many finite AW^*-factors. In particular, such an algebra is orthoseparable.

12B. Notation as in Exercise 10. Assume, in addition, that Z is infinite-dimensional (equivalently, $\mathscr{X}$ is infinite) and orthoseparable. (For example, let Z be any infinite-dimensional commutative von Neumann algebra on a separable Hilbert space.) Choose (as we may) σ to be a nonisolated point of $\mathscr{X}$. Then none of the conditions in (iii), (x) of Exercise 10 hold.

Application: Z 'the' algebra of diagonal operators on a separable, infinite-dimensional Hilbert space (relative to a fixed orthonormal basis) [cf. **37**, p. 29]. Then $\mathscr{X}=\beta\mathbb{N}$ (the Stone-Čech compactification of the discrete space $\mathbb{N}$ of positive integers), and the nonisolated points σ are the points of $\beta\mathbb{N}-\mathbb{N}$.

Chapter 8

The Regular Ring of a Finite Baer $*$-Ring

The present chapter is based on [**8**].

§ 46. Preliminaries

At the outset we assume that *A is a finite Baer $*$-ring satisfying* $\mathrm{LP}\sim\mathrm{RP}$ (cf. the remarks at the beginning of Chapter 7); at this level of generality, the questions outweigh the answers, but it is surprising how far one can get. Gradually the hypotheses are strengthened, the main results being obtained for a somewhat restricted class of finite Baer $*$-rings (including all finite AW^*-algebras). As an application, it is shown in the next chapter that the algebra of all $n\times n$ matrices over any AW^*-algebra (finite or not) is also an AW^*-algebra [§62, Cor. 1 of Th. 1].

Before motivating the chapter informally, we record for convenient reference some consequences of the above assumption:

Proposition 1. *If A is a finite Baer $*$-ring satisfying* $\mathrm{LP}\sim\mathrm{RP}$, *then*

(1) *A satisfies the parallelogram law* (P),

(2) *A has* GC,

(3) *A has a unique dimension function D, and D is completely additive,*

(4) $D(e\cup f)+D(e\cap f)=D(e)+D(f)$ *for every pair of projections e, f in A, and*

(5) *the relations* $x,y\in A$, $yx=1$ *imply* $xy=1$.

Proof. (1), (2) See [§14, Cor. 2 of Prop. 7].

(3) This is true for any finite Baer $*$-ring with GC [§33, Th. 1].

(4) This follows at once from the parallelogram law and the properties of D.

(5) If $yx=1$ it is clear that $R(\{x\})=\{0\}$, thus $\mathrm{RP}(x)=1$; then $\mathrm{LP}(x)\sim\mathrm{RP}(x)=1$ implies, by finiteness, that $\mathrm{LP}(x)=1$. Then $(1-xy)x=x-xyx=x-x=0$, and $1-xy=0$ results from $\mathrm{LP}(x)=1$. ∎

As noted above, $yx=1$ implies that $\mathrm{RP}(x)=1$ (and that x is invertible). The converse is false; $\mathrm{RP}(x)=1$ does not imply that x is invertible (cf. Exercise 1). In a von Neumann algebra, $\mathrm{RP}(x)=1$ means that x is injective; in general, one can hope to get an inverse for x only if one is willing to accept an unbounded operator.

Returning to abstract rings, the 'rings with invertiblity' par excellence are the regular rings: a ring C is said to be regular if, for each $x \in C$, there exists $y \in C$ with $xyx = x$. It follows that $e = yx$ and $f = xy$ are idempotents; $xe = x = fx$ shows that they are trying to act like a 'right-idempotent' and a 'left-idempotent' for x; and y acts like an 'inverse' for x relative to the idempotents e and f. We defer until Section 51 a closer look at regularity; for the present it suffices, by way of motivation, to report the following fact: Each finite von Neumann algebra $\mathscr{A}$ may be enlarged, in a canonical and minimal way, to a regular ring $\mathscr{C}$; the new elements of $\mathscr{C}$ are certain unbounded operators 'affiliated' with $\mathscr{A}$, these being assumed densely defined and closed; the algebraic operations in $\mathscr{C}$ are the usual ones followed by the closure operation (e.g., the 'product' of two operators in $\mathscr{C}$ is the closure of the usual composition product).

Consider now the given finite Baer $*$-ring A. Can A be enlarged to a regular Baer $*$-ring C? In general, we do not know. Following the exposition in **[8]**, it is shown in the next section that a $*$-ring $\mathbf{C}$ containing A can be constructed, along the lines of the classical construction for von Neumann algebras; and in Section 48 it is shown that $\mathbf{C}$ has no new projections (i.e., none not already in A) and that $\mathbf{C}$ is also a finite Baer $*$-ring satisfying LP $\sim$ RP. Gradually strengthening the hypotheses on A, we reach regularity of $\mathbf{C}$ in Section 52; the hypotheses on A are regrettably restrictive, but at least they are inherited by $\mathbf{C}$ (Section 53) and they are satisfied by the finite AW^*-algebras. {In the wake of all this technology, it should be remembered that some rather unpropertied finite Baer $*$-rings—for example, the ring of 2×2 matrices over the field of three elements—are already regular.}

Why consider $\mathbf{C}$ at all? The answer lies in the fact that the ring of $n \times n$ matrices over a regular ring is known to be regular. Putting aside its intrinsic algebraic interest, the justification for considering $\mathbf{C}$ is that it is needed in the proof that the algebra of $n \times n$ matrices over an AW^*-algebra is also an AW^*-algebra **[5]**; the latter result is exposed in the next chapter (see Section 62).

Remarks. There is a related circle of ideas that should be mentioned here. The projection lattice of A is complemented (with a canonical complementation $e \mapsto e' = 1 - e$) and modular [§ 34, Prop. 1]. Suppose, in addition, that for some integer n, $n \geq 4$, there exist n orthogonal equivalent projections $e_1, \ldots, e_n$ with sum 1. (This is equivalent to assuming that $A = B_n$ for some $*$-ring B [§ 16, Prop. 1], or that the order of every homogeneous summand of A is divisible by n [§ 42, Exer. 2].) The e_i are pairwise perspective [§ 17, Exer. 12, (vi)]. Summarizing:

the projections of A form a complemented modular lattice of 'order' ≥ 4 in the sense of von Neumann [**71**, p. 93]. It follows that there exists a regular ring R, unique up to isomorphism, such that the projection lattice of A is isomorphic to the lattice $\mathscr{R}$ of principal right ideals of R (this is von Neumann's coordinatization theorem [**71**, p. 208, Th. 14.1]). The mapping $e \mapsto e' = 1-e$ is an order-reversing involution in the projection lattice of A; it induces a like mapping in the lattice $\mathscr{R}$, denoted, say, $(xR) \mapsto (xR)'$. It follows that the ring R possesses an involution $x \mapsto x^*$ such that $(xR)'$ is the right-annihilator of Rx^* (order ≥ 3 is sufficient for this result [**71**, p. 113, Th. 4.3]). Since $e \cap e' = e(1-e) = 0$, it follows that $(xR) \cap (xR)' = (0)$ for all $x \in R$, therefore the involution of R is proper [**71**, p. 114, Th. 4.5]. Regular rings with proper involution are called $*$-regular; we look into such matters in Section 51, but mention here that such a ring is a Rickart $*$-ring. Thus, R is a regular Rickart $*$-ring; its projection lattice is identifiable with the lattice $\mathscr{R}$ in the obvious way, hence is isomorphic with the projection lattice of A, hence is complete; thus R is a regular Baer $*$-ring.

To summarize: Assuming A has an order $n, n \geq 4$, there exists a regular Baer $*$-ring R such that A and R have isomorphic projection lattices. This is achieved without extra hypotheses on A (other than finiteness, LP $\sim$ RP, and the order condition). However, the only connection between A and R is their common projection lattice. Whether A can in general be embedded as a subring of R is not known (presumably, one has to be able to 'reconstitute' the elements of A and R in terms of their projections via some sort of 'spectral theory').

How does this compare with what is accomplished in the present chapter? We do reach a regular Baer $*$-ring $\mathbf{C}$ with the same projection lattice as A. (1) The hypothesis on 'order' is dropped (though we essentially steer clear of abelian summands by assuming partial isometries to be addable); (2) we must impose a number of rather severe axioms on A; (3) A is visibly a $*$-subring of $\mathbf{C}$ from the start.

Exercises

1A. Let $\mathscr{H}$ be a separable, infinite-dimensional Hilbert space, let $\xi_1, \xi_2, \xi_3, \ldots$ be an orthonormal basis of $\mathscr{H}$, let T be the operator defined by $T\xi_n = (1/n)\xi_n$, and let $\mathscr{A} = \{T\}''$ be the bicommutant of T, calculated in $\mathscr{L}(\mathscr{H})$. Then $\mathscr{A}$ is a commutative (hence finite) von Neumann algebra, $\mathrm{RP}(T) = I$, but T is not invertible.

2A. If e is any projection in A, then eAe is also a finite Baer $*$-ring satisfying LP $\sim$ RP.

3A, D. If B is a $*$-subring of A such that $B = B''$, then B is also a finite Baer $*$-ring.

Problem: Does B satisfy LP $\sim$ RP? Does it have GC?

The answers are affirmative if A satisfies the (EP)-axiom and the (SR)-axiom.

§ 47. Construction of the Ring C

We assume in this section that A is a finite Baer $*$-ring satisfying LP$\sim$RP [cf. § 46, Prop. 1]. This hypothesis is adequate for the construction of **C**, and for the development of many of its properties (Section 48), but we are able to show that **C** is regular only under added hypotheses (Section 52). Definitions and terminology are motivated by the von Neumann algebra case; the construction follows verbatim the reference [**4**], to which the reader is referred for heuristic discussions of the operatorial motivation.

Definition 1. A *strongly dense domain* (SDD) in A is a sequence of projections (e_n) such that $e_n \uparrow 1$; that is, $e_1 \le e_2 \le e_3 \le \cdots$ and $\sup e_n = 1$. {Thinking heuristically in terms of linear operators, the ranges of the e_n are an increasing sequence of closed linear subspaces whose union is a dense linear subspace. We caution the reader that this picture is adequate for the von Neumann algebra case but may be incorrect in the AW^*-case, where the calculation of suprema is not tied down to vectors.}

Lemma 1. *If* (e_n) *and* (f_n) *are* SDD, *then* $(e_n \cap f_n)$ *is an* SDD.

Proof. Setting $g_n = e_n \cap f_n$ and $g = \sup g_n$, we have $g_n \uparrow g$. For all n,

$$1 - g \le 1 - g_n = (1 - e_n) \cup (1 - f_n),$$

therefore

$$(*) \qquad D(1-g) \le D(1-e_n) + D(1-f_n)$$

[§ 46, Prop. 1, (4)]. Since $D(1-e_n) \downarrow 0$ and $D(1-f_n) \downarrow 0$ [§ 33, Th. 2], it follows from $(*)$ that $D(1-g)=0$ [cf. § 27, Lemma 1, (ii)], thus $g=1$. ∎

Lemma 1 extends at once to finitely many SDD $(e_n), (f_n), \ldots, (k_n)$: the term-by-term infimum $(e_n \cap f_n \cap \ldots \cap k_n)$ is also an SDD. Among other things, SDD are a technique for inferring properties 'in the limit'; the following is a sample:

Lemma 2. *Let* (e_n) *be an* SDD *and let* $x \in A$.

(1) *If* $e_n x e_n = 0$ *for all* n, *then* $x = 0$.

(2) *If* $e_n x e_n$ *is self-adjoint for all* n, *then* x *is self-adjoint.*

Proof. (1) Fix an index m. For all $n \ge m$, $e_n(x e_m) = (e_n x e_n) e_m = 0$; it follows from $\sup\{e_n : n \ge m\} = 1$ that $x e_m = 0$. Since m is arbitrary, $x = 0$.

(2) For all n, $e_n(x^* - x)e_n = (e_n x e_n)^* - (e_n x e_n) = 0$, therefore $x^* - x = 0$ by (1). ∎

Definition 2. An *operator with closure* (OWC) is a pair of sequences (x_n, e_n) with $x_n \in A$ and (e_n) an SDD, such that $m<n$ implies $x_n e_m = x_m e_m$ and $x_n^* e_m = x_m^* e_m$. {Heuristically, one can think of (x_n, e_n) as a linear operator whose restriction to the range of e_n is $x_n e_n$; the condition $(x_n e_n) e_m = x_n e_m = x_m e_m$ for $m<n$ means that the $x_n e_n$ are 'strung together coherently', and similarly for the $x_n^* e_n$.}

Lemma 3. *If* (x_n, e_n) *and* (y_n, f_n) *are* OWC, *then so are* (x_n^*, e_n) *and* $(x_n + y_n, e_n \cap f_n)$. *If, in addition, A is a $*$-algebra over an involutive field F, then* $(\lambda x_n, e_n)$ *is an* OWC *for each* $\lambda \in F$.

Proof. Set $g_n = e_n \cap f_n$. If $m<n$ then

$$\begin{aligned}(x_n + y_n) g_m &= x_n g_m + y_n g_m = x_n(e_m g_m) + y_n(f_m g_m)\\ &= (x_n e_m) g_m + (y_n f_m) g_m = (x_m e_m) g_m + (y_m f_m) g_m\\ &= (x_m + y_m) g_m,\end{aligned}$$

and similarly $(x_n + y_n)^* g_m = (x_m + y_m)^* g_m$; since (g_n) is an SDD (Lemma 1) we conclude that $(x_n + y_n, g_n)$ is an OWC. The other two assertions of the lemma are obvious. ∎

The discussion of products requires another concept:

Definition 3. If $x \in A$ and e is a projection in A, we write $x^{-1}(e)$ for the largest projection g such that $(1-e)xg = 0$, that is, $exg = xg$; thus $x^{-1}(e) = 1 - \mathrm{RP}[(1-e)x]$. {Thinking heuristically in terms of linear operators, the range of the projection $x^{-1}(e)$ is the largest subspace mapped by x into the range of e; identifying subspaces and projections, $x^{-1}(e)$ is the inverse image of e under x (hence the notation).}

Lemma 4. *If* $x \in A$ *and* e *is any projection, then* $e \precsim x^{-1}(e)$.

Proof. $1 - x^{-1}(e) = \mathrm{RP}[(1-e)x] \sim \mathrm{LP}[(1-e)x] \le 1-e$, therefore

$$e \le 1 - \mathrm{LP}[(1-e)x] \sim 1 - \mathrm{RP}[(1-e)x] = x^{-1}(e)$$

by finiteness [§ 17, Prop. 4, (ii)]. ∎

Lemma 5. *Suppose* (e_n) *is an* SDD *and* (x_n) *is a sequence in A such that* $m<n$ *implies* $x_n e_m = x_m e_m$. *Let* (f_n) *be any* SDD *and define* $g_n = e_n \cap x_n^{-1}(f_n)$. *Then* (g_n) *is an* SDD.

Proof. Let $h_n = x_n^{-1}(f_n)$; thus $g_n = e_n \cap h_n$ and h_n is the largest projection such that

$$(*) \qquad (1-f_n) x_n h_n = 0.$$

Note that $g_n \uparrow$. {Proof: If $m<n$ then

$$x_n g_m = x_n e_m g_m = x_m e_m g_m = x_m g_m = x_m h_m g_m,$$

hence

$$(1-f_n)x_n g_m = (1-f_n)(1-f_m)x_m h_m g_m = 0$$

by (∗), thus $g_m \leq h_n$ by the maximality of h_n; also $g_m \leq e_m \leq e_n$, thus $g_m \leq e_n \cap h_n = g_n$.} It is to be shown that $g_n \uparrow 1$. Since $D(h_n) \geq D(f_n)$ by Lemma 4, the relation $1-g_n = (1-e_n) \cup (1-h_n)$ yields

$$D(1-g_n) \leq D(1-e_n) + D(1-h_n) \leq D(1-e_n) + D(1-f_n),$$

therefore $D(1-g_n) \downarrow 0$ (cf. the proof of Lemma 1). ∎

This leads to a natural notion of product for OWC:

Lemma 6. *If* (x_n, e_n) *and* (y_n, f_n) *are* OWC, *and if*

$$k_n = [f_n \cap y_n^{-1}(e_n)] \cap [e_n \cap (x_n^*)^{-1}(f_n)],$$

then $(x_n y_n, k_n)$ *is an* OWC.

Proof. (k_n) is an SDD by Lemmas 5 and 1. Clearly $(1-e_n)y_n k_n = (1-f_n)x_n^* k_n = 0$ (cf. Definition 3), thus

$$(*) \qquad e_n y_n k_n = y_n k_n, \qquad f_n x_n^* k_n = x_n^* k_n.$$

If $m < n$ then $(x_n y_n)k_m = (x_m y_m)k_m$ results from the computation (quote the first equation of (∗) at the appropriate steps)

$$\begin{aligned}(x_n y_n)k_m &= x_n y_n f_m k_m = x_n y_m f_m k_m = x_n(y_m k_m) \\ &= x_n(e_m y_m k_m) = x_m e_m y_m k_m = x_m y_m k_m,\end{aligned}$$

and $(x_n y_n)^* k_m = (x_m y_m)^* k_m$ results similarly from the second equation of (∗). ∎

If $x \in A$ and (e_n) is any SDD, clearly $(x, 1)$ and (x, e_n) deserve to be regarded as 'equal' (they agree, so to speak, on a dense linear subspace); this calls for an equivalence relation:

Definition 4. We say that the OWC (x_n, e_n), (y_n, f_n) are *equivalent*, written $(x_n, e_n) \equiv (y_n, f_n)$, if there exists an SDD (g_n) such that $x_n g_n = y_n g_n$ and $x_n^* g_n = y_n^* g_n$ for all n. (Cf. [§ 48, Prop. 2].) The equivalence is said to be *implemented* via the SDD (g_n).

Lemma 7. *The relation* $\equiv$ *defined above is an equivalence relation in the set of all* OWC.

Proof. Reflexivity and symmetry are obvious. If $(x_n, e_n) \equiv (y_n, f_n)$ via (h_n), and if $(y_n, f_n) \equiv (z_n, g_n)$ via (k_n), then $(x_n, e_n) \equiv (z_n, g_n)$ via $(h_n \cap k_n)$. ∎

The following lemma is useful in the manipulation of representatives of equivalence classes:

Lemma 8. (i) *If* (x_n, e_n) *is an* OWC *and* (g_n) *is any* SDD, *then* $(x_n, e_n \cap g_n)$ *is also an* OWC *and* $(x_n, e_n) \equiv (x_n, e_n \cap g_n)$.

(ii) *Suppose* $(x_n, e_n) \equiv (y_n f_n)$ *via an* SDD (g_n). *Set* $h_n = e_n \cap f_n \cap g_n$. *Then* (x_n, h_n), (y_n, h_n) *are* OWC, *and* $(x_n, h_n) \equiv (y_n, h_n)$ *via* (h_n).

Proof. Routine. ▮

Definition 5. We write $[x_n, e_n]$ for the equivalence class of the OWC (x_n, e_n) with respect to the equivalence relation $\equiv$ defined above. The set of all equivalence classes is denoted $\mathbf{C}$, and its elements are called *closed operators* (CO); $[x_n, e_n]$ is the CO *determined* by the OWC (x_n, e_n). {In our heuristic model, the passage from (x_n, e_n) to $[x_n, e_n]$ corresponds to forming the closure of the linear operator (x_n, e_n), proper care having been taken—via Definition 2—to ensure that the adjoint operator is also densely defined.} We denote the elements of $\mathbf{C}$ by boldface letters $\mathbf{x}, \mathbf{y}, \mathbf{z}, \ldots$. If $x \in A$ we write $\overline{x} = [x, 1]$ for the CO determined by the pair $(x, 1)$ of constant sequences; we write $\overline{A} = \{\overline{x} : x \in A\}$ for the image of A in $\mathbf{C}$ under the injective (Lemma 2) mapping $x \mapsto \overline{x}$, and, more generally, $\overline{S}$ for the image of a subset $S \subset A$.

Lemmas 3 and 6 suggest algebraic operations for OWC:

$$(x_n, e_n) + (y_n, f_n) = (x_n + y_n, e_n \cap f_n),$$

$$(x_n, e_n)^* = (x_n^*, e_n),$$

$$(x_n, e_n)(y_n, f_n) = (x_n y_n, k_n)$$

and, when relevant,

$$\lambda(x_n, e_n) = (\lambda x_n, e_n).$$

Algebraic operations in $\mathbf{C}$ are defined by passage to quotients modulo the equivalence relation $\equiv$; the details are as follows.

If $(x_n, e_n) \equiv (x_n', e_n')$ via (g_n), and $(y_n, f_n) \equiv (y_n', f_n')$ via (h_n), then

$$(x_n + y_n, e_n \cap f_n) \equiv (x_n' + y_n', e_n' \cap f_n')$$

via $(g_n \cap h_n)$. It is trivial that $(x_n^*, e_n) \equiv (x_n'^*, e_n')$ via (g_n), and, when relevant, $(\lambda x_n, e_n) \equiv (\lambda x_n', e_n')$ via (g_n). This paves the way for the following definitions:

Definition 6. If $\mathbf{x}, \mathbf{y} \in \mathbf{C}$, say $\mathbf{x} = [x_n, e_n]$, $\mathbf{y} = [y_n, f_n]$, define

$$\mathbf{x} + \mathbf{y} = [x_n + y_n, e_n \cap f_n],$$

$$\mathbf{x}^* = [x_n^*, e_n];$$

when A is a $*$-algebra over an involutive field F, define

$$\lambda \mathbf{x} = [\lambda x_n, e_n]$$

for all $\lambda \in F$. Define

$$\mathbf{x}\mathbf{y} = [x_n y_n, k_n],$$

where (k_n) is the SDD given in Lemma 6; this is well-defined, by the following lemma:

Lemma 9. *Suppose* $(x_n, e_n) \equiv (x'_n, e'_n)$ *and* $(y_n, f_n) \equiv (y'_n, f'_n)$ *and suppose* (k_n), (k'_n) *are* SDD *such that* $(x_n y_n, k_n)$ *and* $(x'_n y'_n, k'_n)$ *are* OWC. *Then* $(x_n y_n, k_n) \equiv (x'_n y'_n, k'_n)$.

Proof. Let (g_n) be an SDD implementing $(x_n, e_n) \equiv (x'_n, e'_n)$; we can suppose, in addition, that (x_n, g_n) and (x'_n, g_n) are OWC (Lemma 8). Similarly, let (h_n) be an SDD implementing $(y_n, f_n) \equiv (y'_n, f'_n)$, such that (y_n, h_n) and (y'_n, h_n) are OWC.

Changing notation, we can suppose that $(x_n, e_n) \equiv (x'_n, e_n)$ via (e_n), $(y_n, f_n) \equiv (y'_n, f_n)$ via (f_n), and that (k_n), (k'_n) are SDD such that $(x_n y_n, k_n)$ and $(x'_n y'_n, k'_n)$ are OWC; we are to show that $(x_n y_n, k_n) \equiv (x'_n y'_n, k'_n)$. Define

$$g_n = f_n \cap y_n^{-1}(e_n);$$

then (g_n) is an SDD (Lemma 5) and $(1-e_n) y_n g_n = 0$ (because $g_n \leq y_n^{-1}(e_n)$), that is, $e_n y_n g_n = y_n g_n$; it follows that

(i) $$(x_n y_n) g_n = (x'_n y'_n) g_n.$$

{Proof: $x_n y_n g_n = x_n e_n y_n g_n = x'_n e_n y_n g_n = x'_n y_n g_n = x'_n y_n f_n g_n = x'_n y'_n f_n g_n = x'_n y'_n g_n$.} Similarly, defining

$$h_n = e_n \cap (x_n^*)^{-1}(f_n),$$

(h_n) is an SDD such that

(ii) $$(y_n^* x_n^*) h_n = (y_n'^* x_n'^*) h_n.$$

Setting $d_n = g_n \cap h_n$, (i) and (ii) yield

$$(x_n y_n) d_n = (x'_n y'_n) d_n, \qquad (x_n y_n)^* d_n = (x'_n y'_n)^* d_n,$$

thus $(x_n y_n, k_n) \equiv (x'_n y'_n, k'_n)$ via (d_n). ∎

Theorem 1. *Let A be a finite Baer $*$-ring satisfying* LP $\sim$ RP. *Define* **C**, *and the operations in* **C**, *as indicated above. Then* (1) **C** *is a $*$-ring with unity* $\overline{1}$ *(and, if A is a $*$-algebra over an involutive field F, then so is* **C**), *and* (2) *the mapping* $x \mapsto \overline{x}$ *is a $*$-isomorphism of A onto a $*$-subring* $\overline{A}$ *of* **C**.

Proof. (1) We illustrate the routine proof by verifying the associative law $(\mathbf{x}\mathbf{y})\mathbf{z} = \mathbf{x}(\mathbf{y}\mathbf{z})$. Say

$$\mathbf{x} = [x_n, e_n], \qquad \mathbf{y} = [y_n, f_n], \qquad \mathbf{z} = [z_n, g_n].$$

Then $\mathbf{x}\mathbf{y}=[x_n y_n, h_n]$ for suitable (h_n), hence $(\mathbf{x}\mathbf{y})\mathbf{z}=[(x_n y_n)z_n, k_n]$ for suitable (k_n). Similarly, $\mathbf{x}(\mathbf{y}\mathbf{z})=[x_n(y_n z_n), k_n']$ for suitable (k_n'). Since $(x_n y_n z_n, k_n)\equiv(x_n y_n z_n, k_n')$ (via any SDD), we conclude that $(\mathbf{x}\mathbf{y})\mathbf{z}=\mathbf{x}(\mathbf{y}\mathbf{z})$.

(2) For example, to show that $(x+y)^{-}=\bar{x}+\bar{y}$, define $\bar{x}+\bar{y}$ by applying Definition 6 to the representatives $(x,1)$, $(y,1)$ of the equivalence classes $\bar{x}, \bar{y}$:

$$\bar{x}+\bar{y}=[x+y, 1\cap 1]=(x+y)^{-}. \quad \blacksquare$$

We write 1 for the unity element of $\mathbf{C}$, that is, we identify 1 with $\bar{1}$. But in general we refrain from identifying x with $\bar{x}$; there are conceptual advantages to maintaining the distinction between A and $\bar{A}$ (as we do throughout this chapter) until the properties of $\mathbf{C}$ have been fully developed. The question of when $\bar{A}=\mathbf{C}$ is discussed in Sections 51 and 53.

In the next section we develop some properties of $\mathbf{C}$ that require no further hypotheses on A.

Exercises

1A. Let A be a finite Baer $*$-ring satisfying $\mathrm{LP}\sim\mathrm{RP}$ and let e be a projection in A (thus eAe is also a finite Baer $*$-ring satisfying $\mathrm{LP}\sim\mathrm{RP}$).

(i) $\bar{e}\mathbf{C}\bar{e}$ coincides with the set of all $\mathbf{x}\in\mathbf{C}$ such that $\mathbf{x}=[x_n, e_n]$ for suitable $x_n\in eAe$.

(ii) Writing $\mathbf{C}_{eAe}$ for the ring accruing to eAe via Theorem 1, we have

$$\mathbf{C}_{eAe}=\bar{e}\mathbf{C}\bar{e}$$

in the sense that there exists a natural $*$-isomorphism between the two rings.

2A. Lemma 2 holds with (e_n) replaced by any increasingly directed family of projections (e_ι) such that $\sup e_\iota=1$.

3A. Let A be a finite Baer $*$-ring satisfying $\mathrm{LP}\sim\mathrm{RP}$, let Λ be a fixed limit ordinal, and consider well-directed families of projections $(e_\rho)_{\rho<\Lambda}$ with $e_\rho\uparrow 1$.

(i) The entire section can be recast in terms of such families; write $\mathbf{C}_\Lambda$ for the ring given by the analogue of Theorem 1. (In particular, $\mathbf{C}=\mathbf{C}_\omega$, where ω is the first infinite ordinal.)

(ii) $\mathbf{C}_{\alpha+\omega}=\mathbf{C}$ for any ordinal α.

(iii) If $\Lambda=\Omega$ (the first uncountable ordinal) and if A is orthoseparable, then the construction collapses: $\mathbf{C}_\Omega=\bar{A}$.

4A. Let A be a finite Baer $*$-ring satisfying $\mathrm{LP}\sim\mathrm{RP}$ and let (e_n) be an SDD in A. If x,y are elements of A such that, for all n, $e_n x e_n$ commutes with $e_n y e_n$, then $xy=yx$. In particular, if $e_n x e_n$ is normal for all n, then x is also normal.

§ 48. First Properties of C

As in the preceding section, A is a finite Baer $*$-ring satisfying $\mathrm{LP}\sim\mathrm{RP}$, and $\mathbf{C}$ is the ring constructed there. We show in this section that $\mathbf{C}$ has no new projections (they are all of the form $\bar{e}$, e a projection in A) and that **C** is itself a finite Baer $*$-ring satisfying LP$\sim$RP (and a little more).

The first proposition, which confirms the heuristic guide in [§ 47, Def. 2], is useful in lifting properties from A to **C**:

Proposition 1. *If* $\mathbf{x} \in \mathbf{C}$, $\mathbf{x} = [x_n, e_n]$, *then*

$$\mathbf{x}\bar{e}_m = \overline{x_m e_m}, \qquad \bar{e}_m \mathbf{x} = \overline{e_m x_m}$$

for all m.

Proof. Fix an index m. Define an SDD (f_n) by the trivial choices $f_n = 0$ for $n < m$ and $f_n = 1$ for $n \geq m$. For all n, one has

$$(x_n e_m) f_n = (x_m e_m) f_n, \qquad (x_n e_m)^* f_n = (x_m e_m)^* f_n;$$

indeed, for $n \geq m$ these reduce to $x_n e_m = x_m e_m$, and for $n < m$ they reduce to $0 = 0$. Since $\mathbf{x}\bar{e}_m = [x_n e_m, g_n]$ for suitable SDD (g_n), and the foregoing relations show that $(x_n e_m, g_n) \equiv (x_m e_m, 1)$ via (f_n), we have $[x_n e_m, g_n] = [x_m e_m, 1]$, that is, $\mathbf{x}\bar{e}_m = (x_m e_m)^-$. It follows that $\mathbf{x}^* \bar{e}_m = [x_n^*, e_n] \bar{e}_m = (x_m^* e_m)^-$, thus $\bar{e}_m \mathbf{x} = (e_m x_m)^-$. ∎

In the next proposition, we simplify the task of verifying equivalence of OWC; so to speak, in testing for equivalence, the adjoints take care of themselves:

Proposition 2. *If* $\mathbf{x} = [x_n, e_n]$, $\mathbf{y} = [y_n, f_n]$ *and if* (g_n) *is an* SDD *such that*

$$x_n g_n = y_n g_n \quad \textit{for all } n,$$

then $\mathbf{x} = \mathbf{y}$. *In fact, it suffices to assume that* $h_n x_n g_n = h_n y_n g_n$ *for a pair of* SDD (g_n), (h_n).

Proof. We prove the second assertion. Note first that if (k_n) is any SDD, then (k_n, k_n) is an OWC such that $(k_n, k_n) \equiv (1, 1)$ via (k_n), thus $[k_n, k_n] = 1$. Then

$$\mathbf{x} = 1\mathbf{x}1 = [h_n, h_n][x_n, e_n][g_n, g_n] = [h_n x_n g_n, r_n]$$

for a suitable SDD (r_n), and similarly $\mathbf{y} = [h_n y_n g_n, s_n]$ for suitable (s_n); since $(h_n x_n g_n, r_n) \equiv (h_n x_n g_n, s_n) = (h_n y_n g_n, s_n)$ (via any SDD) we conclude that $\mathbf{x} = \mathbf{y}$. ∎

Lemma. *If* $\mathbf{x} \in \mathbf{C}$ *then there exists a projection* $f \in A$ *such that* (1) $\bar{f}\mathbf{x} = \mathbf{x}$, *and* (2) $\mathbf{y}\mathbf{x} = 0$ *if and only if* $\mathbf{y}\bar{f} = 0$.

Proof. Say $\mathbf{x} = [x_n, e_n]$. Define $f_n = \mathrm{LP}(x_n e_n)$ and let $f = \sup f_n$; thus f is the smallest projection in A such that

$$(*) \qquad f x_n e_n = x_n e_n \quad \text{for all } n.$$

Since $\bar{f}\mathbf{x}=[fx_n, g_n]$ for suitable (g_n), $\bar{f}\mathbf{x}=\mathbf{x}$ results from $(*)$ and Proposition 2.

Suppose $\mathbf{y}\in\mathbf{C}$ with $\mathbf{yx}=0$. Say $\mathbf{y}=[y_n, h_n]$. For all m and n, we have, by Proposition 1,

$$0=\bar{h}_n\mathbf{y}\mathbf{x}\bar{e}_m=(h_n y_n x_m e_m)^{-},$$

thus $(h_n y_n)(x_m e_m)=0$ and therefore $(h_n y_n) f_m=0$; since m is arbitrary, $h_n y_n f=0$, and since n is arbitrary, $\mathbf{y}\bar{f}=0$ by Proposition 2. Conversely, $\mathbf{y}\bar{f}=0$ implies $\mathbf{yx}=\mathbf{y}(\bar{f}\mathbf{x})=0$. ∎

Theorem 1. *$\mathbf{C}$ has no new projections; that is, if $\mathbf{e}\in\mathbf{C}$ is a projection then $\mathbf{e}=\bar{e}$ with e a (unique) projection in A.*

Proof. By the lemma, there exists a projection e in A such that $\mathbf{C}(1-\bar{e})=L(\{\mathbf{e}\})$; but $L(\{\mathbf{e}\})=\mathbf{C}(1-\mathbf{e})$, therefore $1-\bar{e}=1-\mathbf{e}$. ∎

Corollary 1. *$\mathbf{C}$ is a Baer $*$-ring.*

Proof. Since $\mathbf{C}$ has no new projections, its projections form a complete lattice (isomorphic to the projection lattice of A via $e\mapsto\bar{e}$); moreover, the lemma to the theorem shows that $\mathbf{C}$ is a Rickart $*$-ring, therefore $\mathbf{C}$ is a Baer $*$-ring [§4, Prop. 1]. ∎

We remark that if $x\in A$ and $e=\mathrm{RP}(x)$, then $\mathrm{RP}(\bar{x})$ exists (Corollary 1), and it follows from Theorem 1 that $\mathrm{RP}(\bar{x})=\bar{e}$; thus $\mathrm{RP}(\bar{x})=(\mathrm{RP}(x))^{-}$ for all x in A.

Corollary 2. *$\mathbf{C}$ has* GC *and satisfies the parallelogram law* (P).

Proof. Since A has these properties [§ 46, Prop. 1] and $\mathbf{C}$ has no new projections, $\mathbf{C}$ inherits the properties via the embedding $x\mapsto\bar{x}$. ∎

Corollary 2 will be superseded by the next theorem.

Lemma. *If (x_n) is a sequence in A and (e_n) is an* SDD *such that $x_n e_m=x_m e_m$ whenever $m<n$, and if $f_n=\mathrm{LP}(x_n e_n)$, then $f_n\uparrow$.*

Proof. If $m<n$, then $f_n(x_m e_m)=f_n(x_n e_m)=(f_n x_n e_n)e_m=(x_n e_n)e_m=x_m e_m$, therefore $f_m\le f_n$. ∎

Theorem 2. *If $\mathbf{x}\in\mathbf{C}$ then $\mathrm{LP}(\mathbf{x})\sim\mathrm{RP}(\mathbf{x})$, via a partial isometry of the form $\bar{w}$, w a partial isometry in A.*

Proof. Say $\mathbf{x}=[x_n, e_n]$. Adopt the notation of the lemma, and set $f=\sup f_n$; thus $f_n\uparrow f$, and we have $\bar{f}=\mathrm{LP}(\mathbf{x})$ by the proof of the lemma to Theorem 1.

Write $\mathrm{RP}(\mathbf{x})=\bar{e}$, e a projection in A (Theorem 1), and let $h=1-e$. Thus $\mathbf{x}\bar{h}=0$, and, citing [§ 47, Def. 6], we have

$$0=[x_n, e_n][h, 1]=[x_n h, k_n], \tag{1}$$

where

$$k_n = [1 \cap h^{-1}(e_n)] \cap [e_n \cap (x_n^*)^{-1}(1)$$
$$= 1 \cap h^{-1}(e_n) \cap e_n \cap 1 = e_n \cap h^{-1}(e_n).$$

In particular,

(2) $$(1-e_n)h k_n = 0$$

by the definition of $h^{-1}(e_n)$. Applying Proposition 1 to (1), we infer that $x_n h k_n = 0$ for all n; since $h k_n = e_n h k_n$ by (2), it follows that

(3) $$x_n e_n h k_n = 0.$$

Defining $g_n = \mathrm{RP}(x_n e_n)$, it results from (3) that $g_n h k_n = 0$, therefore $g_n \mathrm{LP}(h k_n) = 0$; citing [§ 3, Prop. 7], we have

$$0 = g_n \mathrm{LP}(h k_n) = g_n[h - h \cap (1-k_n)],$$

thus $g_n \le (1-h) + h \cap (1-k_n)$ and therefore

(4) $$D(g_n) \le 1 - D(h) + D(1-k_n).$$

Since $g_n = \mathrm{RP}(x_n e_n) \sim \mathrm{LP}(x_n e_n) = f_n$, it results from (4) that

$$D(f_n) \le 1 - D(h) + D(1-k_n),$$

therefore

(5) $$D(h) \le D(1-f_n) + D(1-k_n);$$

since $D(1-f_n) \downarrow D(1-f)$ and $D(1-k_n) \downarrow 0$ [§ 33, Th. 2], we infer from (5) that $D(h) \le D(1-f)$ (cf. [§ 27, Lemma 1, (ii)]), that is, $D(f) \le D(1-h)$, therefore $f \precsim 1 - h = e$.

We have shown that $f \precsim e$, where $\bar{f} = \mathrm{LP}(\mathbf{x})$ and $\bar{e} = \mathrm{RP}(\mathbf{x})$. Dually, $e \precsim f$, therefore $e \sim f$ by the Schröder-Bernstein theorem (which is trivial in a finite ring). ∎

Corollary 1. $\mathbf{C}$ *is finite; in fact,* $\mathbf{y}\mathbf{x} = 1$ *implies* $\mathbf{x}\mathbf{y} = 1$.

Proof. Assuming $\mathbf{x}^*\mathbf{x} = 1$, we are to show that the projection $\mathbf{f} = \mathbf{x}\mathbf{x}^*$ is also 1. Write $\mathbf{f} = \bar{f}$, f a projection in A (Theorem 1). Since $\mathrm{LP}(\mathbf{x}) = \mathbf{f}$ and $\mathrm{RP}(\mathbf{x}) = 1$ [cf. § 3, Prop. 4], it results from Theorem 2 that $f \sim 1$, therefore $f = 1$ by the finiteness of A. The second assertion then follows from Theorem 2 on applying [§ 46, Prop. 1, (5)] to $\mathbf{C}$. ∎

Corollary 2. *If* $\mathbf{e}, \mathbf{f}$ *are projections in* $\mathbf{C}$, *say* $\mathbf{e} = \bar{e}$ *and* $\mathbf{f} = \bar{f}$, *then* $\mathbf{e} \sim \mathbf{f}$ *in* $\mathbf{C}$ *if and only if* $e \sim f$ *in* A.

Proof. If $\mathbf{w}$ is a partial isometry in $\mathbf{C}$ such that $\mathbf{w}^*\mathbf{w} = \mathbf{e}$ and $\mathbf{w}\mathbf{w}^* = \mathbf{f}$, then $\mathrm{LP}(\mathbf{w}) = \mathbf{f}$ and $\mathrm{RP}(\mathbf{w}) = \mathbf{e}$, therefore $e \sim f$ in A by the theorem. The converse is obvious. ∎

Note that the central projections of $\mathbf{C}$ are the projections $\overline{h}$, with h a central projection in A (Theorem 1 and [§ 47, Def. 6]); in view of Corollary 2, A and $\mathbf{C}$ have the 'same structure' in the sense of Section 15.

The final proposition is for application later on (cf. [§ 51, Prop. 1]):

Proposition 3. *If* $\mathbf{x}=[x_n, e_n]$ *and the* x_n *are all invertible in* A, *then* $\mathbf{x}$ *is invertible in* $\mathbf{C}$ *and* $\mathbf{x}^{-1}=[x_n^{-1}, h_n]$ *for a suitable* SDD (h_n).

Proof. Setting $f_n=\mathrm{LP}(x_n e_n)$, we have $f_n\uparrow$ by the lemma to Theorem 2; since $f_n\sim \mathrm{RP}(x_n e_n)=e_n$ (by the invertibility of x_n) and since $e_n\uparrow 1$, it follows that $f_n\uparrow 1$. {Proof: $D(\sup f_n)=\sup D(f_n)=\sup D(e_n)=1$.} Set $y_n=x_n^{-1}$. If $m<n$ then $y_n f_m=y_m f_m$; for,

$$y_n x_m e_m = y_n x_n e_m = 1\, e_m = y_m x_m e_m,$$

$(y_n-y_m)x_m e_m=0$, therefore $(y_n-y_m)f_m=0$. Similarly, defining $g_n=\mathrm{LP}(x_n^* e_n)$, we have $g_n\uparrow 1$ and $y_n^* g_m=y_m^* g_m$ for $m<n$. Setting $h_n=f_n\cap g_n$, it follows that (y_n, h_n) is an OWC. Let $\mathbf{y}=[y_n, h_n]$. Since $x_n y_n=y_n x_n=1$ for all n, it follows that $\mathbf{x}\mathbf{y}=\mathbf{y}\mathbf{x}=1$. ∎

Exercises

It is assumed in the following exercises that A is a finite Baer $*$-ring satisfying LP$\sim$RP, with center Z, and $\mathbf{C}$ is the ring constructed in Section 47.

1A. (i) If $\mathbf{x}=[x_n, e_n]$ and $x_n\in Z$ for all n, then $\mathbf{x}$ is central in $\mathbf{C}$.

(ii) If $x\in A$, then $\overline{x}$ is central in $\mathbf{C}$ iff $x\in Z$.

(iii) The central projections in $\mathbf{C}$ are the projections $\overline{h}$, h a central projection in A.

2A. If $\mathbf{e}$ is a projection in $\mathbf{C}$, say $\mathbf{e}=\overline{e}$, then $\mathbf{e}$ is abelian relative to $\mathbf{C}$ iff e is abelian relative to A.

3A. If $(x_n, e_n)\equiv(y_n, f_n)$, then $x_n(e_n\cap f_n)=y_n(e_n\cap f_n)$ and $x_n^*(e_n\cap f_n)=y_n^*(e_n\cap f_n)$ for all n.

4A. If $\mathbf{x}=[x_n, e_n]$ and if, for each n, x_n is invertible in e_nAe_n, then $\mathbf{x}$ is invertible in $\mathbf{C}$. More precisely, if $y_n\in e_nAe_n$ with $x_ny_n=y_nx_n=e_n$, then $\mathbf{x}^{-1}=[y_n, e_n]$.

5A. If (f_n) is any orthogonal sequence (finite or infinite) of projections in A and if, for each n, $a_n\in f_nAf_n$, then there exists $\mathbf{x}\in\mathbf{C}$ such that $\mathbf{x}\overline{f}_n=\overline{f}_n\mathbf{x}=\overline{a}_n$ for all n.

6A. Assume, in addition, that A is orthoseparable [cf. §33, Th. 4] and that A satisfies the (EP)-axiom. Suppose $a\in A$ with $\mathrm{RP}(a)=1$.

(i) There exists an SDD (e_n) such that $e_n\in\{a^*a\}''$ and a^*ae_n is invertible in e_nAe_n for all n.

(ii) $\overline{a}$ is invertible in $\mathbf{C}$.

7C. The results of this (and the preceding) section are valid for a finite Baer $*$-ring satisfying the (EP)-axiom and the (SR)-axiom.

8A. If A satisfies the (EP)-axiom, then $\mathbf{C}$ is strongly semisimple; it follows that for every positive integer n, the matrix ring $\mathbf{C}_n$ is strongly semisimple.

9A. The entire section can be recast in terms of well-directed families $(e_\rho)_{\rho<\Lambda}$, Λ a fixed limit ordinal [cf. §47, Exer. 3].

§ 49. C Has no New Partial Isometries

As in the preceding two sections, A is a finite Baer $*$-ring satisfying LP $\sim$ RP. To fulfill the promise in the section heading, we assume, in addition, that *partial isometries in A are addable* (as is automatically the case when A has no abelian summand, or when A is an AW^*-algebra [§ 20, Th. 1]).

Theorem 1. $\mathbf{C}$ *has no new unitaries.*

Proof. Let $\mathbf{u} \in \mathbf{C}$ be unitary; we are to show that $\mathbf{u} = \bar{u}$ with $u \in A$ unitary. {The following argument uses only the hypothesis $\mathbf{u}^* \mathbf{u} = 1$, hence it shows anew that $\mathbf{C}$ is finite [cf. § 48, Cor. 1 of Th. 2].} In view of [§ 47, Lemma 8], we can suppose that $\mathbf{u} = [x_n, e_n]$, $\mathbf{u}^* \mathbf{u} = [x_n^* x_n, e_n]$ and $(x_n^* x_n, e_n) \equiv (1, e_n)$ via (e_n), thus

$$(*) \qquad x_n^* x_n e_n = e_n \quad \text{for all } n.$$

Defining $w_n = x_n e_n$, we have $w_n^* w_n = e_n$ by $(*)$. Setting $f_n = w_n w_n^* = \mathrm{LP}(w_n)$, it follows (as in the proof of [§ 48, Prop. 3]) that $f_n \uparrow 1$. Note that

$$w_n e_m = w_m \quad \text{whenever } m < n.$$

Set $e_0 = f_0 = w_0 = 0$ and define

$$v_n = w_n(e_n - e_{n-1}) = w_n - w_{n-1} \qquad (n = 1, 2, 3, \ldots).$$

Elementary computations yield $v_n^* v_n = e_n - e_{n-1}$ and $v_n v_n^* = f_n - f_{n-1}$ $(n = 1, 2, 3, \ldots)$, thus the v_n are partial isometries with orthogonal initial projections and orthogonal final projections; by the addability hypothesis, there exists $u \in A$ such that

$$u^* u = \sup(e_n - e_{n-1}) = 1, \qquad u u^* = \sup(f_n - f_{n-1}) = 1$$

and $u(e_n - e_{n-1}) = v_n = w_n - w_{n-1}$ for $n = 1, 2, 3, \ldots$. Then

$$u e_n = \sum_1^n u(e_i - e_{i-1}) = \sum_1^n (w_i - w_{i-1}) = w_n = x_n e_n,$$

therefore $\bar{u} = \mathbf{u}$ [§ 48, Prop. 2]. ∎

Corollary 1. $\mathbf{C}$ *has no new partial isometries.*

Proof. Let $\mathbf{w}$ be a partial isometry in $\mathbf{C}$, say $\mathbf{w}^* \mathbf{w} = \mathbf{e}$, $\mathbf{w} \mathbf{w}^* = \mathbf{f}$. Since $\mathbf{C}$ is finite and has GC [§ 48], it follows from $\mathbf{e} \sim \mathbf{f}$ that $1 - \mathbf{e} \sim 1 - \mathbf{f}$ [§ 17, Prop. 4]. Say $\mathbf{v}^* \mathbf{v} = 1 - \mathbf{e}$, $\mathbf{v} \mathbf{v}^* = 1 - \mathbf{f}$. Then $\mathbf{u} = \mathbf{w} + \mathbf{v}$ is unitary and $\mathbf{u} \mathbf{e} = \mathbf{w}$. Say $\mathbf{u} = \bar{u}$ (Theorem 1) and $\mathbf{e} = \bar{e}$ [§ 48, Th. 1]; then $\mathbf{w} = \bar{u} \bar{e} = (ue)^{-}$, where $w = ue$ is a partial isometry in A. ∎

Corollary 2. *Partial isometries in* $\mathbf{C}$ *are addable.*

Proof. Immediate from Corollary 1 and the assumed addability of partial isometries in A. ∎

Exercises

1A. The results in this section may be generalized to well-directed families [cf. §48, Exer. 9].

2C. The results in this section hold for a finite Baer $*$-ring satisfying the (EP)-axiom and the (SR)-axiom, in which partial isometries are addable (as is the case when there is no abelian summand).

§ 50. Positivity in C

The minimal hypotheses for the rest of the chapter are the following two:

(1°) *A is a finite Baer $*$-ring satisfying the* (EP)-*axiom and the* (UPSR)-*axiom.*

(2°) *Partial isometries in A are addable* (as is the case when A has no abelian summand, or when A is an AW^*-algebra [§20, Th. 1]).

It follows that A also satisfies LP $\sim$ RP [§20, Cor. of Th. 3], therefore the results of Sections 46–49 are applicable to A. Moreover, a strong form of polar decomposition holds for A [§21, Prop. 2].

Our objective in this section is to initiate the study of positivity in **C** (which culminates in Section 53, under much heavier hypotheses on A). For the present we need only the following added assumptions (which will be superseded by the conditions (3°) and (4°) of the next section):

(A) 2 *is invertible in A.*

(B) *If $x, y \in A$ and $x^*x + y^*y = 0$, then, $x = y = 0$.*

Let us review the notion of positivity in an arbitrary $*$-ring B. We say that $x \in B$ is positive, written $x \geq 0$, if $x = y_1^* y_1 + \cdots + y_m^* y_m$ for suitable elements $y_1, \ldots, y_m \in B$ [§13, Def. 8]. The self-adjoint elements of B may be ordered by writing $x \leq y$ (or $y \geq x$) in case $y - x \geq 0$. It is immediate that the relation is reflexive ($x \leq x$) and transitive ($x \leq y$ and $y \leq z$ imply $x \leq z$), but the property of antisymmetry ($x \leq y$ and $y \leq x$ imply $x = y$) may fail in general. In order that antisymmetry hold (and hence that the self-adjoints be partially ordered by this relation), it is sufficient that B have the following property: $x_1^* x_1 + \cdots + x_m^* x_m = 0$ implies $x_1 = \cdots = x_m = 0$ (for any m) [cf. §13, Exer. 9]. We note, for use later on, that if x and y are self-adjoint elements such that $x \leq y$, then $t^* x t \leq t^* y t$ for every $t \in B$.

In view of the (UPSR)-axiom, the notion of positivity in A is simplified: $x \geq 0$ if and only if $x = y^* y$ for some y (indeed, for some $y \geq 0$). It follows at once that (B) extends to finitely many terms: if $x_1^* x_1 + \cdots$

$+x_m^* x_m = 0$, then $x_1 = \cdots = x_m = 0$. In particular, the ordering of self-adjoints in A is antisymmetric. Though we know nothing of square roots in **C**, the ordering of its self-adjoints is antisymmetric too:

Proposition 1. *If* $\mathbf{x}_1, \ldots, \mathbf{x}_m \in \mathbf{C}$ *and* $\mathbf{x}_1^* \mathbf{x}_1 + \cdots + \mathbf{x}_m^* \mathbf{x}_m = 0$, *then* $\mathbf{x}_1 = \cdots = \mathbf{x}_m = 0$.

Proof. We illustrate the proof with $m=2$. Suppose $\mathbf{x}^*\mathbf{x} + \mathbf{y}^*\mathbf{y} = 0$. Say $\mathbf{x} = [x_n, e_n]$, $\mathbf{y} = [y_n, f_n]$; then $[x_n^* x_n + y_n^* y_n, g_n] = 0$ for suitable (g_n). For a suitable SDD (h_n), $(x_n^* x_n + y_n^* y_n) h_n = 0$, therefore

$$(x_n h_n)^* (x_n h_n) + (y_n h_n)^* (y_n h_n) = 0$$

for all n; in view of (B) we have $x_n h_n = y_n h_n = 0$ for all n, hence $\mathbf{x} = \mathbf{y} = 0$ [§48, Prop. 2]. ∎

Another application of (B):

Lemma. *If* $x, y \in A$ *then* $\mathrm{RP}(x^* x + y^* y) = \mathrm{RP}(x) \cup \mathrm{RP}(y)$.

Proof. Let $g = \mathrm{RP}(x^* x + y^* y)$. On the one hand,

$$g \leq \mathrm{RP}(x^* x) \cup \mathrm{RP}(y^* y) = \mathrm{RP}(x) \cup \mathrm{RP}(y)$$

[cf. §22, Lemma to Prop. 2]. On the other hand

$$\begin{aligned} 0 &= (1-g)(x^* x + y^* y)(1-g) \\ &= [x(1-g)]^* [x(1-g)] + [y(1-g)]^* [y(1-g)], \end{aligned}$$

therefore $x(1-g) = y(1-g) = 0$ by (B). Thus $xg = x$, $yg = y$, and so $\mathrm{RP}(x) \leq g$, $\mathrm{RP}(y) \leq g$.

{We remark that, in view of Proposition 1, the same argument yields

$$\mathrm{RP}(\mathbf{x}_1^* \mathbf{x}_1 + \cdots + \mathbf{x}_m^* \mathbf{x}_m) = \mathrm{RP}(\mathbf{x}_1) \cup \cdots \cup \mathrm{RP}(\mathbf{x}_m)$$

for any $\mathbf{x}_1, \ldots, \mathbf{x}_m \in \mathbf{C}$.} ∎

Theorem 1. *If* $x \in A$ *and* (e_n) *is an* SDD *such that* $e_n x e_n \geq 0$ *for all* n, *then* $x \geq 0$.

Proof. We know that $x^* = x$ [§47, Lemma 2]. Let $e = \mathrm{RP}(x) = \mathrm{LP}(x)$ and set $y = x + (1-e)$. Since $x = eye = e^* ye$, it would suffice to show that $y \geq 0$. Since the sum of positives is positive, clearly $e_n y e_n \geq 0$ for all n; moreover, $y^* y = x^* x + 1 - e$, therefore $\mathrm{RP}(y) = e \cup (1-e) = 1$ by the lemma.

Changing notation, we can suppose that $\mathrm{RP}(x) = 1$. Let r be the unique positive square root of $x^* x = x^2$, and write $x = ur$ with $u^* u = uu^* = 1$ [§21, Prop. 2]; it will suffice to show that $u = 1$.

Note that $u^* = u$. {Proof: $r(u^* - u)r = (ur)^* r - r(ur) = x^* r - rx = xr - rx = 0$ (because $r \in \{x^2\}''$ [§13, Def. 10]), therefore $u^* - u = 0$ results from $\mathrm{RP}(r) = 1$.} Thus u is a symmetry [§13, Def. 6]; in view of (A), the formula $g=(1/2)(1+u)$ defines a projection such that $u=2g-1$. It will suffice to show that $g=1$.

Write $r=s^2$ with $s\geq 0$, $s\in\{r\}''$. Since $ur=ru$ (because $x=ru$ with x, r and u self-adjoint), it follows that $su=us$, thus

$$(*)\qquad x=ur=us^2=sus=2sgs-s^2\,.$$

Since $e_n x e_n \geq 0$ we may write $e_n x e_n = t_n^* t_n$ for suitable t_n (e.g., with $t_n \geq 0$); then $(*)$ yields

$$2(gse_n)^*(gse_n)=(se_n)^*(se_n)+t_n^* t_n\,.$$

Citing the lemma, it results that

$$\mathrm{RP}(gse_n)=\mathrm{RP}(se_n)\cup \mathrm{RP}(t_n)\geq \mathrm{RP}(se_n);$$

but $\mathrm{RP}(gse_n)\leq \mathrm{RP}(se_n)$ trivially, thus

$$(**)\qquad \mathrm{RP}(gse_n)=\mathrm{RP}(se_n)\,.$$

But $\mathrm{RP}(s)=\mathrm{RP}(r)=\mathrm{RP}(x)=1$ implies that $\mathrm{RP}(se_n)=e_n$ [cf. §23, Lemma to Prop. 1]; citing $(**)$, we have

$$e_n=\mathrm{RP}(gse_n)\sim \mathrm{LP}(gse_n)\leq g\,,$$

thus $D(e_n)\leq D(g)$ for all n; since $D(e_n)\uparrow 1$ it results that $D(g)=1$, $g=1$. ∎

An important consequence of Theorem 1 is that the notions of positivity in A and $\mathbf{C}$ are consistent:

Corollary. *If* $\mathbf{x}=\overline{x}$, $x\in A$, *then* $\mathbf{x}\geq 0$ *in* $\mathbf{C}$ *if and only if* $x\geq 0$ *in* A.

Proof. Suppose $\mathbf{x}\geq 0$ in $\mathbf{C}$, say $\mathbf{x}=\mathbf{y}_1^*\mathbf{y}_1+\cdots+\mathbf{y}_m^*\mathbf{y}_m$. It is clear from the definition of the operations in $\mathbf{C}$ that we can write $\mathbf{x}=[x_n, e_n]$ with $x_n\geq 0$ for all n. Citing [§48, Prop. 1], we have

$$\overline{e_n}\,\overline{x}\,\overline{e_n}=\overline{e_n}\,\mathbf{x}\,\overline{e_n}=\overline{e_n}\,\overline{x_n e_n}\,,$$

thus $e_n x e_n = e_n x_n e_n \geq 0$ for all n, therefore $x\geq 0$ in A by Theorem 1. The converse is obvious. ∎

Exercises

1A. The results of this section hold with sequences replaced by well-directed families [cf. §49, Exer. 1].

2A. If A satisfies (1°), (2°) and (B), and if, in addition, A is orthoseparable, then, for every $a\in A$, $(1+a^*a)^-$ is invertible in $\mathbf{C}$ (in particular, 2 is invertible in $\mathbf{C}$).

3C. Theorem 1 fails when (UPSR) is weakened to (SR).

4A. Let B be a finite Baer $*$-ring satisfying (1°), (2°), (A) and (B), and let (e_ρ) be a well-directed family of projections in B with $e_\rho\uparrow e$. Then, for each $x\in B$, $x^* e_\rho x \uparrow x^* e x$ (interpreted in the obvious way relative to the partial ordering of self-adjoints described earlier in the section).

§ 51. Cayley Transform

Minimal hypotheses for the rest of the chapter: *we assume, in addition to* (1°), (2°) *of the preceding section, that*

(3°) *A is symmetric, that is,* $1+x^*x$ *is invertible in A, for all* $x \in A$, *and*

(4°) *A contains a central element i such that* $i^2=-1$ *and* $i^*=-i$.

The results of Sections 46–50 are applicable to A by virtue of the following lemma:

Lemma. *A satisfies conditions* (A) *and* (B) *of Section 50.*

Proof. $2=1+1^*1$ is invertible in A by (3°). Suppose $x, y \in A$ and $x^*x+y^*y=0$. Write $x^*x=r^2$, $y^*y=s^2$ with r and s self-adjoint, $r \in \{r^2\}''$, $s \in \{s^2\}''$. Since $r^2=-s^2$, obviously $r \in \{s^2\}'$, therefore $rs=sr$. Then $(r+is)^*(r+is)=r^2+s^2=0$, therefore $r+is=0$; taking adjoint, $r-is=0$, hence $2r=0$, $r=0$, $s=0$. Thus $x=y=0$. ∎

Remarks. 1. In the presence of (4°), and the availability of square roots, (3°) is clearly equivalent to the invertibility of $x+i$ in A for all self-adjoints $x \in A$. {But (3°) can be formulated in any $*$-ring with unity.}

2. We also write i for the corresponding element of $\mathbf{C}$, that is, we identify i with $\bar{i}$; clearly i is also central in $\mathbf{C}$.

3. Since 2 is invertible in A, it follows from (4°) that every $x \in A$ has a unique *Cartesian decomposition* $x=y+iz$, y and z self-adjoint; explicitly,

$$y=(1/2)(x+x^*), \qquad z=(1/2i)(x-x^*).$$

Similarly, every $\mathbf{x} \in \mathbf{C}$ has a unique Cartesian decomposition.

4. It follows from (1°) and (2°) that A may be written as a direct sum $A=B \oplus C$, where C satisfies (4°) and every element in the center of B is self-adjoint (see Exercise 1). Thus, in assuming that A satisfies (4°), we are setting aside the 'purely real' part B.

Property (3°) lifts to $\mathbf{C}$:

Proposition 1. *For all* $\mathbf{x} \in \mathbf{C}$, $1+\mathbf{x}^*\mathbf{x}$ *is invertible in* $\mathbf{C}$.

Proof. Say $\mathbf{x}=[x_n, e_n]$. Then $1+\mathbf{x}^*\mathbf{x}=[1+x_n^*x_n, f_n]$ for suitable (f_n). By (3°), $1+x_n^*x_n$ is invertible in A for all n, therefore $1+\mathbf{x}^*\mathbf{x}$ is invertible in $\mathbf{C}$ and

$$(1+\mathbf{x}^*\mathbf{x})^{-1}=[(1+x_n^*x_n)^{-1}, h_n]$$

for suitable (h_n) [§48, Prop. 3]. ∎

Corollary. *If* $\mathbf{x} \in \mathbf{C}$, $\mathbf{x}^*=\mathbf{x}$, *then* $\mathbf{x}+i$ *is invertible in* $\mathbf{C}$.

Proof. $(\mathbf{x}-i)(\mathbf{x}+i)=(\mathbf{x}+i)(\mathbf{x}-i)=1+\mathbf{x}^2$ is invertible in $\mathbf{C}$ by the proposition. ∎

The corollary sets the stage for the Cayley transform; the following proposition (and its proof) is valid in any $*$-ring B with unity, possessing a central element i with $i^2 = -1$ and $i^* = -i$, such that $x+i$ is invertible for every self-adjoint element x in B (note that $2=(1+i)(1-i)$ is invertible in such a ring):

Proposition 2. *The formulas*

$$\mathbf{u}=(\mathbf{x}-i)(\mathbf{x}+i)^{-1},$$
$$\mathbf{x}=i(1+\mathbf{u})(1-\mathbf{u})^{-1}$$

define mutually inverse bijections between the set of all self-adjoint elements $\mathbf{x}$ *and the set of all unitary elements* $\mathbf{u}$ *such that* $1-\mathbf{u}$ *is invertible in* $\mathbf{C}$. *If* $\mathbf{x}$ *and* $\mathbf{u}$ *are so paired, then* $\{\mathbf{x}\}'=\{\mathbf{u}\}'$, $\{\mathbf{x}\}''=\{\mathbf{u}\}''$ (*the commutants are computed in* $\mathbf{C}$). *We call* $\mathbf{u}$ *the Cayley transform of* $\mathbf{x}$.

Proof. Suppose $\mathbf{x}^*=\mathbf{x}$. Defining $\mathbf{u}=(\mathbf{x}-i)(\mathbf{x}+i)^{-1}$, we have

$$\mathbf{u}^*=(\mathbf{x}-i)^{-1}(\mathbf{x}+i)=\mathbf{u}^{-1},$$

thus $\mathbf{u}$ is unitary; moreover,

$$\begin{aligned}1-\mathbf{u}&=(\mathbf{x}+i)(\mathbf{x}+i)^{-1}-(\mathbf{x}-i)(\mathbf{x}+i)^{-1}\\&=[(\mathbf{x}+i)-(\mathbf{x}-i)](\mathbf{x}+i)^{-1}=2i(\mathbf{x}+i)^{-1}\end{aligned}$$

thus $1-\mathbf{u}$ is invertible, and we recover $\mathbf{x}$ by the calculation

$$\begin{aligned}i(1+\mathbf{u})(1-\mathbf{u})^{-1}&=i(1+\mathbf{u})(1/2i)(\mathbf{x}+i)\\&=(1/2)[(\mathbf{x}+i)+\mathbf{u}(\mathbf{x}+i)]\\&=(1/2)[(\mathbf{x}+i)+(\mathbf{x}-i)]=\mathbf{x}.\end{aligned}$$

Conversely, assuming $\mathbf{u}\in\mathbf{C}$ is unitary and $1-\mathbf{u}$ is invertible, define $\mathbf{x}=i(1+\mathbf{u})(1-\mathbf{u})^{-1}=i(1-\mathbf{u})^{-1}(1+\mathbf{u})$. Then

$$\mathbf{x}(1-\mathbf{u})=i(1+\mathbf{u}). \tag{1}$$

Also $(1-\mathbf{u})\mathbf{x}=i(1+\mathbf{u})$; taking adjoints, $\mathbf{x}^*(1-\mathbf{u}^*)=-i(1+\mathbf{u}^*)$, and right-multiplication by $\mathbf{u}$ yields $\mathbf{x}^*(\mathbf{u}-1)=-i(\mathbf{u}+1)$, thus

$$\mathbf{x}^*(1-\mathbf{u})=i(1+\mathbf{u}). \tag{2}$$

From (1) and (2) we have $\mathbf{x}(1-\mathbf{u})=\mathbf{x}^*(1-\mathbf{u})$; since $1-\mathbf{u}$ is invertible, we conclude that $\mathbf{x}=\mathbf{x}^*$. Moreover, (1) may be rewritten $(\mathbf{x}+i)\mathbf{u}=\mathbf{x}-i$, therefore $(\mathbf{x}-i)(\mathbf{x}+i)^{-1}=\mathbf{u}$. ∎

Remarks. 1. With notations as in Proposition 2, write $\mathbf{u}=\overline{u}$ with $u\in A$ unitary [§49, Th. 1]. Since $1-\mathbf{u}$ is invertible, $\mathrm{RP}(1-\mathbf{u})=1$, therefore $\mathrm{RP}(1-u)=1$ (see the remark following [§48, Cor. 1 of Th. 1]). It is shown in the next section (under an extra hypothesis on A) that, conversely, if $u \in A$ is any unitary with $\mathrm{RP}(1-u)=1$, then $\bar{u}$ is the Cayley transform of some self-adjoint element $\mathbf{x}$ of $\mathbf{C}$.

2. If $x \in A$ then $\{x\}'$ denotes the commutant of x in A, whereas $\{\bar{x}\}'$ is the commutant of $\bar{x}$ in $\mathbf{C}$; there can be no confusion as long as we refrain from identifying x with $\bar{x}$.

3. Proposition 2 is also valid with A in place of $\mathbf{C}$.

Our principal goal (attained in the next section, but only under an additional hypothesis on A) is to prove that $\mathbf{C}$ is regular in the sense of the following definition:

Definition 1. A ring B is said to be *regular* if, for each $x \in B$, there exists $y \in B$ with $xyx = x$. A *$*$-regular* ring is a regular ring with proper involution. (Cf. Exercises 5, 6.)

Proposition 3. *If B is a $*$-ring with unity, the following conditions are equivalent:*

(a) *B is $*$-regular.*

(b) *for each $x \in B$, there exists a projection e such that $Bx = Be$;*

(c) *B is regular and is a Rickart $*$-ring.*

Proof. (a) implies (b): If $x \in B$, by hypothesis there exists $y \in B$ with $xyx = x$. Then $g = xy$ is an idempotent with $gx = x$, therefore $xB \subset gB$; on the other hand, $gB = xyB \subset xB$, thus $xB = gB$. Then

$$L(\{x\}) = L(xB) = L(gB) = L(\{g\}) = B(1-g),$$

therefore

$$R(L(\{x\})) = R(B(1-g)) = R(\{1-g\}) = gB = xB.$$

Forgetting about g, we have shown that

$$R(L(\{x\})) = xB \quad \text{for all } x.$$

In particular,

$$R(L(\{xx^*\})) = xx^*B \quad \text{for all } x.$$

Since $L(\{x\}) = L(\{xx^*\})$ by properness of the involution, we infer that

$$(*) \qquad xB = xx^*B \quad \text{for all } x.$$

Let $x \in B$. We seek a projection e such that $Bx = Be$. By $(*)$, we have $x = xx^*z$ for a suitable element z. Setting $e = x^*z$, we have

$$e = x^*z = (xx^*z)^*z = z^*xx^*z = (x^*z)^*(x^*z) = e^*e,$$

thus e is a projection. Moreover, $x = xx^*z = xe$ shows that $Bx \subset Be$; since $e = e^* = z^*x$, we have also $Be \subset Bx$, thus $Bx = Be$.

(b) implies (c): Let $x \in B$. By hypothesis, $Bx = Be$, e a projection. Then

$$R(\{x\}) = R(Bx) = R(Be) = R(\{e\}) = (1-e)B.$$

This shows that B is a Rickart $*$-ring and that $e = \mathrm{RP}(x)$. Since $Bx = Be$, $e = yx$ for a suitable element y; then $xyx = xe = x$, thus B is regular.

(c) implies (a): Every Rickart $*$-ring has proper involution [§ 3, Prop. 2]. ∎

Proposition 4. *Let B be a $*$-regular ring with unity, and let $x \in B$, $e = \mathrm{RP}(x)$, $f = \mathrm{LP}(x)$.*

(i) *There exists a unique element y such that $y \in eBf$, $yx = e$, $xy = f$.*

(ii) *In particular, if $e = f$ (as is the case when x is normal), then x is invertible in eBe.*

(iii) *If x is self-adjoint [normal] then y is also self-adjoint [normal].*

Proof. As noted in the proof of Proposition 3, $Bx = Be$. Choose $y \in B$ with $yx = e$. Then also

$$(eyf)x = ey(fx) = eyx = ee = e;$$

replacing y by eyf, we can suppose $y \in eBf$. Then

$$(xy - f)x = xyx - fx = xe - fx = x - x = 0,$$

therefore $(xy - f)f = 0$, thus $xy - f = 0$. If also $z \in eBf$ satisfies $zx = e$, then

$$y = ey = zxy = zf = z,$$

which proves the asserted uniqueness.

If x is normal, that is, if $x^*x = xx^*$, then $e = f$ results from properness of the involution [cf. § 5, Lemma 5]. It is elementary that an invertible normal [self-adjoint] element in a $*$-ring with unity has normal [self-adjoint] inverse. ∎

In extending A to $\mathbf{C}$, we are reaching for a regular ring; when A is already regular, the extension collapses:

Theorem 1. *If A is regular then $\bar{A} = \mathbf{C}$.*

Proof. In view of the Cartesian decomposition, it suffices to show that if $\mathbf{x} \in \mathbf{C}$, $\mathbf{x}^* = \mathbf{x}$, then $\mathbf{x} \in \bar{A}$. Let $\mathbf{u} = \bar{u}$ be the Cayley transform of $\mathbf{x}$. As remarked following Proposition 2, $\mathrm{LP}(1-u) = \mathrm{RP}(1-u) = 1$; by Proposition 4, $1-u$ has an inverse b in A, hence $\mathbf{x} = i(1+\mathbf{u})(1-\mathbf{u})^{-1} = \bar{x}$, where $x = i(1+u)b$. ∎

This has some surprising consequences:

Corollary 1. *If A is regular and (x_n, e_n) is an* OWC, *then there exists $x \in A$ such that $xe_n = x_n e_n$ and $e_n x = e_n x_n$ for all n.*

Proof. Let $\mathbf{x} = [x_n, e_n]$, write $\mathbf{x} = \bar{x}$ by Theorem 1, and quote [§ 48, Prop. 1]. ∎

Corollary 2. *If A is regular, (f_n) is a sequence of orthogonal projections, and $a_n \in f_n A f_n$ for all n, then there exists $x \in A$ such that $f_n x = x f_n = a_n$ for all n.*

Proof. If the sequence (f_n) is finite, let x be the sum of the a_n. Otherwise, let $f = \sup f_n$ and define

$$x_n = a_1 + \cdots + a_n, \qquad e_n = f_1 + \cdots + f_n + (1-f);$$

then (x_n, e_n) is an OWC, and Corollary 1 provides a suitable element x. ∎

Corollary 3. *Suppose that* (1) *for each $x \in A$ there exists a positive integer k such that $x^* x \leq k1$, and* (2) *A contains an infinite sequence (f_n) of nonzero orthogonal projections. Then A is not regular.*

Proof. Assume to the contrary that A is regular. Setting $a_n = n f_n$, Corollary 2 provides an element $x \in A$ such that $f_n x = x f_n = n f_n$ for all n. Choose k as in (1); then $f_n x^* x f_n \leq k f_n$, thus

$$(*) \qquad n^2 f_n \leq k f_n \quad \text{for all } n.$$

Fix m with $m^2 > k$; thus $m^2 1 - k1 = 1 + \cdots + 1$ ($m^2 - k$ summands) is ≥ 0, and it is invertible by (1°) and (4°). Then $m^2 f_m \geq k f_m$; in view of $(*)$ this yields $(m^2 1 - k1) f_m = 0$ (cf. the remarks preceding [§ 50, Prop. 1]), hence, by the invertibility of $m^2 1 - k1$, the contradiction $f_m = 0$. {In connection with (2), see Exercise 11.} ∎

To put it another way, if A is regular and satisfies (2), then it can't satisfy (1). We return to such matters in the discussion of boundedness in Section 54.

Exercises

1A. Let A be a Baer $*$-ring satisfying the (EP)-axiom, and assume that partial isometries in A are addable. Then there exists a unique central projection h such that (i) hA satisfies (4°), and (ii) every element in the center of $(1-h)A$ is self-adjoint. (A $*$-ring of the latter kind is called *purely real.*)

2D. *Problem:* How does one proceed in the purely real case (see Exercise 1), i.e., in the absence of (4°)?

3A. If (A_ι) is a family of rings satisfying (1°)–(4°), then the complete direct product of the A_ι also satisfies them [cf. § 1, Exer. 13].

4A. Let B be a $*$-ring with unity, possessing a central element i with $i^2 = -1$ and $i^* = -i$, such that $x + i$ is invertible for each self-adjoint element x. (Briefly, we say that *B is a $*$-ring with Cayley transform.*) Suppose that for each $x \geq 0$ there exists a unique $r \geq 0$ with $x = r^2$. Then B satisfies the (UPSR)-axiom.

5A. Let B be a ring with unity, $x \in B$. The following conditions are equivalent: (a) there exists $y \in B$ with $xyx = x$; (b) there exists an idempotent $e \in B$ with

$Bx=Be$; (c) there exists an idempotent $f \in B$ with $xB=fB$. Thus the validity of (b) for each $x \in B$ is equivalent to regularity.

6A. The following conditions on a $*$-ring B are equivalent: (a) B is $*$-regular; (b) the involution of B is proper, and, for each $x \in B$, there exists a projection $e \in B$ with $Bx=Be$; (c) B is regular and is a weakly Rickart $*$-ring.

7A. Proposition 4 holds for any $*$-regular ring.

8A. If B is a $*$-regular ring and I is any ideal in B, then I is strict [cf. §22, Exer. 1] and B/I is $*$-regular.

9A. If B is a $*$-regular ring with unity, in which $x^*x+y^*y=0$ implies $x=y=0$, then B is symmetric.

10A. Assume: (1) C is a $*$-ring with unity and proper involution, (2) B is a $*$-subring of C, (3) B contains all the projections of C, and (4) for every $x \in C$, $1+x^*x$ is invertible and $(1+x^*x)^{-1} \in B$. Then B is a Rickart $*$-ring iff C is a Rickart $*$-ring. In particular, B is a Baer $*$-ring iff C is a Baer $*$-ring.

11A. (i) Condition (2) of Corollary 3 can fail only if A is the direct sum of finitely many factors of Type I.

(ii) An integral domain is regular if and only if it is a field.

(iii) A finite Baer $*$-factor of Type I, satisfying $\mathrm{LP} \sim \mathrm{RP}$, need not be regular.

12C. Let B be a ring with unity and let n be a positive integer. Then B_n is regular iff B is regular.

13C. Let B be a $*$-regular ring and let n be a positive integer. If the relations $x_1, \ldots, x_n \in B$, $x_1^*x_1+\cdots+x_n^*x_n=0$ imply $x_1=\cdots=x_n=0$, then B_n is also $*$-regular.

14A. The following conditions on a $*$-regular ring B are equivalent: (a) B satisfies $\mathrm{LP} \sim \mathrm{RP}$; (b) if e, f are projections that are algebraically equivalent [§1, Exer. 6] then $e \sim f$.

15C. Let B be a $*$-regular ring satisfying $\mathrm{LP} \sim \mathrm{RP}$ (cf. Exercise 14).

(i) *If* $x_1, \ldots, x_n$ are elements of B such that $x_1^*x_1+\cdots+x_n^*x_n=0$, then $\mathrm{RP}(x_1), \ldots, \mathrm{RP}(x_n)$ are central abelian projections.

(ii) If B has no abelian direct summand, then B_n is $*$-regular for all n.

16A. The results of this section generalize to well-directed families (but the statement of Corollary 2 must be left in sequential form).

17C. In a regular Baer $*$-ring, (i) $yx=1$ implies $xy=1$ (in particular, the ring is finite), and (ii) the projection lattice is a continuous geometry.

18A. For any $\mathbf{x} \in \mathbf{C}$, the element $\mathbf{y}=(1+\mathbf{x}^*\mathbf{x})^{-1}$ satisfies $\mathbf{y}^2 \leq 1$.

19C. Except for Corollary 3, the results of the present section hold with (UPSR) weakened to (SR).

§ 52. Regularity of C

To proceed further, it is necessary to augment the hypotheses (1°)–(4°) of the preceding section with some more spectral theory. The appropriate axiom is as follows:

Definition 1. We say that A satisfies the *unitary spectral axiom* (briefly, the (US)-*axiom*) if, for each unitary $u \in A$ with $\mathrm{RP}(1-u)=1$, there exists a sequence of projections $e_n \in \{u\}''$ such that $e_n \uparrow 1$ and $(1-u)e_n$ is invertible in $e_n A e_n$ for all n.

The following proposition assesses the strength of the (US)-axiom:

Proposition 1. (i) *The* (US)-*axiom is implied by the hypothesis* (1°) *when A is orthoseparable (equivalently, the center of A is orthoseparable).*

(ii) *Every Rickart C^*-algebra (in particular, every AW^*-algebra) satisfies the* (US)-*axiom.*

Proof. (i) We show, more generally, that if A is any orthoseparable Baer $*$-ring satisfying the (WEP)-axiom, then A satisfies the (US)-axiom.

Suppose $u \in A$ is unitary with $\mathrm{RP}(1-u)=1$. Set $a=1-u$. Obviously $\{a\}'=\{u\}'$; moreover, $xu=ux$ iff $u^*x=xu^*$ iff $x^*u=ux^*$, thus $\{u\}'$ is a $*$-subring. It follows that $\{a\}''=\{u\}''$ is a commutative $*$-subring of A.

Let (f_ι) be a maximal orthogonal family of nonzero projections such that, for each ι, there exists $b_\iota \in \{a\}''$ with $ab_\iota = b_\iota a = f_\iota$ (hence af_ι is invertible in $f_\iota A f_\iota$, with inverse $b_\iota f_\iota$). We assert that $\sup f_\iota = 1$. Let $g = 1 - \sup f_\iota$, note that $g \in \{a\}''$ [§4, Prop. 7], and assume to the contrary that $g \neq 0$. Since $\mathrm{RP}(a)=1$, it follows that $ag \neq 0$. By the (WEP)-axiom, there exists $c \in \{ga^*ag\}'' = \{ga^*a\}'' \subset \{a\}''$ with $(ga^*a)c^*c = f$, f a nonzero projection. Thus

$$(*) \qquad f = (ga^*c^*c)a = a(ga^*c^*c).$$

Clearly $f \leq g$, therefore f is orthogonal to every f_ι; setting $b = ga^*c^*c \in \{a\}''$, $(*)$ shows that maximality is contradicted.

By orthoseparability, the family (f_ι) is countable; write it as a (possibly finite) sequence (f_n). Define $e_n = f_1 + \cdots + f_n$ (if there are only finitely many n, then $e_n = 1$ for sufficiently large n). Then $e_n \uparrow 1$ and $e_n a = a e_n = \sum_1^n a f_k$ is invertible in $e_n A e_n$.

The foregoing applies, in particular, if A satisfies the hypothesis (1°); since A then has GC [§ 46, Prop. 1], the finiteness of A yields the parenthetical criterion for orthoseparability [§ 33, Th. 4].

(ii) This is easy spectral theory [cf. § 8, Prop. 1]. ∎

We assume for the rest of the section that, in addition to the hypotheses (1°)–(4°) of the preceding section,

(5°) *A satisfies the* (US)-*axiom.*

We can now characterize the unitaries that occur as Cayley transforms (see also Exercise 3):

Proposition 2. *If* $\mathbf{u} \in \mathbf{C}$ *is unitary, then* $1-\mathbf{u}$ *is invertible in* $\mathbf{C}$ *if and only if* $\mathrm{RP}(1-\mathbf{u})=1$ *(equivalently, writing* $\mathbf{u}=\bar{u}$ *with* $u \in A$ *unitary,* $\mathrm{RP}(1-u)=1$*). Thus, the Cayley transform pairs the self-adjoints* $\mathbf{x}$ *of* $\mathbf{C}$ *with the unitaries* u *in* A *such that* $\mathrm{RP}(1-u)=1$.

Proof. The 'only if' part is trivial. Conversely, suppose $\mathrm{RP}(1-\mathbf{u})=1$. Writing $\mathbf{u}=\bar{u}$ with $u \in A$ unitary, we have $\mathrm{RP}(1-u)=1$ (remark following [§ 48, Cor. 1 of Th. 1]). By the (US)-axiom, there exists an SDD (e_n) such that $e_n \in \{u\}''$ and $(1-u)e_n$ has an inverse y_n in $e_n A e_n$, thus

$$(*) \qquad (1-u)y_n = y_n(1-u) = e_n .$$

If $m<n$, it follows from the uniqueness of inverses that $y_n e_m = e_m y_n = y_m$, therefore (y_n, e_n) is an OWC; setting $\mathbf{y}=[y_n, e_n]$, $(*)$ yields $(1-\mathbf{u})\mathbf{y} = \mathbf{y}(1-\mathbf{u})=1$ [§ 48, Prop. 2], thus $1-\mathbf{u}$ is invertible in $\mathbf{C}$. ∎

At this point, one could develop the spectral theory in $\mathbf{C}$; we defer this until the next section, preferring instead to take the shortest path to regularity. The following formulation of the (US)-axiom will be more convenient:

Lemma 1. *If* $u \in A$ *is unitary and* $e=\mathrm{RP}(1-u)$, *then there exists a sequence of projections* $e_n \in \{u\}''$ *such that* $e_n \uparrow e$ *and* $(1-u)e_n$ *is invertible in* $e_n A e_n$.

Proof. Note that $e \in \{1-u\}''=\{u\}''$ [§ 3, Prop. 10]; $1-e$ is the largest projection such that $(1-u)(1-e)=0$, that is, $u(1-e)=1-e$ (hence also $u^*(1-e)=1-e$). So to speak, '$u=1$ on $1-e$', and Proposition 2 is not applicable directly when $1-e \neq 0$; we remedy this by considering the unitary element

$$v = ue-(1-e).$$

Obviously $v \in \{u\}''$. We show that $\mathrm{RP}(1-v)=1$. Let $g=1-\mathrm{RP}(1-v)$; thus $g \in \{1-v\}''=\{v\}'' \subset \{u\}''$, and $(1-v)g=0$. We have

$$\begin{aligned} 1-v &= [e+(1-e)]-[ue-(1-e)] \\ &= (1-u)e+2(1-e) = 1-u+2(1-e), \end{aligned}$$

multiplication by e yields $e(1-v)=e(1-u)=1-u$, and multiplying this by g we have

$$(1-u)g = e(1-v)g = e0 = 0,$$

hence $eg=0$ (recall that $e=\mathrm{RP}(1-u)$). It follows that

$$0=(1-v)g=(1-u)g+2(1-e)g=0+2g-2eg=2g,$$

therefore $g=0$ (recall that 2 is invertible in A). Thus $\mathrm{RP}(1-v)=1$. Note also (from the definition of v) that $ve=ue$, thus $(1-v)e=(1-u)e=1-u$.

By the (US)-axiom, choose a sequence of projections $g_n \in \{v\}'' \subset \{u\}''$ with $g_n \uparrow 1$ and $(1-v)g_n$ invertible in $g_n A g_n$. Then $eg_n \in \{u\}''$, $eg_n \uparrow e$, and $(1-u)eg_n = (1-v)eg_n$ is invertible in $eg_n A eg_n$, thus the sequence $e_n = eg_n$ meets all requirements. ∎

This is the key to constructing the 'relative inverses' needed for regularity.

Lemma 2. *If* $\mathbf{x} \in \mathbf{C}$, $\mathbf{x}^* = \mathbf{x}$, *and if* $\mathbf{e} = \mathrm{RP}(\mathbf{x})$, *then* $\mathbf{x}$ *is invertible in* $\mathbf{eCe}$.

Proof. Let $\mathbf{u} = \overline{u}$ be the Cayley transform of $\mathbf{x}$. Since $\mathbf{x} = i(1-\mathbf{u})^{-1}(1+\mathbf{u})$, it is clear that $\mathrm{RP}(\mathbf{x}) = \mathrm{RP}(1+\mathbf{u})$; thus, writing $\mathbf{e} = \overline{e}$, e a projection in A, we have $e = \mathrm{RP}(1+u)$. Set $v = -u$; thus v is unitary and $\mathrm{RP}(1-v) = e$. By Lemma 1, there exists a sequence of projections $e_n \in \{v\}'' = \{u\}''$ such that $e_n \uparrow e$ and $(1-v)e_n$ has an inverse z_n in $e_n A e_n$. Thus

$$(*) \qquad (1+u)z_n = z_n(1+u) = e_n \quad \text{for all } n.$$

As argued in the proof of Proposition 2, $z_n e_m = z_m$ when $m < n$. Setting $f_n = e_n + (1-e)$, it is routine to verify that (z_n, f_n) is an OWC, with $z_n f_m = z_m$ when $m < n$; define $\mathbf{z} = [z_n, f_n]$. From $(*)$ we see that

$$(1+u)z_n f_n = e_n f_n = e_n = ef_n;$$

it follows that $(1+\mathbf{u})\mathbf{z} = \overline{e}$ and, since the z_n commute with u, also $\mathbf{z}(1+\mathbf{u}) = \overline{e}$. Define $\mathbf{y} = -i(1-\mathbf{u})\mathbf{z}$; then

$$\mathbf{xy} = [i(1+\mathbf{u})(1-\mathbf{u})^{-1}][-i(1-\mathbf{u})\mathbf{z}] = (1+\mathbf{u})\mathbf{z} = \overline{e}$$

and similarly $\mathbf{yx} = \overline{e}$. Note that $\mathbf{z} \in \mathbf{eCe}$; indeed, $z_n e = (z_n e_n)e = z_n e_n = z_n$ and similarly $ez_n = z_n$, thus $\mathbf{z}\overline{e} = \mathbf{z} = \overline{e}\mathbf{z}$. It then follows from the defining formula for $\mathbf{y}$ that $\mathbf{y} \in \mathbf{eCe}$. Thus $\mathbf{x}$ is invertible in $\mathbf{eCe}$, with inverse $\mathbf{y}$. (Moreover, $\mathbf{y}^* = \mathbf{y}$ results from $\mathbf{x}^* = \mathbf{x}$ and the uniqueness of inverses.) ∎

Theorem 1. $\mathbf{C}$ *is* $*$*-regular.*

Proof. Let $\mathbf{x} \in \mathbf{C}$, $\mathbf{e} = \mathrm{RP}(\mathbf{x}) = \mathrm{RP}(\mathbf{x}^*\mathbf{x})$; it will suffice to show that $\mathbf{Cx} = \mathbf{Ce}$ [§ 51, Prop. 3]. The inclusion $\mathbf{Cx} \subset \mathbf{Ce}$ results from $\mathbf{x} = \mathbf{xe}$. Applying Lemma 2 to the self-adjoint element $\mathbf{x}^*\mathbf{x}$, there exists an element $\mathbf{y}$ such that $\mathbf{e} = \mathbf{y}(\mathbf{x}^*\mathbf{x})$, thus $\mathbf{Ce} \subset \mathbf{Cx}$. ∎

Definition 2. We call $\mathbf{C}$ *the regular ring of* A.

The regular ring is characterized as follows (assuming A satisfies (1°)–(5°); see also Exercises 4 and 5):

Proposition 3. *Suppose* D *is a* $*$*-ring with unity, such that* (1) D *is regular,* (2) A *is a* $*$*-subring of* D, (3) *every unitary element of* D *belongs* *to* A, (4) *the relations* $x, y \in D$, $x^*x + y^*y = 0$ *imply* $x = y = 0$, *and* (5) *the element* i *of* A *is also central in* D. *Then* D *is* $*$*-isomorphic to* $\mathbf{C}$ (*via an extension of the embedding* $a \mapsto \bar{a}$ *of* A *into* $\mathbf{C}$).

Proof. It follows from (1) and (4) that D is $*$-regular [§ 51, Def. 1]. By (3), D and A have the same unity element. If e is any projection in D, then the symmetry $u = 2e - 1$ is in A by (3), therefore $e = (1/2)(1+u) \in A$; thus D contains no new projections. Since D is a Rickart $*$-ring [§ 51, Prop. 3] with complete projection lattice, D is a Baer $*$-ring [§ 4, Prop. 1]. For any $x \in D$, $\mathrm{RP}(1 + x^*x) = 1$ [§ 50, Lemma], therefore $1 + x^*x$ is invertible in D by $*$-regularity [§ 51, Prop. 4]. It follows that the Cayley transform is operative in D (remarks preceding [§ 51, Prop. 2]).

Let $x \in D$. We assert that there exists an SDD (e_n) such that $xe_n \in A$ and $x^*e_n \in A$ for all n. By the Cartesian decomposition (and [§ 47, Lemma 1]) we can suppose $x^* = x$. Let u be the Cayley transform of x; by (3), $u \in A$. Since $1 - u$ is invertible in D, we have $\mathrm{RP}(1-u) = 1$ (in D or in A—it's the same). Adopt the notations in the proof of Proposition 2, in particular, $(1-u)y_n = e_n$; writing $(1-u)^{-1}$ for the inverse of $1-u$ in D, we have $y_n = (1-u)^{-1}e_n$, therefore $xe_n = i(1+u)(1-u)^{-1}e_n = i(1+u)y_n \in A$ for all n.

Each $x \in D$ determines an element $\varphi(x)$ of $\mathbf{C}$ as follows. By the preceding, we can choose an SDD (e_n) such that $xe_n \in A$ and $x^*e_n \in A$ for all n. Applying [§ 47, Lemma 5] to the sequence (xe_n) in A, we know that $g_n = e_n \cap (xe_n)^{-1}(e_n)$ is an SDD; since $(1-e_n)xe_ng_n = 0$, we have

$$(*) \qquad xg_n = xe_ng_n = (e_nxe_n)g_n \quad \text{for all } n.$$

Similarly, defining $h_n = e_n \cap (x^*e_n)^{-1}(e_n)$, (h_n) is an SDD such that

$$(**) \qquad x^*h_n = x^*e_nh_n = (e_nx^*e_n)h_n \quad \text{for all } n.$$

Setting $x_n = e_nxe_n$ and $k_n = g_n \cap h_n$, it is clear from $(*)$ and $(**)$ that (x_n, k_n) is an OWC. We propose to define $\varphi(x) = [x_n, k_n]$, hence must check that the indicated element of $\mathbf{C}$ is unambiguously determined. Suppose also (e_n') is an SDD such that $xe_n' \in A$ and $x^*e_n' \in A$ for all n; applying the foregoing construction, we arrive at an OWC (x_n', k_n'), where $x_n' = e_n'xe_n'$. Setting $f_n = k_n \cap k_n'$, it is elementary that $f_nx_nf_n = f_nxf_n = f_nx_n'f_n$, therefore $[x_n, k_n] = [x_n', k_n']$ by [§ 48, Prop. 2].

It is routine to check that $\varphi: D \to \mathbf{C}$ is a $*$-monomorphism. Finally, φ is surjective. For, suppose $\mathbf{x} \in \mathbf{C}$, $\mathbf{x}^* = \mathbf{x}$. If $\mathbf{u} = \bar{u}$ is the Cayley transform of $\mathbf{x}$, then $\mathrm{LP}(1-u) = \mathrm{RP}(1-u) = 1$; hence $1-u$ is invertible in D, the formula $x = i(1+u)(1-u)^{-1}$ defines a self-adjoint element of D whose Cayley transform is also u, and a straightforward argument shows that $\varphi(x) = \mathbf{x}$. ∎

Proposition 1 yields two special cases that are worth noting explicitly:

Corollary. *If A satisfies* (1°)–(4°) *and is orthoseparable, or if A is a finite AW^*-algebra, then A has a unique regular ring in the sense of Definition 2 and Proposition 3.*

Proposition 3 yields a remarkable application of regularity:

Theorem 2. *If (h_α) is an orthogonal family of central projections in A with $\sup h_\alpha = 1$, then $\mathbf{C}$ is $*$-isomorphic to the complete direct product of the $\overline{h}_\alpha \mathbf{C}$, via the mapping $\mathbf{x} \mapsto (\overline{h}_\alpha \mathbf{x})$.*

Proof. {We remark that it suffices to assume that A satisfies (1°)–(4°) and that $\mathbf{C}$ is regular (cf. Exercise 3).}

Write $\mathbf{h}_\alpha = \overline{h}_\alpha$, $\mathbf{C}_\alpha = \mathbf{h}_\alpha \mathbf{C}$, and let D be the complete direct product of the $\mathbf{C}_\alpha$ [§ 1, Exer. 13]. Write 1 for the unity element $(\mathbf{h}_\alpha)$ of D.

Define a mapping $\varphi: \mathbf{C} \to D$ by $\varphi(\mathbf{x}) = (\mathbf{h}_\alpha \mathbf{x})$. It is clear that φ is a $*$-monomorphism, with $\varphi(1) = 1$. It remains to show that φ is surjective.

Note that $\varphi(i) = (\mathbf{h}_\alpha i)$ is a central, skew-adjoint unitary element of D; that is, writing $\varphi(i) = i$, D satisfies (4°). If $(\mathbf{x}_\alpha) \in D$ is self-adjoint, then $\mathbf{x}_\alpha$ is self-adjoint for all α, $\mathbf{x}_\alpha + \mathbf{h}_\alpha i$ has an inverse $\mathbf{y}_\alpha$ in $\mathbf{C}_\alpha$, therefore $(\mathbf{x}_\alpha) + i = (\mathbf{x}_\alpha + \mathbf{h}_\alpha i)$ has inverse $(\mathbf{y}_\alpha)$ in D. It follows that the Cayley transform is operative in D (remarks preceding [§ 51, Prop. 2]).

Since $2 = (1+i)(1-i)$ is invertible, every element of D has a Cartesian decomposition; thus, to show that φ is surjective, it will suffice to show that its range contains every self-adjoint element of D.

Suppose $(\mathbf{x}_\alpha) \in D$ is self-adjoint. Let $(\mathbf{u}_\alpha)$ be the Cayley transform of $(\mathbf{x}_\alpha)$. In particular, $(\mathbf{u}_\alpha)$ is unitary, therefore $\mathbf{u}_\alpha$ is unitary in $\mathbf{C}_\alpha$ for each α. Write $\mathbf{u}_\alpha = \overline{u}_\alpha$ with $u_\alpha \in h_\alpha A$ [§49, Cor. 1], thus $u_\alpha^* u_\alpha = u_\alpha u_\alpha^* = h_\alpha$. By addability, there exists $u \in A$ with $h_\alpha u = u_\alpha$ for all α (hence u is unitary), thus $\varphi(\overline{u}) = (\mathbf{u}_\alpha)$. Observe that $\mathrm{RP}(1-\overline{u}) = 1$, that is, $\mathrm{RP}(1-u) = 1$. {Proof: Let $g = 1 - \mathrm{RP}(1-u)$. We have $(1-\overline{u})\overline{g} = 0$, hence $0 = \varphi(1-\overline{u})\varphi(\overline{g}) = (1-(\mathbf{u}_\alpha))\varphi(\overline{g})$; since $1-(\mathbf{u}_\alpha)$ is invertible in D (by the theory of the Cayley transform), we conclude that $\varphi(\overline{g}) = 0$, therefore $\overline{g} = 0$ by the injectivity of φ. Thus $g = 0$ as asserted.} Since $\mathrm{LP}(1-\overline{u}) = \mathrm{RP}(1-\overline{u}) = 1$ and $\mathbf{C}$ is $*$-regular, $1-\overline{u}$ is invertible in $\mathbf{C}$ [§51, Prop. 4]; defining

$$\mathbf{x} = i(1+\overline{u})(1-\overline{u})^{-1},$$

we have $\mathbf{x}^* = \mathbf{x}$. Since $\mathbf{x}(1-\overline{u}) = i(1+\overline{u})$, application of φ yields $\varphi(\mathbf{x})(1-\varphi(\overline{u})) = i(1+\varphi(\overline{u}))$, thus

$$\varphi(\mathbf{x})(1-(\mathbf{u}_\alpha)) = i(1+(\mathbf{u}_\alpha)),$$

therefore $\varphi(\mathbf{x}) = i(1+(\mathbf{u}_\alpha))(1-(\mathbf{u}_\alpha))^{-1} = (\mathbf{x}_\alpha)$. ∎

Exercises

1A. Let B be a Rickart $*$-ring in which 2 is invertible. If B satisfies the (US)-axiom and e is any projection in B, then eBe also satisfies the (US)-axiom.

2A. If (A_ι) is a family of rings satisfying (1°)–(5°), then the complete direct product of the A_ι also satisfies (1°)–(5°).

3C. Suppose that (1°′) A is a finite Baer $*$-ring satisfying the (EP)-axiom and the (SR)-axiom (hence **C** exists [§48, Exer. 7]), and that (2°)–(4°) hold. The following conditions are equivalent: (a) **C** is regular; (b) if $u \in A$ is unitary and $\mathrm{RP}(1-u)=1$, then $1-\bar{u}$ is invertible in **C**; (c) if $u \in A$ is unitary and $e=\mathrm{RP}(1-u)$, then $1-\bar{u}$ is invertible in $\bar{e}\mathbf{C}\bar{e}$.

4C. Suppose that (1°′) A is a finite Baer $*$-ring satisfying the (EP)-axiom and the (SR)-axiom, and that (2°)–(5°) hold. Then A has a unique regular ring in the sense of Definition 2 and Proposition 3. {Note: When A has no abelian summand, the hypothesis (2°) is redundant; when A is orthoseparable (equivalently, the center of A is orthoseparable), the hypothesis (5°) is redundant.}

5C. Let A be a Baer $*$-factor of Type II_1, satisfying the (EP)-axiom and the (SR)-axiom, in which every element of the form $1+x^*x$ is invertible, and possessing a central element z such that $z^* \neq z$. Then A has a unique regular ring in the sense of Definition 2 and Proposition 3.

6A. Every $*$-regular ring with unity trivially satisfies (5°).

7C. Suppose A satisfies (1°)–(4°) and is regular. If (h_α) is an orthogonal family of central projections in A with $\sup h_\alpha = 1$, then A is $*$-isomorphic to the complete direct product of the $h_\alpha A$, via the mapping $x \mapsto (h_\alpha x)$. The same result holds with (1°) weakened to (1°′) (see Exercise 3).

8C. (i) Suppose that (1°′) A is a finite Baer $*$-ring satisfying the (EP)-axiom and the (SR)-axiom, (2°′) A has no abelian summand, and that A satisfies (3°)–(5°). Then $\mathbf{C}_n$ is $*$-regular for all n.

(ii) In particular, if A is a nonabelian, finite Baer $*$-factor satisfying the (EP)-axiom and the (SR)-axiom, if A is symmetric, and if A possesses a central element z such that $z^* \neq z$, then $\mathbf{C}_n$ is $*$-regular for all n.

9A. If A is a complete Boolean algebra with the identity involution [cf. §3, Exer. 14], then A satisfies (1°), (2°) and (4°) trivially and (5°) vacuously (but not (3°)), and is a commutative regular ring.

10D. *Problems:* (i) Can the regularity of **C** be reached with fewer axioms? (ii) Can orthoseparability be omitted in the Corollary following Proposition 3? (iii) What about the purely real case [cf. §51, Exer. 1]?

§ 53. Spectral Theory in C

As in the preceding section, we assume that A satisfies the hypotheses (1°)–(5°). In this section we exploit (5°) to show that all properties hypothesized for A lift to **C**.

A key consequence of the (US)-axiom (5°) is that a self-adjoint element of **C** can be represented in a form suitable for 'spectral theory':

Proposition 1. *If* $\mathbf{x}\in\mathbf{C}$, $\mathbf{x}^*=\mathbf{x}$, *and if* $\mathbf{u}=\overline{u}$ *is the Cayley transform of* $\mathbf{x}$, *then one can write* $\mathbf{x}=[x_n, e_n]$ *with* $x_n, e_n\in\{u\}''$, $x_n^*=x_n$, $x_n e_n=x_n$.

Proof. Adopt the notations in the proof of [§52, Prop. 2] (we know that $\mathrm{RP}(1-u)=1$ by the trivial half of that proposition). By elementary algebra [§1, Exer. 12],

$$y_n\in\{(1-u)e_n, ((1-u)e_n)^*\}''\subset\{u\}''.$$

In e_nAe_n, ue_n is unitary and y_n is the inverse of $e_n-ue_n=(1-u)e_n$; defining

$$x_n=i(e_n+ue_n)y_n,$$

we have $x_n^*=x_n$ by elementary algebra (see the proof of [§51, Prop. 2]). The formula

$$(*)\qquad x_n=i(1+u)y_n$$

shows that $x_n\in\{u\}''$; also, $m<n$ implies $y_n e_m=y_m$, therefore $x_n e_m=x_m$. Since $x_n^*=x_n$, it follows that (x_n, e_n) is an OWC, and $(*)$ yields

$$[x_n, e_n]=i(1+\mathbf{u})\mathbf{y}=i(1+\mathbf{u})(1-\mathbf{u})^{-1}=\mathbf{x}. \quad \blacksquare$$

The next proposition is a substitute for the assertion that 'functions' of a self-adjoint element $\mathbf{x}$ lie in $\{\mathbf{x}\}''$:

Proposition 2. *If* $\mathbf{x}\in\mathbf{C}$, $\mathbf{x}^*=\mathbf{x}$, *if* $\mathbf{u}=\overline{u}$ *is the Cayley transform of* $\mathbf{x}$, *and if* $\mathbf{y}=[y_n, f_n]$ *with* $y_n\in\{u\}''$ *for all n, then* $\mathbf{y}\in\{\mathbf{x}\}''$.

Proof. Assuming $\mathbf{z}\in\{\mathbf{x}\}'$, it is to be shown that $\mathbf{y}\mathbf{z}=\mathbf{z}\mathbf{y}$. Since $\{\mathbf{x}\}'$ is a $*$-subring of $\mathbf{C}$, we can suppose (by the Cartesian decomposition) that $\mathbf{z}^*=\mathbf{z}$. Let $\mathbf{w}=\overline{w}$ be the Cayley transform of $\mathbf{z}$. Since $\mathbf{x}\mathbf{z}=\mathbf{z}\mathbf{x}$, we infer that $\mathbf{u}\mathbf{w}=\mathbf{w}\mathbf{u}$ [§51, Prop. 2], thus $uw=wu$. Since $y_n\in\{u\}''$, it follows that $wy_n=y_nw$ for all n, therefore $\mathbf{w}\mathbf{y}=\mathbf{y}\mathbf{w}$; since $\{\mathbf{w}\}'=\{\mathbf{z}\}'$, it follows that $\mathbf{z}\mathbf{y}=\mathbf{y}\mathbf{z}$. $\blacksquare$

The following proposition well illustrates the power of the (US)-axiom and the advantages of unique positive square roots:

Proposition 3. *If* $\mathbf{x}\in\mathbf{C}$, *then* $\mathbf{x}\geq 0$ *if and only if* $\mathbf{x}=\mathbf{y}^*\mathbf{y}$ *for some* $\mathbf{y}\in\mathbf{C}$. *In fact,* $\mathbf{C}$ *satisfies the* (UPSR)*-axiom.*

Proof. Suppose $\mathbf{x}\geq 0$ (that is, $\mathbf{x}=\mathbf{y}_1^*\mathbf{y}_1+\cdots+\mathbf{y}_m^*\mathbf{y}_m$ for suitable $\mathbf{y}_1,\ldots,\mathbf{y}_m\in\mathbf{C}$). Write $\mathbf{x}=[x_n, e_n]$ as in the statement of Proposition 1. Citing [§48, Prop. 1] we have $\overline{x}_n=(e_nx_ne_n)^-=\overline{e}_n\mathbf{x}\overline{e}_n\geq 0$ in $\mathbf{C}$, therefore $x_n\geq 0$ in A [§50, Cor. of Th. 1].

Let r_n be the unique positive square root of x_n; in particular, $r_n\in\{x_n\}''\subset\{u\}''$ [§13, Def. 10]. Thus, the x_n, e_n, r_n all belong to the commutative

*-subring $\{u\}''$. It follows from the uniqueness of positive square roots that

(1) $$r_n e_m = r_m \quad \text{when } m<n;$$

for, $r_n e_m = e_m r_n e_m \geq 0$ and $(r_n e_m)^2 = r_n^2 e_m = x_n e_m = x_m$. From (1) we see that (r_n, e_n) is an OWC; defining $\mathbf{r} = [r_n, e_n]$, it results from $r_n^2 = x_n$ that

(2) $$\mathbf{x} = \mathbf{r}^2.$$

Since $r_n \in \{u\}''$, we have

(3) $$\mathbf{r} \in \{\mathbf{x}\}''$$

by Proposition 2. Next, we note that

(4) $$\mathbf{r} \geq 0;$$

for, if s_n is the unique positive square root of r_n, then the above argument shows that $s_n e_m = s_m$ when $m<n$, thus $\mathbf{s} = [s_n, e_n]$ is a self-adjoint element with $\mathbf{r} = \mathbf{s}^2$.

In view of (2), (3), (4), it remains only to prove uniqueness: assuming that $\mathbf{t} \geq 0$ and $\mathbf{x} = \mathbf{t}^2$, it is to be shown that $\mathbf{t} = \mathbf{r}$. Obviously $\mathbf{t}\mathbf{x} = \mathbf{x}\mathbf{t}$; since $\mathbf{r} \in \{\mathbf{x}\}''$, we have $\mathbf{r}\mathbf{t} = \mathbf{t}\mathbf{r}$. Then

$$(\mathbf{r}+\mathbf{t})(\mathbf{r}-\mathbf{t}) = \mathbf{r}^2 - \mathbf{t}^2 = \mathbf{x} - \mathbf{x} = 0;$$

setting $\mathbf{y} = \mathbf{r} - \mathbf{t}$, it follows that

(5) $$0 = \mathbf{y}(\mathbf{r}+\mathbf{t})\mathbf{y} = \mathbf{y}^*\mathbf{r}\mathbf{y} + \mathbf{y}^*\mathbf{t}\mathbf{y}.$$

Since $\mathbf{y}^*\mathbf{r}\mathbf{y} \geq 0$ and $\mathbf{y}^*\mathbf{t}\mathbf{y} \geq 0$, it results from (5) that $\mathbf{y}\mathbf{r}\mathbf{y} = \mathbf{y}\mathbf{t}\mathbf{y} = 0$ [§50, Prop. 1]; then

$$\mathbf{y}^3 = \mathbf{y}(\mathbf{r}-\mathbf{t})\mathbf{y} = 0,$$

hence $\mathbf{y}^4 = 0$, and $\mathbf{y} = 0$ results from the properness of the involution. ∎

Definition 1. If $\mathbf{x} \in \mathbf{C}$, $\mathbf{x} \geq 0$, we write $\mathbf{x}^{\frac{1}{2}}$ for the unique positive square root of $\mathbf{x}$ given by Proposition 3. We know, in addition, that $\mathbf{x}^{\frac{1}{2}} \in \{\mathbf{x}\}''$.

Lemma. *If $\mathbf{x} \in \mathbf{C}$, $\mathbf{x} \neq 0$, then there exists $a \in A$ such that $\mathbf{x}\overline{a} = \overline{f}$, f a nonzero projection.*

Proof. Say $\mathbf{x} = [x_n, e_n]$. Since $\mathbf{x} \neq 0$, there exists an index m such that $x_m e_m \neq 0$ [§48, Prop. 2]. By the (EP)-axiom in A, there exists $b^* = b \in \{e_m x_m^* x_m e_m\}''$ such that $(e_m x_m^* x_m e_m) b^2 = e$, e a nonzero projection. Thus

$$b e_m x_m^* x_m e_m b = e;$$

setting $w = x_m e_m b$, we have $w^* w = e$. Define $f = w w^*$. Citing [§48, Prop. 1], we have

$$\overline{w} = (x_m e_m)^- \overline{b} = \mathbf{x} \overline{e}_m \overline{b},$$

hence $\overline{f} = \overline{w}\,\overline{w}^* = \mathbf{x}(e_m b w^*)^-$; let $a = e_m b w^*$. ∎

Proposition 4. **C** *satisfies the* (EP)*-axiom. In fact, given any* $\mathbf{z} \in \mathbf{C}$, $\mathbf{z} \neq 0$, *there exists* $y \in A$ *such that* $\overline{y} \in \{\mathbf{z}^* \mathbf{z}\}''$, $y \geq 0$, *and* $\mathbf{z}^* \mathbf{z} \overline{y}^2 = \overline{f}$, $\overline{f}$ *a nonzero projection.*

Proof. Set $\mathbf{x} = (\mathbf{z}^* \mathbf{z})^{\frac{1}{2}}$, let $\mathbf{u} = \overline{u}$ be the Cayley transform of $\mathbf{x}$, and write $\mathbf{x} \doteq [x_n, e_n]$ as in Proposition 1. Adopt the notation in the proof of the lemma. In particular, $a = e_m b w^*$, where

$$b \in \{e_m x_m^* x_m e_m\}'' = \{x_m^2\}'' \subset \{u\}'';$$

then $w = x_m e_m b$ also belongs to $\{u\}''$, thus $a \in \{u\}''$. Set $y = (a^* a)^{\frac{1}{2}} \in \{a^* a\}'' \subset \{u\}''$. Then $\overline{a}, \overline{y} \in \{\mathbf{x}\}''$ by Proposition 2, and the relation $f = f^* f$ yields

$$\overline{f} = \overline{a}^* \mathbf{x}^* \mathbf{x} \overline{a} = \mathbf{x}^2 (a^* a)^- = \mathbf{z}^* \mathbf{z} \overline{y}^2. \quad ∎$$

Proposition 4 completes the proof that **C** has all the properties hypothesized for A—plus regularity:

Theorem 1. *Assume A satisfies* (1°)–(5°).

(i) **C** *satisfies* (1°)–(5°) *and is regular.*

(ii) $\overline{A} = \mathbf{C}$ *if and only if A is regular.*

(iii) *In particular,* **C** *is its own regular ring, thus the operation* $A \mapsto \mathbf{C}$ *is idempotent.*

Proof. (i) **C** is a finite Baer $*$-ring [§48] satisfying the (EP)-axiom (Proposition 4) and the (UPSR)-axiom (Proposition 3), thus it satisfies (1°); **C** also satisfies (2°)–(4°) (see [§49, Cor. 2 of Th. 1] and [§51, Prop. 1]), it inherits (5°) from A since it has no new unitaries [§49, Th. 1] or projections [§48, Th. 1], and it is regular [§52, Th. 1].

(ii) If A is regular, then $\overline{A} = \mathbf{C}$ [§51, Th. 1]; the converse is immediate from the regularity of **C**.

(iii) It follows from (i) that **C** has a regular ring **D** in the sense of [§52, Def. 2]; but $\overline{\mathbf{C}} = \mathbf{D}$ by (ii), thus **C** is its own regular ring. ∎

It follows that the properties that accrue to A in virtue of (1°)–(5°) also accrue to **C**; for example, **C** admits a strong form of polar decomposition:

Proposition 5. *If* $\mathbf{x} \in \mathbf{C}$ *one can write* $\mathbf{x} = \overline{w}\mathbf{r}$ *with* $\mathbf{r} = (\mathbf{x}^* \mathbf{x})^{\frac{1}{2}}$ *and w a partial isometry in A such that* $\overline{w}^* \overline{w} = \mathrm{RP}(\mathbf{x})$, $\overline{w}\,\overline{w}^* = \mathrm{LP}(\mathbf{x})$.

Proof. Since $\mathbf{C}$ satisfies (1°) and (2°), we can apply [§21, Prop. 2] to it; the only partial isometries available in $\mathbf{C}$ are of the form $\overline{w}$ [§49, Cor. 1 of Th. 1]. ∎

Proposition 6. *If* $\mathbf{x} \in \mathbf{C}$, *then* $\mathbf{x} \geq 0$ *if and only if one can write* $\mathbf{x} = [x_n, e_n]$ *with* $x_n \geq 0$ *for all n.*

Proof. The 'only if' part is trivial. Conversely, suppose $\mathbf{x} = [x_n, e_n]$ with $x_n \geq 0$ (or merely $e_n x_n e_n \geq 0$) for all n. Citing [§48, Prop. 1], we have $\overline{e}_n \mathbf{x} \overline{e}_n = (e_n x_n e_n)^- \geq 0$ for all n, therefore $\mathbf{x} \geq 0$ by [§50, Th. 1] (which is applicable to $\mathbf{C}$ since $\mathbf{C}$ also satisfies (1°)–(4°)). ∎

The final result of the section bears on the notion of boundedness introduced in the next section (as do Exercises 3 and 5):

Proposition 7. *Let* $\mathbf{x}, \mathbf{y} \in \mathbf{C}$. *Then* $\mathbf{x}^* \mathbf{x} \leq \mathbf{y}^* \mathbf{y}$ *if and only if* $\mathbf{x} = \mathbf{w} \mathbf{y}$ *with* $\mathbf{w}^* \mathbf{w} \leq 1$.

Proof. If $\mathbf{x} = \mathbf{w}\mathbf{y}$ with $\mathbf{w}^*\mathbf{w} \leq 1$, then $\mathbf{x}^*\mathbf{x} = \mathbf{y}^*\mathbf{w}^*\mathbf{w}\mathbf{y} \leq \mathbf{y}^* 1 \mathbf{y}$.

Conversely, suppose $\mathbf{x}^*\mathbf{x} \leq \mathbf{y}^*\mathbf{y}$. Set $\mathbf{f} = \mathrm{RP}(\mathbf{y})$. It is elementary that $\mathbf{x}\mathbf{f} = \mathbf{x}$ (see the parenthetical remark at the end of the proof of [§50, Lemma]). By Proposition 5, we may write $\mathbf{y} = \mathbf{v}\mathbf{s}$ with $\mathbf{s} = (\mathbf{y}^*\mathbf{y})^{\frac{1}{2}}$ and $\mathbf{v}^*\mathbf{v} = \mathbf{f} = \mathrm{RP}(\mathbf{y}) = \mathrm{RP}(\mathbf{s})$, $\mathbf{v}\mathbf{v}^* = \mathrm{LP}(\mathbf{y})$. Let $\mathbf{t}$ be the inverse of $\mathbf{s}$ in $\mathbf{f}\mathbf{C}\mathbf{f}$ [§51, Prop. 4], and define $\mathbf{w} = \mathbf{x}\mathbf{t}\mathbf{v}^*$. Then

$$\mathbf{w}\mathbf{y} = (\mathbf{x}\mathbf{t}\mathbf{v}^*)(\mathbf{v}\mathbf{s}) = \mathbf{x}\mathbf{t}\mathbf{f}\mathbf{s} = \mathbf{x}\mathbf{t}\mathbf{s} = \mathbf{x}\mathbf{f} = \mathbf{x}$$

and

$$\begin{aligned} \mathbf{w}^*\mathbf{w} &= (\mathbf{t}\mathbf{v}^*)^*\mathbf{x}^*\mathbf{x}(\mathbf{t}\mathbf{v}^*) \leq (\mathbf{t}\mathbf{v}^*)^*\mathbf{y}^*\mathbf{y}(\mathbf{t}\mathbf{v}^*) \\ &= \mathbf{v}\mathbf{t}\mathbf{s}^2\mathbf{t}\mathbf{v}^* = \mathbf{v}\mathbf{f}\mathbf{v}^* = \mathbf{v}\mathbf{v}^* = \mathrm{LP}(\mathbf{y}) \leq 1. \end{aligned}$$ ∎

Exercises

In the following exercises, it is assumed that A satisfies (1°)–(5°).

1A. If $\mathbf{x} \in \mathbf{C}$, $\mathbf{x} \geq 0$, $\mathrm{RP}(\mathbf{x}) = \mathbf{e}$, and if $\mathbf{y}$ is the inverse of $\mathbf{x}$ in $\mathbf{e}\mathbf{C}\mathbf{e}$, then $\mathbf{y} \geq 0$.

2A. If $\mathbf{x}, \mathbf{y} \in \mathbf{C}$, $0 \leq \mathbf{x} \leq \mathbf{y}$, and if $\mathbf{x}$ is invertible, then $\mathbf{y}$ is invertible and $0 \leq \mathbf{y}^{-1} \leq \mathbf{x}^{-1}$.

3A. (i) If $\mathbf{y} \in \mathbf{C}$, $\mathbf{y} \geq 0$, then $\mathbf{y} \leq 1$ if and only if $\mathbf{y}^2 \leq 1$. (ii) In particular, $0 \leq (1 + \mathbf{x}^*\mathbf{x})^{-1} \leq 1$ for all $\mathbf{x} \in \mathbf{C}$. (iii) If $\mathbf{x}^* = \mathbf{x}$ and $\mathbf{y} = (\mathbf{x} + i)^{-1}$, then $\mathbf{y}^*\mathbf{y} \leq 1$.

4A. If $\mathbf{x} \in \mathbf{C}$, one can write $\mathbf{x} = \mathbf{u}\mathbf{r}$ with $\mathbf{r} \geq 0$ and $\mathbf{u} = \overline{u}$ unitary.

5A. If $\mathbf{x} \in \mathbf{C}$ and $\mathbf{z}$ is in the center of $\mathbf{C}$, then $\mathbf{x}^*\mathbf{x} \leq \mathbf{z}$ iff $\mathbf{x}\mathbf{x}^* \leq \mathbf{z}$.

6A. Suppose A has the property that the relations $a \in A$, $aa^* \leq a^*a$ imply $aa^* = a^*a$ (as is the case when A is a finite von Neumann algebra). Then the relations $\mathbf{x} \in \mathbf{C}$, $\mathbf{x}\mathbf{x}^* \leq \mathbf{x}^*\mathbf{x}$ imply $\mathbf{x}\mathbf{x}^* = \mathbf{x}^*\mathbf{x}$.

§ 54. C Has no New Bounded Elements

We assume, as in Sections 52 and 53, that A satisfies the hypotheses (1°)–(5°), and will shortly add another.

The notion of boundedness alluded to in the section heading can be defined in an abstract $*$-ring:

Definition 1. An element x of a $*$-ring B with unity is said to be *bounded* if there exists a positive integer k such that $\mathbf{x}^*\mathbf{x} \leq k1$ (in the sense of the ordering described in Section 50). The set of all such elements x is denoted B_0; it is called the *bounded subring* of B, in view of the following proposition:

Proposition 1. (i) *If B is any $*$-ring with unity, then the set B_0 of all bounded elements in B is a subring of B.*

(ii) *B_0 contains all partial isometries (in particular, all projections) of B.*

(iii) *If $n1$ is invertible in B for every positive integer n, then B_0 is a $*$-subring of B.*

Proof. (i) Suppose $x, y \in B_0$. Say $x^*x \leq m1$, $y^*y \leq n1$, m and n positive integers. The boundedness of $x-y$ and xy result from the computations

$$(x-y)^*(x-y) \leq (x+y)^*(x+y) + (x-y)^*(x-y)$$
$$= 2x^*x + 2y^*y \leq 2m1 + 2n1 = (2m+2n)1$$

and

$$(xy)^*(xy) = y^*(x^*x)y \leq y^*(m1)y = m(y^*y) \leq (mn)1,$$

where the last inequality is justified by the identity

$$(mn)1 - m(y^*y) = m(n1 - y^*y).$$

(ii) If $w^*w = e$, e a projection, then $1 - w^*w = (1-e)^*(1-e) \geq 0$, thus $w \in B_0$.

(iii) We observe first that if $ny \geq 0$, n a positive integer, then $y \geq 0$. {Proof: Obviously $n^2y = n(ny) \geq 0$. Say $n^2y = y_1^*y_1 + \cdots + y_m^*y_m$. Write $c = (n1)^{-1}$; then c is self-adjoint and central, thus $y = c(n^2y)c = (y_1c)^*(y_1c) + \cdots + (y_mc)^*(y_mc) \geq 0$.}

Assuming $x \in B_0$, we are to show that $x^* \in B_0$. Say $x^*x \leq m1$. Then

$$0 \leq (xx^* - m1)^2 + x(m1 - x^*x)x^*$$
$$= (xx^*)^2 - 2m(xx^*) + m^2 1 + m(xx^*) - (xx^*)^2$$
$$= m^2 1 - m(xx^*) = m(m1 - xx^*),$$

therefore $m1 - xx^* \geq 0$ by the preceding observation. Thus $xx^* \leq m1$, $x^* \in B_0$. ∎

In particular, A_0 is a $*$-subring of A, $\mathbf{C}_0$ is a $*$-subring of $\mathbf{C}$; obviously $\overline{A}_0 \subset \mathbf{C}_0$; and A_0, $\mathbf{C}_0$ are themselves Baer $*$-rings [cf. §4, Exer. 6]. The notion of boundedness has occured tacitly before (cf. [§51, Cor. 3 of Th. 1], [§53, Prop. 7]).

The claim in the section heading is that $\overline{A}_0 = \mathbf{C}_0$; to validate it, we require an additional axiom on A:

Definition 2. We say that A satisfies the *positive sums axiom* (briefly, the (PS)-*axiom*) provided that if (f_n) is an orthogonal sequence of projections in A with $\sup f_n = 1$, and if, for each n, $a_n \in f_n A f_n$ with $0 \le a_n \le 1$, then there exists $a \in A$ such that $a f_n = a_n$ for all n. {Informally, '$a = \sum a_n$'.}

In addition to (1°)–(5°), we assume for the rest of the section that

(6°) *A satisfies the* (PS)-*axiom.*

Remarks. Assume the notations of Definition 2.

1. The conditions '$a_n \in f_n A f_n$ and $0 \le a_n \le 1$' are equivalent to '$0 \le a_n \le f_n$'.
2. The elements a_n, f_m all commute: $a_n f_n = f_n a_n = a_n$, and $a_n f_m = f_m a_n = a_m a_n = 0$ when $m \ne n$.
3. In an AW^*-algebra (or even a Rickart C^*-algebra) the construction of a is easy spectral theory; in fact, writing the commutant of the set of all f_n as a C*-sum [§10, Prop. 3], one need only assume that the $a_n \in f_n A f_n$ are bounded in norm.
4. The element a is unique since $\sup f_n = 1$. Moreover, $0 \le a \le 1$. {Proof: Setting $e_n = f_1 + \cdots + f_n$, we have $e_n \uparrow 1$ and $e_n a e_n = a_1 + \cdots + a_n \ge 0$ for all n, therefore $a \ge 0$ [§50, Th. 1]; also $e_n(1-a)e_n = (f_1 - a_1) + \cdots + (f_n - a_n) \ge 0$ for all n, thus $a \le 1$.}
5. The condition $\sup f_n = 1$ can be dropped, by adjoining $1 - \sup f_n$ to the sequence (but then a need not be unique).
6. The point of the (PS)-axiom is that $a \in A$; one can always construct an $\mathbf{a} \in \mathbf{C}$ in the obvious way (cf. the proof of [§51, Cor. 2 of Th. 1]).

Lemma. *Let* $\mathbf{x} \in \mathbf{C}$. *In order that* $0 \le \mathbf{x} \le 1$, *it is necessary and sufficient that* $\mathbf{x} = \overline{a}$ *for some* $a \in A$ *with* $0 \le a \le 1$.

Proof. Suppose $0 \le \mathbf{x} \le 1$. Let $\mathbf{u} = \overline{u}$ be the Cayley transform of $\mathbf{x}$ and write $\mathbf{x} = [x_n, e_n]$ as in [§53, Prop. 1]. Then $0 \le \overline{e}_n \mathbf{x} \overline{e}_n \le \overline{e}_n$; since $\overline{e}_n \mathbf{x} \overline{e}_n = (e_n x_n e_n)^- = \overline{x}_n$, we have $0 \le x_n \le e_n$ [§50, Cor. of Th. 1], therefore

$$0 \le x_n \le 1 \quad \text{for all } n. \tag{1}$$

Note that

$$x_1 \le x_2 \le x_3 \le \cdots; \tag{2}$$

for, if $m<n$ then, since $x_n^{\frac{1}{2}} \in \{x_n\}'' \subset \{u\}''$, we have

$$x_m = x_n e_m = x_n^{\frac{1}{2}} e_m x_n^{\frac{1}{2}} \leq x_n^{\frac{1}{2}} 1 x_n^{\frac{1}{2}} = x_n .$$

Set $e_0 = x_0 = 0$ and define

$$f_n = e_n - e_{n-1}, \qquad a_n = x_n - x_{n-1} .$$

From (1) and (2) we have $0 \leq a_n \leq 1$, and it is easy to see that $a_n \in f_n A f_n$. The (PS)-axiom yields $a \in A$ with

$$a f_n = a_n = x_n - x_{n-1} \quad \text{for all } n,$$

therefore

$$a e_n = \sum_1^n a f_j = \sum_1^n (x_j - x_{j-1}) = x_n ,$$

thus $\bar{a} = \mathbf{x}$. Moreover, $0 \leq a \leq 1$ by Remark 4 above.

Conversely, if $a \in A$ and $0 \leq a \leq 1$, then $0 \leq \bar{a} \leq 1$ by the trivial half of [§ 50, Cor. of Th. 1]. ∎

Theorem 1. *If* $\mathbf{x} \in \mathbf{C}$, *then* $\mathbf{x}$ *is bounded in* $\mathbf{C}$ *if and only if* $\mathbf{x} = \bar{a}$ *with* a *bounded in* A. *More precisely,* $\mathbf{x}^* \mathbf{x} \leq k1$ *(k a positive integer) if and only if* $\mathbf{x} = \bar{a}$ *with* $a^* a \leq k1$.

Proof. Suppose $\mathbf{x}^* \mathbf{x} \leq k1$, k a positive integer. By the (UPSR)-axiom, we may write $k1 = s^2$ with $s \geq 0$, $s \in \{k1\}''$; clearly s is central in A. Moreover, $s^2 = 1 + (k-1)1$ is invertible in A by symmetry, thus $t = s^{-1}$ is central in A. (Informally, $t = k^{-\frac{1}{2}} 1$.) Setting $\mathbf{y} = \bar{t}\mathbf{x}$, we have $\mathbf{y}^* \mathbf{y} \leq 1$. By the lemma, there exists $b \in A$, $0 \leq b \leq 1$, such that

(1) $$\mathbf{y}^* \mathbf{y} = \bar{b} .$$

Write $b = c^2$ with $c \in A$, $c \geq 0$. Then (1) yields

(2) $$\mathbf{x}^* \mathbf{x} = \bar{s} \mathbf{y}^* \mathbf{y} \bar{s} = \bar{s} \bar{b} \bar{s} = (\overline{sc})^2 .$$

Write $\mathbf{x} = \bar{w} \mathbf{r}$ with $\mathbf{r} = (\mathbf{x}^* \mathbf{x})^{\frac{1}{2}}$ and $w \in A$ a partial isometry [§ 53, Prop. 5]. Since $sc = s^{\frac{1}{2}} c s^{\frac{1}{2}} \geq 0$, it results from (2), and the uniqueness of positive square roots in $\mathbf{C}$ [§ 53, Prop. 3], that $\mathbf{r} = (sc)^-$. Then $\mathbf{x} = \bar{w} \mathbf{r} = \bar{a}$ with $a = wsc$. Moreover, $(a^* a)^- = \mathbf{x}^* \mathbf{x} \leq k1$, therefore $a^* a \leq k1$ by [§ 50, Cor. of Th. 1].

Conversely, if $a \in A$ and $a^* a \leq k1$, then $\bar{a}^* \bar{a} \leq k1$ by the trivial half of [§ 50, Cor. of Th. 1]. ∎

Corollary 1. *If* $\mathbf{x}_1, \ldots, \mathbf{x}_m \in \mathbf{C}$ *and*

$$\mathbf{x}_1^* \mathbf{x}_1 + \cdots + \mathbf{x}_m^* \mathbf{x}_m = 1 ,$$

then $\mathbf{x}_j = \bar{a}_j$ *for suitable* $a_j \in A$ *with* $a_j^* a_j \leq 1$.

Proof. Since $1-\mathbf{x}_j^*\mathbf{x}_j=\sum_{k\neq j}\mathbf{x}_k^*\mathbf{x}_k$, we have $0\leq\mathbf{x}_j^*\mathbf{x}_j\leq 1$; quote the lemma. ∎

Corollary 2. *If every $a\in A$ is bounded (for example, if A is a finite AW^*-algebra), then $\overline{A}$ is the bounded subring of* $\mathbf{C}$.

Proof. Immediate from the theorem. ∎

We conclude the chapter with a result on boundedness in quotient rings (see also Exercise 6):

Theorem 2. *If I is a restricted ideal of A, then the bounded subring of A/I is the canonical image of the bounded subring of A, thus $A_0/A_0\cap I$ is naturally $*$-isomorphic to $(A/I)_0$.*

Proof. Write $x\mapsto\tilde{x}$ for the canonical mapping $A\to A/I$. Recall that A/I is a Rickart $*$-ring [§ 23, Prop. 1].

If a is bounded in A, it is obvious that $\tilde{a}$ is bounded in A/I.

Conversely, suppose $\tilde{x}$ is a bounded element of A/I; we seek $a\in A_0$ with $\tilde{x}=\tilde{a}$. Say $\tilde{x}^*\tilde{x}\leq k1$, k a positive integer. Then

$$k1-\tilde{x}^*\tilde{x}=\tilde{y}_1^*\tilde{y}_1+\cdots+\tilde{y}_m^*\tilde{y}_m$$

for suitable $y_1,\dots,y_m\in A$, thus there exists $c\in I$ with

$$k1-c=x^*x+y_1^*y_1+\cdots+y_m^*y_m. \tag{1}$$

In particular, $k1-c\geq 0$, so we may write $k1-c=r^2$ with $r\geq 0$. In view of (1) we have $x^*x\leq r^2$, therefore $\overline{x}^*\overline{x}\leq\overline{r}^2$ in $\mathbf{C}$; it follows that

$$\overline{x}=\mathbf{w}\overline{r} \tag{2}$$

for suitable $\mathbf{w}\in\mathbf{C}$ with $\mathbf{w}^*\mathbf{w}\leq 1$ [§ 53, Prop. 7]. In view of Theorem 1, $\mathbf{w}=\overline{w}$ with $w\in A_0$, thus (2) yields

$$x=wr, \tag{3}$$

and it will suffice to show that $\tilde{r}=\tilde{b}$ with $b\in A_0$. Let $u=\tilde{r}$. We know that $u^*=u$ and $u^2=k1-\tilde{c}=k1$. Set $s=(k^{-\frac{1}{2}}r)^{\sim}$ (cf. the proof of Theorem 1). Then $s^*=s$ and $s^2=k^{-1}u^2=1$, thus s is a symmetry in A/I. Then $(1/2)(1+s)$ is a projection in A/I, therefore $(1/2)(1+s)=\tilde{e}$ for a suitable projection e in A [§ 23, Prop. 1]. Then $s=(2e-1)^{\sim}$, thus

$$u=\tilde{r}=k^{\frac{1}{2}}s=(k^{\frac{1}{2}}(2e-1))^{\sim};$$

writing $b=k^{\frac{1}{2}}(2e-1)$, we have $\tilde{r}=\tilde{b}$ with $b\in A_0$. ∎

Exercises

1A. Assume A satisfies (1°)–(5°). If every element of A is bounded (that is, $A_0 = A$) and if A is not the direct sum of finitely many factors of Type I, then $\bar{A} \subset \mathbf{C}$ properly.

2A. Assume A satisfies (1°)–(4°). (i) For every $\mathbf{x} \in \mathbf{C}$, $(1+\mathbf{x}^*\mathbf{x})^{-1}$ is bounded. (ii) If A also satisfies (5°), and if $\mathbf{x}^* = \mathbf{x}$, then $(\mathbf{x}+i)^{-1}$ is bounded.

3A. If B is a $*$-ring with unity then B_0 is a $*$-subring of B under either of the following hypotheses: (i) x^*x and xx^* are unitarily equivalent, for every $x \in B$; (ii) B is a Rickart $*$-ring with WPD [§ 21, Def. 1].

4A. (i) If A satisfies (1°)–(4°) and is regular, then A satisfies (5°) and (6°).
(ii) If A satisfies (1°)–(5°), then $\mathbf{C}$ satisfies (1°)–(6°).

5A. An example of a ring satisfying (1°)–(6°) that is neither regular nor AW^* can be constructed as follows. If (A_ι) is any family of $*$-rings satisfying (1°)–(6°), then the complete direct product A of the A_ι also satisfies (1°)–(6°). In particular, let (A_ι) be a family of finite AW^*-algebras: if at least one of the A_ι is infinite-dimensional, then A is not regular; if the family is infinite, then A is not an AW^*-algebra.

6A. Assume A satisfies (1°)–(5°). Note that A and $\mathbf{C}$ have the same p-ideals, every ideal of $\mathbf{C}$ is restricted, and the p-ideals are paired bijectively with the ideals of $\mathbf{C}$.

Suppose $\mathscr{I}$ is a p-ideal of A, I is the ideal of A generated by $\mathscr{I}$, and $\mathbf{I}$ the ideal of $\mathbf{C}$ generated by $\bar{\mathscr{I}}$. Explicitly,

$$I = \{x \in A : \mathrm{RP}(x) \in \mathscr{I}\},$$
$$\mathbf{I} = \{\mathbf{x} \in \mathbf{C} : \mathrm{RP}(\mathbf{x}) = \bar{e},\ e \in \mathscr{I}\}.$$

Thus $\mathbf{I} \cap \bar{A} = \bar{I}$.

(i) I is maximal-restricted in A if and only if $\mathbf{I}$ is maximal in $\mathbf{C}$, in which case $\mathbf{C}/\mathbf{I}$ is a regular Baer $*$-factor.

(ii) The bounded subring of $\mathbf{C}/\mathbf{I}$ is the canonical image of the bounded subring of $\mathbf{C}$, thus $\mathbf{C}_0/\mathbf{C}_0 \cap \mathbf{I}$ is naturally $*$-isomorphic to $(\mathbf{C}/\mathbf{I})_0$.

(iii) If, in addition, A satisfies (6°), then $A_0/A_0 \cap I$ is naturally $*$-isomorphic to $(\mathbf{C}/\mathbf{I})_0$.

(iv) If A satisfies (6°) and all elements of A are bounded (as is the case when A is a finite AW^*-algebra), then the natural $*$-monomorphism $x + I \mapsto \bar{x} + \mathbf{I}$ maps A/I onto the bounded subring of $\mathbf{C}/\mathbf{I}$, thus A/I is $*$-isomorphic to $(\mathbf{C}/\mathbf{I})_0$. {In a sense, we may speak of $\mathbf{C}/\mathbf{I}$ as 'the regular ring' of A/I; the rub is that A/I need not satisfy (1°)–(5°); for instance, the (EP)-axiom may fail in A/I [cf. § 45, Exer. 10, (iii)].}

7A. If A is the real AW^*-algebra of bounded, complex sequences that are real at infinity [§ 4, Exer. 14], then A is a commutative Baer $*$-ring that satisfies (1°), (3°), (5°) and (6°), but neither (2°) nor (4°).

8A. A complete Boolean algebra trivially satisfies (6°) [cf. § 52, Exer. 9].

9A. Let A be the $*$-ring of all complex sequences $x = (\lambda_n)$ that are ultimately real (that is, λ_n is real for all sufficiently large n); the ring operations of A are the coordinatewise ones, and the involution is $x^* = (\lambda_n^*)$. Then A is a commutative, regular Baer $*$-ring satisfying the (EP)-axiom and the (UPSR)-axiom; A satisfies (1°), (3°), (5°) and (6°), but neither (2°) nor (4°).

Chapter 9

Matrix Rings over Baer $*$-Rings

The chapter is based on [**9**].

§ 55. Introduction

Let A be a Baer $*$-ring, n a positive integer, A_n the $*$-ring of all $n \times n$ matrices over A (with $*$-transpose as the involution [§ 16]). Is A_n a Baer $*$-ring?

In general, the answer is negative (see the exercises). However, if A is properly infinite, then A_n is $*$-isomorphic to A—thereby killing off the question—under rather weak hypotheses (e.g., orthogonal GC is sufficient [§ 17, Cor. of Th. 1], hence also PC [§ 14, Prop. 5]).

We are thus left with the problem of determining conditions on a finite Baer $*$-ring A that are sufficient to ensure that A_n is a Baer $*$-ring. The problem is largely open. Using the theory of the regular ring developed in the preceding chapter, we show that A_n is a Baer $*$-ring provided that A satisfies conditions (1°)–(6°) of Section 54, (7°) A_n satisfies the parallelogram law (P), and (8°) every orthogonal sequence of projections in A_n has a supremum; the hypotheses (1°)–(8°) are severe but the case of AW^*-algebras is covered (see Section 62).

For the rest of the chapter (excluding the exercises, and excluding [§ 62, Cor. 1 of Th. 1]) A denotes a finite Baer $*$-ring, with center Z, satisfying the conditions (1°)–(6°) of the preceding chapter; for convenient reference, we record them here:

(1°) *A is a finite Baer $*$-ring satisfying the* (EP)*-axiom and the* (UPSR)*-axiom;*

(2°) *partial isometries in A are addable;*

(3°) *$1 + a^*a$ is invertible in A for all $a \in A$;*

(4°) *A contains a central element i such that $i^2 = -1$ and $i^* = -i$;*

(5°) *if $u \in A$ is unitary and* $\mathrm{RP}(1-u) = 1$, *then there exists a sequence of projections $e_k \in \{u\}''$ such that $e_k \uparrow 1$ and $(1-u)e_k$ is invertible in $e_k A e_k$ for all k;*

(6°) *if* f_k *is an orthogonal sequence of projections in* A *with* $\sup f_k = 1$, *and if, for each* k, $a_k \in f_k A f_k$ *with* $0 \leq a_k \leq 1$, *then there exists* $a \in A$ *such that* $a f_k = a_k$ *for all* k.

Remarks. 1. All conditions (1°)–(6°) hold for A a finite AW^*-algebra.

2. When A is orthoseparable, condition (5°) is redundant [§ 52, Prop. 1, (i)].

3. When A is regular, conditions (5°) and (6°) are redundant; for, condition (5°) holds trivially (we may take $e_k = 1$ for all k [§ 51, Prop. 4, (ii)]) and condition (6°) is a consequence of (1°)–(4°) [§ 51, Cor. 2 of Th. 1].

4. Condition (4°) means that A has no 'purely real' part [§ 51, Exer. 1].

5. For a nonregular, non-AW^* example of A, see [§ 54, Exer. 5].

The following notations are also fixed for the rest of the chapter:

C denotes the regular ring of A constructed in the preceding chapter [§ 52, Def. 2] (we drop the boldface, identify $\bar{A}$ with A, and regard A as a $*$-subring of C).

n denotes a fixed positive integer. We write A_n and C_n for the $n \times n$ matrix rings (with $*$-transpose as involution), and regard A_n as a $*$-subring of C_n. For simplicity, we use lower case letters for the elements of all rings in sight; thus $x = (x_{ij})$ denotes a typical element of C_n. It is shown in the next section that C_n contains no partial isometries (projections, unitaries) not already in A_n.

Exercises

1A. If $A = \{0, 1\}$ is the field with two elements, with the identity involution, the $*$-ring A_2 is a regular Baer ring but is not $*$-regular; in particular, A_2 is not a Baer $*$-ring.

2A. If A is the real AW^*-algebra of bounded, complex sequences that are real at infinity [cf. § 54, Exer. 7], then A_2 is not a Rickart $*$-ring. This can be seen in two steps:

(i) View the elements of A_2 as bounded sequences of 2×2 matrices (with the norm of a 2×2 complex matrix defined as the sum of the absolute values of its entries).

(ii) For $k = 1, 2, 3, \ldots$ let e_k be the element of A_2 whose kth coordinate is the projection

$$\begin{pmatrix} \frac{1}{2} & \frac{i}{2} \\ -\frac{i}{2} & \frac{1}{2} \end{pmatrix}$$

and all other coordinates are the zero matrix. Then e_k is an orthogonal sequence of projections in A_2 with no supremum.

3A. If $A=\mathbb{Z}$, the ring of integers with the identity involution, then A_2 is not a Rickart $*$-ring.

4A. If A is the ring of 2×2 matrices over the field of three elements, then A is a Baer $*$-ring but A_2 is not a Rickart $*$-ring.

5A. If A is a Baer $*$-ring in which partial isometries are not addable, then A_n is not a Baer $*$-ring for any $n\geq 2$.

6A. If A is a Baer $*$-ring such that A_2 is a Rickart $*$-ring, does it follow that A_2 is a Baer $*$-ring?

§ 56. Generalities

In this section we derive some properties of A_n and C_n using only the hypotheses (1°)–(6°). The central result is that A_n is a Rickart $*$-ring (Theorem 1); this is the principal dividend of (6°).

The first proposition holds under the weaker hypotheses of Section 50:

Proposition 1. *If* $x_1,\dots,x_m\in C_n$ *and* $x_1^*x_1+\cdots+x_m^*x_m=0$, *then* $x_1=\cdots=x_m=0$. *In particular, the involution of* C_n *is proper.*

Proof. Say $x_k=(x_{ij}^k)$ $(k=1,\dots,m)$. The (j,j) coordinate of the given equation reads

$$\sum_{k=1}^{m}\left(\sum_{i=1}^{n} x_{ij}^{k*}x_{ij}^{k}\right)=0,$$

hence $x_{ij}^k=0$ [§ 50, Prop. 1]. ∎

The next proposition requires only (1°)–(5°) (see also [§ 52, Exer. 8]):

Proposition 2. C_n *is* $*$-*regular, hence is a regular Rickart* $*$-*ring.*

Proof. Since C is regular [§ 52, Th. 1], C_n is regular by a general theorem of von Neumann [§ 51, Exer. 12]. Since the involution of C_n is proper (Proposition 1), the proposition follows from [§ 51, Prop. 3]. ∎

The full force of the hypotheses (1°)–(6°) is used in the next proposition:

Proposition 3. *If* $x_1,\dots,x_m\in C_n$ *and* $x_1^*x_1+\cdots+x_m^*x_m=1$ *(the identity matrix), then* $x_1,\dots,x_m\in A_n$. *In particular,* C_n *has no new projections (unitaries, partial isometries).*

Proof. Say $x_k=(x_{ij}^k)$. The (j,j) coordinate of the given equation reads

$$\sum_{k=1}^{m}\left(\sum_{i=1}^{n} x_{ij}^{k*}x_{ij}^{k}\right)=1,$$

hence $x_{ij}^k\in A$ [§ 54, Cor. 1 of Th. 1].

If $w\in C_n$ is a partial isometry, say $w^*w=e$, then $w\in A_n$ results from the equation $w^*w+(1-e)^*(1-e)=1$. Thus C_n has no new partial isometries (hence no new projections or unitaries). ∎

Theorem 1. *A_n is a Rickart $*$-ring.*

Proof. Since C_n is a Rickart $*$-ring (Proposition 2), the theorem is immediate from the fact that A_n contains all projections of C_n (Proposition 3). ∎

Corollary. *If A is a finite AW^*-algebra, then A_n is a Rickart C^*-algebra. (It is shown in Section 62 that A_n is also an AW^*-algebra.)*

Proof. The $n \times n$ matrix algebra over a C^*-algebra is also a C^*-algebra (e. g., by the Gel'fand-Naĭmark theorem); quote Theorem 1. ∎

Note that the center of A_n consists of all the 'scalar' matrices

$$\operatorname{diag}(z,\dots,z) = \begin{pmatrix} z & & 0 \\ & \ddots & \\ 0 & & z \end{pmatrix}$$

with $z \in Z$ (the center of A), thus A and A_n have $*$-isomorphic centers (as do C and C_n). In particular, A and A_n have the 'same' central projections, identified via

$$h \leftrightarrow \operatorname{diag}(h,\dots,h).$$

These are also the central projections of C and C_n, since (i) A and C have the same projections [§ 48, Th. 1], (ii) A_n and C_n have the same projections (Proposition 3), and (iii) an element $a \in A$ is central in A iff it is central in C [cf. § 47, Def. 6]. If h is a central projection in A and $(x_{ij}) \in C_n$, we write $h(x_{ij}) = (h x_{ij})$ for the product of $\operatorname{diag}(h,\dots,h)$ and (x_{ij}); that is, for purposes of notating products, we identify h with $\operatorname{diag}(h,\dots,h)$.

In view of the coarse structure theory (see [§ 15, Th. 2] and [§ 18, Th. 2]), the effect of the following theorem is to reduce the study of A_n and C_n to the cases that A is homogeneous or of Type II:

Theorem 2. *If (h_α) is an orthogonal family of central projections in A with $\sup h_\alpha = 1$, then C_n is $*$-isomorphic to the complete direct product of the $h_\alpha C_n$, via the mapping $(x_{ij}) \mapsto (h_\alpha(x_{ij}))$.*

Proof. The indicated mapping is clearly a $*$-homomorphism. It is injective: if $(h_\alpha(x_{ij})) = 0$, that is, if $h_\alpha(x_{ij}) = (h_\alpha x_{ij}) = 0$ for all α, then $h_\alpha x_{ij} = 0$ for all α, i, j, hence $x_{ij} = 0$ for all i, j, thus $(x_{ij}) = 0$.

It remains to show that the mapping is surjective. Suppose that for each α we are given a matrix (x_{ij}^α) in $h_\alpha C_n$; thus $x_{ij}^\alpha \in h_\alpha C$ for all α, i, j. Since C is the complete direct product of the $h_\alpha C$ [§ 52, Th. 2], it follows that for each pair i, j there exists $x_{ij} \in C$ such that $h_\alpha x_{ij} = x_{ij}^\alpha$ for all α. Then

$$h_\alpha(x_{ij}) = (h_\alpha x_{ij}) = (x_{ij}^\alpha),$$

thus $(x_{ij})\in C_n$ maps onto the given element $((x_{ij}^{\alpha}))$ of the complete direct product of the $h_\alpha C_n$. ∎

Corollary. *Notation as in Theorem 2. If, for each α, $w_\alpha \in h_\alpha A_n$ is a partial isometry [projection, unitary] then there exists a unique partial isometry [projection, unitary] $w\in A_n$ such that $h_\alpha w = w_\alpha$ for all α.*

Proof. By the theorem, there exists a unique $w\in C_n$ such that $h_\alpha w = w_\alpha$ for all α. Since $w_\alpha w_\alpha^* w_\alpha = w_\alpha$, that is, $h_\alpha(ww^*w-w)=0$ for all α, it follows that $ww^*w=w$, that is, w is a partial isometry. By Proposition 3, $w\in A_n$. If, moreover, the w_α are projections [unitaries] in $h_\alpha A_n$, it is immediate from the theorem that w is also a projection [unitary]. ∎

The center of A_n, being isomorphic to the center Z of A, is a Baer $*$-ring. In addition, there is a smoothly working notion of central cover available in A_n:

Definition 1. If $x=(a_{ij})\in A_n$, we define

$$C(x) = \sup\{C(a_{ij}) : i,j=1,\ldots,n\},$$

where $C(a_{ij})$ is the central cover of a_{ij} relative to A [§ 6, Prop. 3]. We regard $C(x)$ as a projection in A_n as well, and call it the central cover of x; clearly $C(x)$ is the smallest central projection h such that $hx_{ij}=x_{ij}$ for all i and j, that is, $hx=x$. In particular, the present definition is consistent with [§ 6, Def. 1].

Though we do not know that A_n is a Baer $*$-ring, the characterization of 'very orthogonality' valid for Baer $*$-rings holds also in A_n:

Proposition 4. *If $x,y\in A_n$, the following conditions are equivalent:* (a) $C(x)C(y)=0$; (b) $xA_ny=0$.

Proof. Write $B=A_n$. It is obvious that (a) implies (b).

Conversely, suppose $xBy=0$. Then also $(BxB)B(ByB)=0$. Say $x=(a_{ij})$, $y=(b_{ij})$. We assert that $a_{ij}Ab_{rs}=0$ for all indices i,j,r,s; since pre- and post-multiplication of x by suitable matrices will move a_{ij} to the $(1,1)$ position, and similarly for b_{rs}, it clearly suffices to prove that $a_{11}Ab_{11}=0$. Also, pre- and post-multiplying x by the projection $\operatorname{diag}(1,0,\ldots,0)$, we can suppose that $a_{ij}=0$ unless $i=j=1$, and similarly that $b_{ij}=0$ unless $i=j=1$. It follows that, for all $a\in A$, the condition

$$x[\operatorname{diag}(a,0,\ldots,0)]y=0$$

yields $a_{11}ab_{11}=0$; thus $a_{11}Ab_{11}=0$. Summarizing, it has been shown that for all indices i,j,r,s we have $a_{ij}Ab_{rs}=0$, hence $C(a_{ij})C(b_{rs})=0$ [§ 6, Cor. 1 of Prop. 3]. It is immediate from Definition 1 that $C(x)C(y)=0$. ∎

We shall make use of central cover in A_n only for projections; the following elementary proposition is a special case of [§ 6, Prop. 1]:

Proposition 5. *For projections in* A_n, (1) $e \sim f$ *implies* $C(e)=C(f)$; (2) $e \precsim f$ *implies* $C(e) \le C(f)$; (3) $C(he)=hC(e)$ *when* h *is central;* (4) *if* (e_ι) *is a family of projections that possesses a supremum* e, *then* $C(e)=\sup C(e_\iota)$.

Exercises

1C. Let B be any $*$-ring with unity, n a positive integer, B_n the $*$-ring of all $n \times n$ matrices over B [§ 16, Def. 1].

(i) The involution of B_n is proper if and only if B satisfies the condition that $x_1^* x_1 + \cdots + x_n^* x_n = 0$ implies $x_1 = \cdots = x_n = 0$.

(ii) Let M be the set of all $n$ples $x=(x_1, \ldots, x_n)$, $x_i \in B$, regarded as a left B-module in the obvious way; one can identify B_n with the ring of all module endomorphisms $T: M \to M$. Assume that B satisfies the condition in (i). For $x, y \in M$, define

$$[x,y] = x_1 y_1^* + \cdots + x_n y_n^*;$$

thus $[x,y]$ is a B-valued 'inner product' on M such that $[x,x]=0$ implies $x=0$. Writing T^* for the $*$-transpose of the matrix T, one has $[xT,y]=[x,yT^*]$ identically. For a subset N of M, write

$$N^\perp = \{x \in M : [x,y]=0 \text{ for all } y \in N\}.$$

Prove: B_n is a Baer $*$-ring if and only if $M = N + N^\perp$ for every submodule N of M satisfying $N = N^{\perp\perp}$.

(iii) Assume B is an involutive division ring. Then B_n is a Baer $*$-ring if and only if B satisfies the condition that $x_1^* x_1 + \cdots + x_n^* x_n = 0$ implies $x_1 = \cdots = x_n = 0$.

2A. If A satisfies (1°)–(6°) then A_n and C_n are symmetric. In fact, if x is any element of C_n, then $1 + x^* x$ has an inverse in A_n.

3A. Assume A satisfies (1°)–(6°). If $x \in C_n$ then $\mathrm{LP}(x)$ and $\mathrm{RP}(x)$ are algebraically equivalent in C_n [cf. § 1, Exer. 6]. In particular, if e, f are projections in A_n, then $e - e \cap f$ and $e \cup f - f$ are algebraically equivalent in C_n.

4D. Assume A satisfies (1°)–(6°). *Problems:*
(i) Does A_n (equivalently, C_n) satisfy the parallelogram law (P)?
(ii) Does A_n satisfy $\mathrm{LP} \sim \mathrm{RP}$? Does C_n?
(iii) Is A_n (equivalently, C_n) a Baer $*$-ring?

5A. Assume A satisfies (1°)–(5°). The following conditions on A are equivalent:
(a) the relations $x \in C$, $x^* x \le 1$ imply $x \in A$;
(b) every partial isometry in C_2 is in A_2;
(c) every unitary in C_2 is in A_2;
(d) every symmetry in C_2 is in A_2;
(e) every projection in C_2 is in A_2.
In items (b)–(e), one can replace 2 by any positive integer $n \ge 2$; it follows that, assuming (a), A_n is a Rickart $*$-ring for all n.

§ 57. Parallelogram Law and Generalized Comparability

Does the parallelogram law (P) hold in A_n? Though the evidence is favorable [§ 56, Exer. 3], the question is open. From the next section on, we assume (P) outright. In the present section, we note that (i) if A_n possesses a feeble notion of 'square roots', then A_n satisfies (P), and (ii) if A_n satisfies (P) then it has GC.

Definition 1. A weakly Rickart $*$-ring is said to satisfy the *very weak square root axiom* (VWSR) if, for each element x, there exists an element r such that $x^*x = r^*r$ (hence $\mathrm{RP}(x) = \mathrm{RP}(r)$) and $\mathrm{LP}(r) = \mathrm{RP}(r)$.

It is trivial that a Rickart $*$-ring satisfying the (WSR)-axiom [§ 13, Def. 5] also satisfies the (VWSR)-axiom; in particular, every Rickart C^*-algebra satisfies the (VWSR)-axiom.

Proposition 1. *If A_n satisfies the* (VWSR)*-axiom, then A_n satisfies* $\mathrm{LP} \sim \mathrm{RP}$, *therefore A_n satisfies the parallelogram law* (P).

Proof. The second assertion follows from the first [§ 3, Prop. 7]. Thus, if $x \in A_n$, $e = \mathrm{RP}(x)$, $f = \mathrm{LP}(x)$, it will suffice to show that $e \sim f$. Write $x^*x = r^*r$ with $\mathrm{LP}(r) = \mathrm{RP}(r) = \mathrm{RP}(x) = e$. By the $*$-regularity of C_n, we have $C_n x = C_n e = C_n r$ [§ 51, Prop. 3]; it follows from [§ 21, Prop. 3] that there exists a partial isometry $w \in C_n$ such that $x = wr$, $w^*w = e$, $ww^* = f$. Since $w \in A_n$ [§ 56, Prop. 3], this shows that $e \sim f$ in A_n. ∎

In particular, if A is a finite AW^*-algebra, then A_n is a Rickart C^*-algebra [§ 56, Cor. of Th. 1], therefore A_n satisfies (P) by Proposition 1. {But this is not news, since (P) holds in every Rickart C^*-algebra [§ 13, Th. 1]—by a very different proof.}

We denote by $u_1, \ldots, u_n$ the diagonal matrix units in A_n:

$$u_1 = \mathrm{diag}(1, 0, \ldots, 0), \ldots, u_n = \mathrm{diag}(0, \ldots, 0, 1).$$

Thus, the u_i are orthogonal, equivalent projections with sum 1, and A is $*$-isomorphic to $u_1 A_n u_1$ via the correspondence

$$a \leftrightarrow \mathrm{diag}(a, 0, \ldots, 0)$$

[cf. § 16]. It follows that the centers of A, A_n and $u_1 A_n u_1$ are $*$-isomorphic via the correspondences

$$z \leftrightarrow \mathrm{diag}(z, \ldots, z) \leftrightarrow \mathrm{diag}(z, 0, \ldots, 0).$$

Lemma 1. *Suppose A_n satisfies the parallelogram law* (P). *If e is any projection in A_n, there exists an orthogonal decomposition $e = e_1 + \cdots + e_n$ such that $e_i \precsim u_1$ for all i.*

Proof. The proof for $n=3$ illustrates the principle. Since A_3 is a Rickart $*$-ring [§ 56, Th. 1] satisfying (P), we can apply [§ 13, Prop. 5] to the pair e, u_1: there exist orthogonal decompositions

$$e=e_1+f, \qquad u_1=u+v$$

such that $e_1 \sim u$ and $fu_1=0$; thus $e_1 \precsim u_1$ and $f \le 1-u_1=u_2+u_3$. Repeat the argument on the pair f, u_2, obtaining orthogonal decompositions

$$f=e_2+e_3, \qquad u_2=a+b$$

with $e_2 \sim a$ and $e_3 u_2=0$. Then $e_2 \sim a \le u_2 \sim u_1$ shows that $e_2 \precsim u_1$; and $e_3 \le f \le u_2+u_3$, together with $e_3 \le 1-u_2=u_1+u_3$, yields

$$e_3 \le (u_2+u_3)(u_1+u_3)=u_3 \sim u_1 .$$

Thus $e=e_1+e_2+e_3$ with $e_i \precsim u_1$ $(i=1,2,3)$. ∎

Lemma 2. *If e, f are projections in A_n such that $e \precsim u_1$ and $f \precsim u_1$, then e, f are* GC.

Proof. It is no loss of generality to suppose that $e \le u_1$, $f \le u_1$. Since A has GC and $u_1 A_n u_1$ is $*$-isomorphic to A, it follows that e, f are GC in $u_1 A_n u_1$; since the central projections of $u_1 A_n u_1$ are evidently of the form $\operatorname{diag}(h,0,\dots,0)=hu_1$ with h a central projection in A, it is immediate that e, f are GC in A_n. ∎

Proposition 2. *If A_n satisfies the parallelogram law* (P), *then A_n has* GC.

Proof. {The argument shows, more generally, that if A is any Rickart $*$-ring with GC, such that A_n is a Rickart $*$-ring satisfying (P), then A_n has GC.}

Let e, f be any pair of projections in A_n; we seek orthogonal decompositions

$$e=e'+e'', \qquad f=f'+f''$$

with $e' \sim f'$ and e'', f'' very orthogonal [§ 14, Prop. 1]. By the finite additivity of equivalence [§ 1, Prop. 8], we are free to discard equivalent subprojections of e, f.

The proof for $n=3$ illustrates the method. By Lemma 1, we can write

$$e=e_1+e_2+e_3, \qquad f=f_1+f_2+f_3$$

with $e_i \precsim u_1$, $f_i \precsim u_1$ for all i. Applying Lemma 2 to the pair e_1, f_1 and discarding equivalent subprojections, we can suppose e_1, f_1 are very orthogonal. Then applying Lemma 2 to the pair e_1, f_2 and discarding equivalent subprojections, we can suppose e_1, f_2 are very orthogonal too. Thus, by 9 successive applications of Lemma 2, we are reduced to the case that e_i, f_j are very orthogonal for all i, j; but then it is clear that e, f are very orthogonal [cf. § 56, Prop. 5, (4)]. {Explicitly, if h_{ij} is a central projection such that $e_i h_{ij}=e_i$ and $f_j h_{ij}=0$, then, setting

$$h = \inf_j (\sup_i h_{ij}),$$

we have $e_i h=e_i$ for all i, and $f_j h=0$ for all j, therefore $eh=e$ and $fh=0$.} ∎

Combining Propositions 1 and 2, we have:

Corollary 1. *If A_n satisfies the* (VWSR)*-axiom, then A_n satisfies the parallelogram law* (P) *and has* GC.

Another consequence of Proposition 2:

Corollary 2. *If A_n satisfies the parallelogram law* (P), *and e, f are projections in A_n, then the following conditions are equivalent:* (a) $C(e)C(f)\neq 0$; (b) $eA_n f\neq 0$; (c) *e, f are partially comparable, that is, there exist nonzero subprojections $e_0 \leq e$, $f_0 \leq f$ such that $e_0 \sim f_0$.*

Proof. The equivalence of (a) and (b) is noted in [§ 56, Prop. 4]. It is obvious that (c) implies (a) [cf. § 56, Prop. 5]. Since A_n has GC by Proposition 2, it follows that (b) implies (c) [§ 14, Prop. 2]. ∎

Exercises

1A. Assume A satisfies (1°)–(5°). If C_n satisfies the (VWSR)-axiom, then C_n is $*$-regular, satisfies LP $\sim$ RP, satisfies the parallelogram law (P), and has GC.

2A. Assume A satisfies (1°)–(6°), and suppose every element of A is bounded [§ 54, Def. 1]. If C_n satisfies the (VWSR)-axiom, then so does A_n.

3A. Consider the following condition on a $*$-ring B: If $x \in B$ and e is a projection in B such that $x^*x \in eBe$, then there exists $r \in eBe$ with $x^*x=r^*r$. Call this condition (C).

(i) In a weakly Rickart $*$-ring, the (VWSR)-axiom implies condition (C).

(ii) In a finite Rickart $*$-ring satisfying LP $\sim$ RP, condition (C) implies the (VWSR)-axiom.

(iii) In a $*$-regular ring, the (VWSR)-axiom implies LP $\sim$ RP.

(iv) A finite $*$-regular ring satisfies the (VWSR)-axiom if and only if it satisfies LP $\sim$ RP and condition (C).

§ 58. Finiteness

For the rest of the chapter (excluding Corollary 1 in Section 62) we assume, in addition to (1°)–(6°):

(7°) *A_n satisfies the parallelogram law* (P);

(8°) *every sequence of orthogonal projections in A_n has a supremum.*

On the basis of (1°)–(8°) it will be shown that A_n is a finite Baer $*$-ring (see Section 62).

If A is a finite AW^*-algebra, then A_n is a Rickart C^*-algebra [§ 56, Cor. of Th. 1], therefore A_n satisfies (7°) (see the remarks following [§ 57, Prop. 1]) and (8°) (see [§ 8, Lemma 3 to Prop. 1]). Nevertheless, in the general case (7°) and (8°) must be regarded as unwelcome guests, since they impose conditions on A_n (about which we are trying to prove something) rather than directly on A (about which we are willing to assume practically anything).

Proposition 1. *A_n is finite.*

Proof. A_n is a Rickart $*$-ring [§ 56, Th. 1]; it satisfies the parallelogram law (P) by hypothesis, therefore it has GC [§ 57, Prop. 2]. Write $u_1, \dots, u_n$ for the diagonal matrix units (as in Section 57). Since A is finite, so are the isomorphic rings $u_i A_n u_i$; in other words, the u_i are finite projections. Thus $u_1 + \dots + u_n = 1$, where the u_i are finite; in view of (8°), it follows that 1 is a finite projection in A_n [§ 17, Th. 3]. ∎

In view of (8°), the following variant of finiteness is also valid in A_n [§ 17, Prop. 4, (iii)]:

Corollary. *A_n does not contain an infinite sequence of orthogonal, equivalent nonzero projections.*

The corollary is important for certain exhaustion arguments that underlie the next section (cf. the proof of [§ 26, Prop. 16]).

Exercises

1A. Assume that A satisfies (1°)–(6°), and that A_n satisfies the (VWSR)-axiom and (8°). Then the relations $x, y \in A_n$, $yx = 1$ imply $xy = 1$. (Better yet, see Exercise 3.)

2C. Let B be a regular Baer $*$-ring satisfying LP$\sim$RP and let n be any positive integer. The relations $x, y \in B_n$, $yx = 1$ imply $xy = 1$; in particular, B_n is finite.

3C. Suppose that (1°′) A is a finite Baer $*$-ring satisfying the (EP)-axiom and the (SR)-axiom, and that (2°)–(5°) hold. Let C be the regular ring of A [§ 52, Exer. 4] and let n be any positive integer. The relations $x, y \in C_n$, $yx = 1$ imply $xy = 1$; in particular, C_n and A_n are finite.

§ 59. Simple Projections

As in the preceding section, we assume (1°)–(8°). For brevity, we write $B = A_n$.

As an aid to proving that B is a Baer $*$-ring, we develop here the rudiments of the dimension theory of projections in B. The essential idea is to scan certain of the arguments in Chapter 6 with B in mind. The properties of B that are crucial for the following discussion of dimension: (i) B is finite [§58, Prop. 1], (ii) B has GC [§57, Prop. 2], (iii) B admits a notion of central cover ([§56, Prop. 5], [§57, Cor. 2 of Prop. 2]), and (iv) the central projections of B form a complete Boolean algebra, whose Stone representation space we denote by $\mathscr{X}$ (we identify a central projection with the characteristic function of the clopen set in $\mathscr{X}$ to which it corresponds).

Let us scan Section 26 with B in mind. Everything through Proposition 13 holds verbatim for B. In particular, a nonzero projection e is called *simple* iff there exists a central projection h and an orthogonal decomposition $h=e_1+\cdots+e_m$ with $e\sim e_k$ for $k=1,\dots,m$ [cf. §26, Def. 2]; necessarily $h=C(e)$, and the integer m, which is unique by finiteness [cf. §26, Prop. 1], is called the *order* of e; one defines $T(e)=(1/m)h$, regarded as a continuous function on $\mathscr{X}$ [cf. §26, Def. 3]; if m is a power of 2 then e is called *fundamental* [cf. §26, Def. 4]. If e is simple and $e\sim f$, then f is also simple and $T(e)=T(f)$ [cf. §26, Prop. 11]; conversely, if e and f are simple projections with $T(e)=T(f)$, then $e\sim f$ [cf. §26, Prop. 12].

Lemma 1. *Every nonzero projection in B contains a simple projection. When A is of Type* I, *the simple projection can be taken to be abelian; when A is of Type* II, *it can be taken to be fundamental.*

Proof. We note first that the lemma is true in A (i.e., for $n=1$) [§26, Props. 14 and 16]. Write $u_1,\dots,u_n$ for the diagonal matrix units of B.

Let e be any nonzero projection in B. Since $u_1+\cdots+u_n=1$, we can suppose, for example, that $eu_1\neq 0$. Then $C(e)C(u_1)\neq 0$. In view of [§57, Cor. 2, (c) of Prop. 2] we can suppose, without loss of generality, that $e\leq u_1$ [cf. §15, Prop. 7]. Since u_1Bu_1 is $*$-isomorphic to A, it follows from the first line of the proof that there exists a subprojection $e'\leq e$ that is simple in u_1Bu_1; since the central projections of u_1Bu_1 are the projection hu_1, with h a central projection in B (see the discussion in Section 57), and since $h=hu_1+\cdots+hu_n$ with $hu_1\sim\cdots\sim hu_n$, it is clear that e' is also simple in B [cf. §26, Prop. 8]. If, in addition, A is of Type I [Type II] then e' can be taken to be abelian [fundamental]. ∎

Although we do not yet know that the projection lattice of B is complete, it is a good omen that the above process can be continued to exhaustion:

Lemma 2. *Let e be any nonzero projection in B. There exists an orthogonal family* (e_ι) *of simple projections such that* $e=\sup e_\iota$. *If A is of Type* I, *the* e_ι *can be taken to be abelian; if A is of Type* II, *they can be taken to be fundamental.*

Proof. Let (e_ι) be a maximal orthogonal family of projections of the desired kind with $e_\iota \leq e$ for all ι. We assert that $\sup e_\iota$ exists and is equal to e. Thus, assuming f is a projection such that $e_\iota \leq f$ for all ι, it is to be shown that $e \leq f$. Let $x = e(1-f)$ and assume to the contrary that $x \neq 0$. Let $g = \mathrm{LP}(x)$; thus $0 \neq g \leq e$. By Lemma 1, there exists a projection g_0 of the desired kind with $g_0 \leq g$. Then for all ι one has $e_\iota x = e_\iota e(1-f) = e_\iota(1-f) = 0$, therefore $e_\iota g = 0$, hence $e_\iota g_0 = 0$, contradicting maximality. ∎

§ 60. Type II Case

Assuming that (1°)–(8°) hold and that A is of Type II, we show in this section that $B = A_n$ is a Baer $*$-ring.

First, we observe that projections in B may be subdivided at will:

Lemma 1. *If e is any projection in B and m is any positive integer, then e is the sum of m orthogonal, equivalent projections.*

Proof. By [§ 57, Lemma 1 to Prop. 2], we are reduced to the corresponding result for A [§ 19, Th. 1]. ∎

We now scan Section 30 with B in mind. Everything through Proposition 4 holds verbatim for B (the needed notion of central cover, divisibility of projections, and completeness of the lattice of central projections, are available in B); the analogue of [§ 30, Prop. 4] for B:

Lemma 2. *If $(f_\varkappa)_{\varkappa \in K}$ is an orthogonal family of fundamental projections in B, such that the orders of the $f_\varkappa$ are bounded, then there exists an orthogonal family (h_α) of nonzero central projections with $\sup h_\alpha = 1$, such that for each α, the set $\{\varkappa \in K : h_\alpha f_\varkappa \neq 0\}$ is finite.*

Lemma 3. *With notation as in Lemma 2, $\sup f_\varkappa$ exists.*

Proof. For each α, write $K_\alpha = \{\varkappa \in K : h_\alpha f_\varkappa \neq 0\}$ and define

$$g_\alpha = \sum_{\varkappa \in K_\alpha} h_\alpha f_\varkappa = h_\alpha \Big(\sum_{\varkappa \in K_\alpha} f_\varkappa \Big)$$

(these are finite sums). By [§ 56, Cor. of Th. 2], there exists a projection $g \in B$ such that $h_\alpha g = g_\alpha$ for all α. It will be shown that $\sup f_\varkappa$ exists and is equal to g.

Given any $\varkappa \in K$, we assert that $f_\varkappa \leq g$, that is, $f_\varkappa(1-g) = 0$. Given any α, it suffices to show that $h_\alpha f_\varkappa(1-g) = 0$ [§ 56, Th. 2]. If $h_\alpha f_\varkappa = 0$ this is trivial. If $h_\alpha f_\varkappa \neq 0$ then $h_\alpha f_\varkappa \leq g_\alpha$ by the definition of g_α; since $h_\alpha g = g_\alpha$, an elementary computation yields $h_\alpha f_\varkappa(1-g) = 0$. Thus g is an upper bound for the $f_\varkappa$.

Finally, if f is a projection in B with $f_\varkappa \le f$ for all $\varkappa$, it is to be shown that $g \le f$. Fixing α, it is enough to show that $h_\alpha g(1-f)=0$, that is, $g_\alpha = g_\alpha f$. For all $\varkappa$, $h_\alpha f_\varkappa \le h_\alpha f$, therefore

$$\sum_{\varkappa \in K_\alpha} h_\alpha f_\varkappa \le h_\alpha f,$$

that is, $g_\alpha \le h_\alpha f$, thus $g_\alpha = g_\alpha(h_\alpha f) = g_\alpha f$. ∎

Lemma 4. *If* $(e_\iota)_{\iota \in I}$ *is any orthogonal family of projections in* B, *then* $\sup e_\iota$ *exists.*

Proof. For each ι we may express e_ι as the supremum of an orthogonal family of fundamental projections [§ 59, Lemma 2], say

(i) $$e_\iota = \sup\{e_{\iota\varkappa} : \varkappa \in K_\iota\}.$$

Let J be the set of all ordered pairs $(\iota, \varkappa)$ with $\iota \in I$ and $\varkappa \in K_\iota$. For $m = 0, 1, 2, 3, \ldots$ let J_m denote the set of all $(\iota, \varkappa)$ in J for which $e_{\iota\varkappa}$ has order 2^m. Thus

$$J = \bigcup_{m=0}^{\infty} J_m .$$

For fixed m, the projections $e_{\iota\varkappa}$ with $(\iota, \varkappa) \in J_m$ have bounded order $(=2^m)$, hence by Lemma 3 we can form

(ii) $$f_m = \sup\{e_{\iota\varkappa} : (\iota, \varkappa) \in J_m\}.$$

It is routine to check that the f_m are orthogonal. Hence by (8°) we can form

(iii) $$e = \sup\{f_m : m = 0, 1, 2, 3, \ldots\}.$$

It will be shown that $\sup e_\iota$ exists and is equal to e.

Given any ι, let us show that $e_\iota \le e$. If $\varkappa \in K_\iota$ then $e_{\iota\varkappa} \le e$. {Proof: If $e_{\iota\varkappa}$ has order 2^m, then $(\iota, \varkappa) \in J_m$, hence $e_{\iota\varkappa} \le f_m \le e$.} Then $e_\iota \le e$ results from (i).

Finally, if g is any projection in B with $e_\iota \le g$ for all ι, it is to be shown that $e \le g$. Given any m, it will suffice by (iii) to show that $f_m \le g$. For all $(\iota, \varkappa) \in J_m$ we have $e_{\iota\varkappa} \le e_\iota \le g$, therefore $f_m \le g$ by (ii). ∎

Theorem 1. *If* A *satisfies* (1°)–(6°) *and is of Type* II, *and if* n *is a positive integer such that* (7°) *and* (8°) *hold, then* A_n *is a Baer* $*$*-ring.*

Proof. Since $B = A_n$ is a Rickart $*$-ring [§ 56, Th. 1], the theorem is immediate from Lemma 4 [§ 4, Prop. 1]. ∎

§ 61. Type I Case

We assume in this section that (1°)–(8°) hold and that A is of Type I. It is to be shown that A_n (equivalently, C_n) is a Baer $*$-ring. In view of [§ 56, Th. 2] and the decomposition theory of Type I rings [§ 18, Th. 2], we can suppose that A is homogeneous of Type I_r, r a positive integer. {Note that if h is a central projection, then $hA_n=(hA)_n$, $hC_n=(hC)_n$, and hC is the regular ring of hA (cf. [§ 52, Prop. 3] or [§ 47, Exer. 1, (ii)].}

Let $e_1, \dots, e_r$ be orthogonal, equivalent, abelian projections in A with sum 1, and let $u_1, \dots, u_n$ be the diagonal matrix units in $B=A_n$. For $i=1, \dots, n$ and $\nu=1, \dots, r$, let $u_{i\nu}$ be the matrix with e_ν in the (i,i) coordinate and zeros elsewhere:

$$u_{i\nu}=\operatorname{diag}(0, \dots, e_\nu, \dots, 0),$$

with e_ν occurring in the ith place. By hypothesis, $e_\nu A e_\nu$ is an abelian Baer $*$-ring; thus, if e is a subprojection of e_ν, then $e=he_\nu$ for a suitable central projection h of A [§ 15, Prop. 6]. Since $u_{i\nu}Bu_{i\nu}$ is $*$-isomorphic to $e_\nu Ae_\nu$, $u_{i\nu}$ is an abelian projection in B; better yet, the preceding comment shows that if u is a subprojection of $u_{i\nu}$ then $u=hu_{i\nu}$ for a suitable central projection h. Summarizing, the $u_{i\nu}$ are orthogonal, equivalent, abelian projections in B with sum 1, possessing the subprojection property just mentioned.

Lemma 1. *B does not contain $nr+1$ orthogonal, equivalent nonzero projections.*

Proof. Suppose f_μ $(\mu=1, 2, \dots, nr+1)$ are orthogonal, equivalent projections in B; it is to be shown that $f_1=0$. Since $\sum_{i,\nu} u_{i\nu}=1$, it suffices to show that $u_{i\nu}f_1=0$ for all i, ν. Assume to the contrary, for example, that $u_{11}f_1 \neq 0$. Then [§ 57, Cor. 2 of Prop. 2] provides nonzero subprojections $u \leq u_{11}$, $f \leq f_1$ with $u \sim f$. As remarked above, $u=hu_{11}$ with h a central projection. The proof continues as in [§ 18, Prop. 4]. ∎

Lemma 2. *If (e_ι) is any orthogonal family of projections in B, there exists a nonzero central projection h such that $he_\iota=0$ for all but finitely many ι.*

Proof. Assuming to the contrary that no such h exists, one shows that for each positive integer k, there exist distinct indices $\iota_1, \dots, \iota_k$ and nonzero subprojections $g_t \leq e_{\iota_t}$ $(t=1, \dots, k)$ such that $g_1 \sim g_2 \sim \cdots \sim g_k$ (for $k=nr+1$ this contradicts Lemma 1); the argument for this is the same as in [§ 14, Prop. 9]. ∎

This leads to the analogue of [§ 18, Prop. 5] for B:

Lemma 3. *Let (e_ι) be any orthogonal family of projections in B. There exists an orthogonal family (h_α) of nonzero central projections with $\sup h_\alpha=1$, such that for each α, the set $\{\iota: h_\alpha e_\iota \neq 0\}$ is finite.*

Proof. Let (h_α) be a maximal such orthogonal family, set $h_0=1-\sup h_\alpha$, and assume to the contrary that $h_0 \neq 0$. The direct summand h_0A also satisfies (1°)–(6°) and is homogeneous of Type I_r, and $(h_0 A)_n = h_0 B$ also satisfies (7°) and (8°); Lemma 2 applied to $h_0 A$ produces a nonzero central projection $h \le h_0$ that contradicts maximality. ∎

Lemma 4. *With notation as in Lemma 3,* $\sup e_\iota$ *exists.*

Proof. Formally the same as [§ 60, Lemma 3]. ∎

With Lemma 4 in hand, the argument of [§ 60, Th. 1] shows that B is a Baer $*$-ring. Putting the homogeneous pieces back together via [§ 56, Th. 2], we have:

Theorem 1. *If A satisfies* (1°)–(6°) *and is of Type* I, *and if n is a positive integer such that* (7°) *and* (8°) *hold, then A_n is a Baer $*$-ring.*

§ 62. Summary of Results

As noted in Section 55, the situation for properly infinite Baer $*$-rings is essentially trivial. Combining the theorems in Sections 60 and 61, we have:

Theorem 1. *If A is a finite Baer $*$-ring satisfying* (1°)–(6°), *and if n is a positive integer such that* (7°) *and* (8°) *hold, then A_n is a finite Baer $*$-ring with* GC.

Proof. Let h be the central projection such that hA is Type I and $(1-h)A$ is Type II [§ 15, Th. 2]. Since $hA_n = (hA)_n$ and $(1-h)A_n = ((1-h)A)_n$ are Baer $*$-rings by the theorems of the preceding two sections, it follows that $A_n = hA_n + (1-h)A_n$ is also a Baer $*$-ring. ∎

Corollary 1. *If A is any AW^*-algebra and n is any positive integer, then A_n is an AW^*-algebra.*

Proof. At any rate, A_n is a C^*-algebra (by the Gel'fand-Naĭmark theorem), thus it is the Baer $*$-ring property that is at issue. By structure theory [§ 15, Th. 1] we are reduced to the properly infinite and finite cases. The properly infinite case is disposed of at once by the observation that every AW^*-algebra has GC (see, e. g., [§ 14, Cor. 1 of Prop. 7]). If A is a finite AW^*-algebra, then A satisfies (1°)–(6°) and A_n satisfies (7°) and (8°), as noted in Section 58; quote Theorem 1. ∎

Noting Remark 3 in Section 55, we have:

Corollary 2. *If A is a regular Baer $*$-ring satisfying* (1°)–(4°), *and if n is a positive integer such that* (7°) *and* (8°) *hold, then A_n is a regular Baer $*$-ring.*

Properties of A_n are hard to come by; none is too humble to be noted:

Corollary 3. *If A satisfies the hypotheses of Theorem 1, then partial isometries in A_n are addable.*

Proof. If $n=1$ this is the hypothesis (2°). Suppose $n>1$. If h is any central projection then $hA_n=(hA)_n$, thus it is clear that A_n has no abelian direct summand. Quote [§ 20, Th. 1, (ii)]. ∎

Exercises

1C. Assume (1°)–(8°). If $x, y \in C_n$ and $yx=1$, then $xy=1$.

2A. Assume (1°)–(8°). If A is of Type I [Type II] then A_n is also of Type I [Type II].

3C. If $A=Z_r$, Z a commutative AW^*-algebra (in other words, A is a homogeneous AW^*-algebra of Type I_r) then every element of A is unitarily equivalent to an upper triangular element of A.

4C. If A is a finite Baer $*$-ring satisfying the (EP)-axiom and the (SR)-axiom, then C_n is strongly semisimple for all n.

5A. Let Z be a commutative AW^*-algebra, write $Z=C(\mathscr{X})$, $\mathscr{X}$ a Stonian space, fix a point $\sigma \in \mathscr{X}$, and let

$$J=\{c \in Z: c=0 \text{ on a neighborhood of } \sigma\}.$$

Then, for every positive integer n, $(Z/J)_n$ is a finite Baer $*$-factor of Type I_n.

6C. If A is a commutative ring with unity and descending chain condition on annihilators, then the following conditions are equivalent: (a) A_n is a Baer ring [cf. § 4, Exer. 4] for every $n \geq 2$; (b) A_n is a Baer ring for some $n \geq 2$; (c) A is the direct sum of finitely many Prüfer rings.

7D. Assuming (1°)–(8°), does A_n satisfy LP$\sim$RP? Does C_n?

8C. If B is a regular Baer $*$-factor of Type II satisfying LP$\sim$RP, then so is B_n, for every positive integer n.

9C. Let A be a finite Baer $*$-factor of Type II, satisfying the (EP)-axiom and the (UPSR)-axiom, in which every element of the form $1+x^*x$ is invertible, and possessing a central element z such that $z^* \neq z$. Suppose, in addition, that A satisfies the (PS)-axiom. {Thus A satisfies (1°)–(6°).} Let n be any positive integer.

(i) A_n is a finite Baer $*$-factor of Type II.

(ii) A_n satisfies LP$\sim$RP.

Thus the conditions (7°) and (8°) are a consequence of (1°)–(6°) in the factorial Type II case.

Hints, Notes and References

§ 1

Definition 2. The notation is borrowed from [**19**, p. 186].

Definition 5. Introduced by Murray and von Neumann [**67**, Def. 6.1.1]. In [**54**] this is called '$*$-equivalence', the term 'equivalence' being reserved for the concept in Exercise 6.

Proposition 8. The validity of this proposition is the reason for the choice of definition of equivalence. Note that the analogous proposition for unitary equivalence is false. {For example, let $\mathscr{H}$ be a Hilbert space with orthonormal basis $\xi_1, \xi_2, \xi_3, \ldots$ and let P, Q, R be the projections in $\mathscr{L}(\mathscr{H})$ whose ranges are the closed linear subspaces

$$[\xi_1, \xi_3, \xi_5, \ldots], \quad [\xi_3, \xi_5, \xi_7, \ldots], \quad [\xi_2, \xi_4, \xi_6, \ldots],$$

respectively. Then P is unitarily equivalent to Q, but $P+R=I$ is not unitarily equivalent to $Q+R$.} Cf. [§ 17, Exer. 17, (i)].

Proposition 9. [**54**, Th. 24].

Definition 8. [**67**, Def. 6.3.1].

Theorem 1. A theorem of Lebow [**57**].

Exercise 2. (ii), (iii) follow from (i) or from Exercise 1.

Exercise 6. [**54**, Th. 14].

Exercise 7. [**54**, Th. 26].

Exercise 8. Cf. [**54**, Th. 27]. This is worked out in [§ 13, Lemma to Prop. 7].

Exercise 9. [**52**, p. 20, Lemma 2], [**54**, p. 24, Exer. 4]. Hint: The element $x=e+(1-f)$ has inverse $f+(1-e)$.

Exercise 10. [**54**, Th. 15]. Hint: Cf. Proposition 9.

Exercise 12. Used in the proof of [§ 53, Prop. 1].

Exercise 14. Masas are the structural elements in the multiplicity analysis of commutative von Neumann algebras [**81**, Part II, p. 4], [**35**, Ch. III].

Exercises 17–19. [**54**, p. 39, Exer. 10].

Exercise 20. Cf. [**37**, Solution 177].

Exercise 21. No. See [§ 12, Exer. 3].

§ 2

Definition 2. The basic reference for C^*-algebras is the treatise of Dixmier [**24**].

Exercise 3. Cf. [**47**, Lemma 4.9]. Hint: Let $a \in A$, set $z = xx^* ay$, and calculate $z^* z$.

Exercise 5. [**40**, p. 75].

§ 3

Definition 2. Cf. [**74**, Def. 2.1], [**64**, Def. 6.1].

Proposition 2. Cf. [**54**, Th. 21].

Proposition 7. [**47**, Lemma 5.3], [**64**, Lemma 6.4].

Corollary of Proposition 8. [**64**, Lemma 6.2].

Example 2. [**47**, p. 249].

Proposition 10. Cf. [**48**, Lemma 4], [**54**, Th. 20].

Exercise 1. [**52**, p. 20], [**54**, p. 36]. Hint: [§ 1, Exer. 9].

Exercise 3. (i) For the Riesz-Schauder theory see, e. g., Banach's book [**1**, Ch. 10, § 2].

Exercise 6. This is worked out in [§ 51, Prop. 3].

Exercise 7. Cf. [**54**, p. 34, Cor.].

Exercise 8. $C = B \cap S' = (B' \cup S)'$.

Exercise 9. Cf. the proof of [§ 59, Lemma 2].

Exercise 11. Cf. [**51**, p. 532, proof of Th. 2], [**63**, Lemma 2.2], [**54**, p. 43, proof of Th. 28].

Exercise 14. A general reference on Boolean algebras is the book of Halmos [**36**].

Exercise 18. (ii) Cf. [§ 16, Prop. 2].

§ 4

Definition 1. [**52**], [**54**].

Definition 2. [**47**, Th. 2.3], [**49**, p. 853].

Definition 3. When $1 \in B$ this coincides with the concept defined in [**54**, p. 29].

Proposition 6. [**54**, p. 30].

Proposition 7. [**54**, Th. 20].

Definition 4. Consistent with [**48**, p. 462]. (Cf. [§ 7, Exer. 9].)

Definition 5. The concept is due to von Neumann [**68**], the terminology to Dixmier [**23**].

Proposition 9. [**47**, p. 236].

Exercise 3. (i) The last line alludes to the well-known Wedderburn theorem [**42**, p. 183, Th. 1].

Exercise 4. (i) [**54**, Th. 3].

Exercise 5. [**54**, p. 34, Cor.].

Exercise 9. Cf. [**54**, Th. 20].

Exercise 10. This is [§ 6, Prop. 2].

Exercise 11. Better yet, see Exercises 24 and 25.

Exercise 13. Cf. [**7**, Lemmas 2.1, 2.2].

Exercise 14. [**52**, pp. 52–53], [**54**, pp. 103–104].

Exercise 17. Cf. [**54**, Th. 6].

Exercise 20. Cf. [**54**, p. 38, Exer. 8].

Exercise 21. (ii) Cf. [§ 7, Exer. 9].

Exercise 22. By the hypothesis (iv), there exists a Hilbert space $\mathscr{K}$, a von Neumann algebra $\mathscr{B}$ of operators on $\mathscr{K}$, and a $*$-monomorphism $\theta: \mathscr{B} \to \mathscr{L}(\mathscr{H})$ with $\theta(\mathscr{B}) = \mathscr{A}$. From (iii) it is clear that θ is completely additive on projections, that is, if (E_ι) is any orthogonal family of projections in $\mathscr{B}$, then $\theta(\sup E_\iota) = \sup \theta(E_\iota)$ (the latter sup as calculated in $\mathscr{L}(\mathscr{H})$). It follows that θ is normal [**22**, p. 16, Footnote 6], therefore $\mathscr{A} = \theta(\mathscr{B})$ satisfies $\mathscr{A} = \mathscr{A}''$ in $\mathscr{L}(\mathscr{H})$ (cf. [**22**, Cor. 4 of Th. 3], [**23**, Ch. I, § 4, Cor. 2 of Th. 2], [**90**, Lemma 4.4], [**112**, Lemma 1]).

Exercise 23. (a) implies (b): If $\theta: A \to \mathscr{L}(\mathscr{H})$ is a $*$-monomorphism such that $\theta(A) = \theta(A)''$ in $\mathscr{L}(\mathscr{H})$, then $\theta(A)$ is an AW^*-subalgebra of $\mathscr{L}(\mathscr{H})$ (Proposition 8, (iv)).

(b) implies (c) trivially (Exercise 21).

(c) implies (d): Since $\theta(A)$ is AW^*-embedded in $\mathscr{L}(\mathscr{H})$, it follows that θ is completely additive on projections, i.e., if (e_ι) is any orthogonal family of projections in A, then $\theta(\sup e_\iota) = \sup \theta(e_\iota)$ (the latter sup as calculated in $\mathscr{L}(\mathscr{H})$). Each vector $\xi \in \mathscr{H}$ determines a positive linear form φ_ξ on A that is CAP, via the formula

$$\varphi_\xi(x) = (\theta(x)\xi|\xi) \qquad (x \in A),$$

thus the family of linear forms $(\varphi_\xi)_{\xi \in \mathscr{H}}$ is contained in $\mathscr{P}$. If $\varphi_\xi(x) = 0$ for all $\xi \in \mathscr{H}$, then $\theta(x) = 0$, $x = 0$.

(d) implies (e): If $x \neq 0$ then $x^* x \neq 0$.

(e) implies (a): A recent result of G. K. Pedersen [**122**].

{Historical note: The implication (e) $\Rightarrow$ (b) was proved by J. Feldman [**26**, Th. 2]. Explicitly, Feldman showed, assuming (e), that there exists a Hilbert space $\mathscr{H}$ and a $*$-monomorphism $\theta: A \to \mathscr{L}(\mathscr{H})$ such that (i) $\theta(1) = I$, and (ii) $\theta(\sup e_\iota) = \sup \theta(e_\iota)$ for every *increasingly directed* family of projections (e_ι) in A (the latter sup as calculated in $\mathscr{L}(\mathscr{H})$). It follows at once that θ is completely additive on projections (consider the family of finite subsums); in particular, if (E_ι) is an orthogonal family of projections in $\theta(A)$, then $\sup E_\iota$ (as calculated in $\mathscr{L}(\mathscr{H})$) is also in $\theta(A)$. Since $\theta(A)$ is an AW^*-algebra (it is $*$-isomorphic to A), it follows from a result of Kaplansky [cf. § 7, Exer. 9] that $T \in \theta(A)$ implies $\mathrm{RP}(T) \in \theta(A)$ (RP as calculated in $\mathscr{L}(\mathscr{H})$). Thus $\theta(A)$ is AW^*-embedded in $\mathscr{L}(\mathscr{H})$. In particular, RP's in $\theta(A)$ are unambiguous (they are the same whether calculated in the AW^*-algebra $\theta(A)$ or in $\mathscr{L}(\mathscr{H})$) and therefore $\theta(\mathrm{RP}(x)) = \mathrm{RP}(\theta(x))$ for all $x \in A$. In view of [§ 3, Prop. 7], it results that $\theta(e \cup f) = \theta(e) \cup \theta(f)$ for any pair of projections e, f in A, thus θ preserves finite sups; it follows from this, and (ii), that $\theta(\sup e_\iota) = \sup \theta(e_\iota)$ for any family of projections (e_ι) in A (the latter sup as calculated in $\mathscr{L}(\mathscr{H})$). In particular, if (E_ι) is any family of projections in $\theta(A)$, then $\sup E_\iota$ (as calculated in $\mathscr{L}(\mathscr{H})$) is also in $\theta(A)$; thus $\theta(A)$ is an AW^*-subalgebra of $\mathscr{L}(\mathscr{H})$. Cf. Exercise 24.}

Exercise 24. (c) implies (a): Proposition 8, (iv).

(a) implies (b): Exercise 21, (i).

(b) implies (c): Assume $\mathscr{A}$ is AW^*-embedded in $\mathscr{L}(\mathscr{H})$. By Exercise 23, $\mathscr{A}$ is $*$-isomorphic to a von Neumann algebra, therefore $\mathscr{A}=\mathscr{A}''$ in $\mathscr{L}(\mathscr{H})$ by Exercise 22. {Historical note: For A finite [§ 15, Def. 3], this result is due to H. Widom [**90**, Th. 4.4] (see also J. Feldman [**26**, Th. 1]); for $\mathscr{A}$ semifinite, to K. Saitô [**78**, Th. 5.2]; for general $\mathscr{A}$, to G. Pedersen [**122**].}

Exercise 25. The implications (a) $\Rightarrow$ (b) $\Rightarrow$ (c) $\Rightarrow$ (d) are trivial.

(d) implies (e): Cf. [§ 9, Exer. 4].

(e) implies (a): Assuming (e), $\mathscr{A}$ is an AW^*-algebra by a result of Kaplansky [cf. § 7, Exer. 9]. By hypothesis, $\mathscr{A}$ is AW^*-embedded in $\mathscr{L}(\mathscr{H})$ in the sense of Exercise 21. By Exercise 24, $\mathscr{A}=\mathscr{A}''$ in $\mathscr{L}(\mathscr{H})$.

Exercise 26. The result is due to R. V. Kadison [**112**, Lemma 1], by a proof more direct than the following.

(a) implies (b): $\sup\mathscr{S}$ is adherent to $\mathscr{S}$ in the strong operator topology.

(b) implies (a): It follows from (b) and easy spectral theory that $T\in\mathscr{A}$ implies $\mathrm{RP}(T)\in\mathscr{A}$ [**112**, p. 176]. If $(E_\iota)_{\iota\in I}$ is any orthogonal family of projections in $\mathscr{A}$, then the sums $\sum_{\iota\in J} E_\iota$, J a finite subset of I, are also in $\mathscr{A}$, therefore $\sup E_\iota\in\mathscr{A}$ by (b). It follows from Exercise 25 that $\mathscr{A}=\mathscr{A}''$ in $\mathscr{L}(\mathscr{H})$.

Exercise 27. (a) implies (b) trivially (see Definition 3).

(b) implies (a): The result is due to G. K. Pedersen [**122**]. Note that if e,f are projections in B, then $e\cup f$ (as calculated in A) is also in B [§ 5, Prop. 7]. Let $(f_\varkappa)$ be any family of projections in B and let $f=\sup f_\varkappa$ (as calculated in A); it is to be shown that $f\in B$. We can suppose $f\neq 0$. Let (e_ι) be a maximal orthogonal family of nonzero projections in B such that $e_\iota\le f$ for all ι (start with some nonzero $f_\varkappa$ and expand by Zorn's Lemma). Set $e=\sup e_\iota$; by hypothesis, $e\in B$. Obviously $e\le f$. To prove that $f=e$ it is enough to show that, for any $\varkappa$, $f_\varkappa\le e$. If, on the contrary, $f_\varkappa-f_\varkappa e\neq 0$, then $f_\varkappa\cup e-e=\mathrm{RP}(f_\varkappa-f_\varkappa e)$ is a nonzero subprojection of f that belongs to B and is orthogonal to every e_ι, contradicting maximality.

§ 5

Definition 1. Cf. [**74**, Def. 2.2].

Exercise 3. Cf. [§ 3, Prop. 11].

Exercise 5. Hint: Consider a maximal orthogonal family of nonzero projections (e_ι), show that $\sup e_\iota$ is a unity element for A, and quote [§ 4, Prop. 1].

Exercise 8. Cf. [**3**, Lemma 3.2].

Exercise 9. The desired norm on A_1 can be introduced via either the Gel'fand-Naĭmark theorem [**75**, Th. 4.8.11] or the regular representation [**75**, Lemma 4.1.13].

Exercise 15. Hint: Exercise 14.

§ 6

Definition 1. Cf. [**71**, p. 242, Def. 1.1], [**18**, Lemme 3.1], [**19**, p. 186], [**35**, § 55], [**80**, Def. 3.3], [**54**, p. 9, Def.].

Proposition 2. [**47**, p. 237, Cor. 1], [**54**, Th. 22].

Proposition 3. [**54**, Th. 9].

Corollary 1. Cf. [**80**, Lemma 3.3], [**54**, Th. 13].

Definition 2. The term was introduced by Halmos in the context of multiplicity theory [**35**, p. 93].

Corollary 2. An early special case (in the context of orthoseparable AW^*-algebras) was proved by Rickart [**74**, Th. 5.18].

Definition 3. Cf. [**67**, Def. 3.1.2], [**54**, p. 18, Exer. 7].

Proposition 4. [**54**, Th. 22; p. 18, Exer. 10].

Corollary 2. [**80**, Th. 3.1].

Definition 4. Cf. [**54**, p. 9, Def.].

Proposition 5. Cf. [**18**, Lemme 6.6].

Exercise 6. Cf. [**18**, Lemme 3.1]. Hint: In a C^*-algebra with unity, every element is a linear combination of unitary elements [**18**, p. 250].

Exercise 7. (iii) [**18**, Lemme 3.1], [**80**, Cor. 3.6].

Exercise 8. Hint: Exercise 1.

Exercise 10. Cf. [**74**, Lemma 5.6], [**21**, Lemme 3.10], [**80**, Lemma 5.2].

Exercise 11. Same as [§ 14, Exer. 17].

Exercise 12. Cf. [**54**, p. 18, Exer. 7].

Exercise 14. Cf. [**67**, Ch. IV, Cor. of Th. III]. For a matricial generalization see [§ 16, Exer. 4].

Exercise 15. Cf. [**23**, Ch. I, § 2, Prop. 2].

Exercise 17. In the case of von Neumann algebras, the result is dramatically sharper [**23**, Ch. I, § 2, Prop. 1].

Exercise 18. Cf. [§ 56, Prop. 4].

Exercise 19. Cf. [§ 4, Exer. 3].

§ 7

Theorem 1. [**47**, p. 236].

Proposition 3. [**47**, Lemma 2.1].

Definition 1. [**52**, p. 30], [**54**, p. 89].

Exercise 1. See [**47**, Th. 2.3] and [**49**, p. 853].

Exercise 2. Cf. [**20**, Th. 2], [**22**, p. 16, footnote (6)], [**23**, Ch. I, § 4, Exer. 9], [**94**, p. 214, Cor.]. The word "commutative" can be omitted [§ 4, Exer. 23].

Exercise 3. [**20**, Th. 2].

Exercise 4. Hint: Let T be a Stonian space that is not hyperstonian, and represent $\mathscr{A}=C(T)$ as operators on a Hilbert space $\mathscr{H}$; it follows from Exercises 2 and 3 that the intrinsic projection lattice operations in $\mathscr{A}$ must differ from the operatorial ones.

Exercises 5, 6. Hint: [§ 3, Exer. 15].

Exercise 8. The equivalence of (a) and (b) follows readily from Exercise 7 [cf. **48**, Lemma 2]. In view of (1), (a) and (c) are equivalent by [§ 4, Exer. 27].

Exercise 9. Immediate from Exercise 8.

Exercise 11. [**15**, Cor. 2.3].

§ 8

Theorem 1. Cf. [**74**, Th. 2.10].

Lemma 3. Cf. [**74**, Th. 2.7].

Proposition 4. [**47**, Lemma 2.1].

Exercise 1. Cf. [**47**, p. 249].

§ 9

Proposition 4. [**47**, Lemma 2.1].

Exercise 1. Hint: [§ 5, Exer. 5].

Exercise 2. [**47**, Lemma 2.1].

Exercise 3. [**47**, Lemma 2.2].

Exercise 4. Essentially the same argument as for [§ 7, Exer. 9].

§ 10

Definition 1. [**45**, p. 411].

Proposition 1. [**47**, p. 238].

Proposition 2. [**47**, Lemma 2.5].

Exercise 1. (i) [**44**].
(ii) In fact, $A_n = \mathbb{C}$ already provides an example.

Exercise 2. [**45**, p. 411].

Exercise 3. (ii) Cf. [**75**, Lemma 4.4.6].
(iii) For the case of general C^*-algebras, see [**45**, Th. 6.4], [**75**, Th. 4.8.5], [**24**, Props. 1.3.7, 1.8.1].

§ 11

Theorem 1. [**47**, Lemma 3.1], [**54**, Th. 30].

Lemma 2. [**52**, p. 25, Lemma 4], [**54**, p. 39, Exer. 11].

Proposition 1. Cf. [**52**, p. 27, Lemma 7], [**54**, Th. 38].

Exercise 1. [**54**, Th. 29]. The proof is written out in [§ 16, Prop. 2].

Exercise 5. Hint: A_2 contains two 'orthogonal' equivalent copies of A. Cf. [§ 20, Prop. 2].

§ 12

Proposition 1. Cf. [**47**, Th. 3.2], [**52**, p. 28, Th. 2], [**54**, Th. 41]. This is proved for von Neumann algebras in [**67**, Lemma 6.1.3].

Exercise 3. No. Hint: Look at matrices.

§ 13

Definition 1. [**54**, p. 81].

Proposition 1. Cf. [**18**, proof of Lemme 3.3].

Proposition 2. [**47**, proof of Th. 5.4].

Corollary. The fact that $\mathrm{LP}(T) \sim \mathrm{RP}(T)$ in a von Neumann algebra is proved in [**67**, Lemma 6.2.1]. The parallelogram law is proved in [**67**, proof of Lemma 7.3.4].

Definition 2. [**16**, p. 388], [**17**, p. 7].

Proposition 3. [**86**, Th. 1].

Proposition 5. Cf. [**18**, proof of Th. 6], [**2**, Appendix II, Lemma II.4].

Definition 3. Same references as for Definition 2.

Proposition 6. Cf. [**16**, p. 391, No. 3]. For Rickart C^*-algebras, the result is due to Kaplansky [cf. **2**, Appendix II, Lemma II.7 and p. 86].

Definition 4. [**52**, p. 37], [**54**, p. 90].

Lemma. Cf. [**54**, Th. 27].

Proposition 7. Cf. [**52**, p. 15, Th. 2; p. 37], [**54**, Th. 60].

Theorem 1. The result is due to Kaplansky [cf. **2**, Appendix I, Th. I.19].

Proposition 8. Better yet, see Exercise 5.

Lemma to Theorem 2. Cf. [**54**, proof of Th. 62].

Theorem 3. For von Neumann algebras, see [**16**, p. 390]; for AW^*-algebras, see [**2**, Appendix II, Th. II.12]. See also Exercise 6.

Theorem 4. Better yet, see Exercise 7.

Exercise 5. Cf. [**54**, Th. 60].

Exercise 6. Hint: Revise the proof of Theorem 3 in the light of Exercise 5.

Exercise 7. [**54**, Th. 62].

Exercise 8. Hint: Revise Theorem 5 in the light of Exercises 6 and 7.

Exercise 10. Cf. the proof of [§ 53, Prop. 3].

Exercise 11. Cf. [§ 51, Lemma].

Exercise 14. Hint: Dropping down to $(e \cup f)A(e \cup f)$, the parallelogram law yields $1-e \sim f$; cf. [§ 20, Prop. 2] (in particular, we get addability of the partial isometries here too).

Exercise 15. Consider $w+w^*$, where w is a partial isometry implementing $e \sim f$.

Exercise 18. Cf. [§ 1, Exer. 18].

Exercise 19. A theorem of A. Brown [**12**].

§ 14

Definition 1. Cf. [**71**, p. 268, Th. 2.7].

Proposition 1. Cf. [**71**, p. 265, Th. 2.1], [**59**, p. 87, Satz 1.1].

Definition 3. PC corresponds to Axiom E of [**54**, p. 41].

Proposition 3. Cf. [**74**, Th. 5.2], [**47**, Lemma 3.3], [**54**, Th. 58].

Proposition 4. Cf. [**67**, proof of Lemma 6.2.3], [**18**, proof of Th. 6], [**47**, proof of Th. 5.6], [**54**, p. 87].

Proposition 5. Cf. [**71**, p. 264, Lemma 2.1], [**47**, Lemma 3.4], [**54**, Th. 35].

Proposition 6. Cf. [**18**, proof of Th. 6].

Corollary 1. [**47**, Th. 5.6]. Comparability in von Neumann factors goes back to [**67**, Lemma 6.2.3]. For generalized comparability in arbitrary von Neumann algebras, see [**18**, Th. 6]. For generalized comparability in continuous geometries, see [**71**, p. 265, Th. 2.1; p. 268, Th. 2.7], [**59**, p. 87, Satz 1.1; p. 89, Satz 1.2].

Theorem 1. Better yet, see Exercise 5.

Proposition 8. Cf. [**47**, Lemma 6.1], [**54**, Th. 55].

Proposition 9. Cf. [**47**, proof of Lemma 4.11], [**54**, proof of Th. 48].

Exercise 1. [**54**, pp. 43–44].

Exercise 2. This is [§ 20, Th. 2].

Exercise 3. Hint: Proposition 4.

Exercise 4. Hint: Proposition 5, [§ 20, Prop. 3], Proposition 4.

Exercise 5. [**54**, Ths. 57, 62].

Exercise 6. Hint: [§ 6, Prop. 4].

Exercise 8. Hint: [§ 13, Prop. 1].

Exercise 9. Hint: It is enough to show PC; cf. Exercise 8.

Exercise 11. Cf. [**71**, p. 269, Def. 2.4], [**59**, p. 97, Satz 4.1], [**61**, p. 225, Remark].

Exercise 12. [**10**, Th. 2]. Hint: Exercise 5 and [§ 13, Exers. 8, 15].

Exercise 13. Same as [§ 20, Exer. 2].

Exercise 14. Hint: Exercise 2. For a simpler proof in the finite case, cf. [**59**, p. 100, Satz 4.4].

Exercise 15. Same as [§ 6, Exer. 10].

Exercise 16. Same proof as for [§ 57, Prop. 2].

Exercise 17. Cf. [**47**, Th. 6.6, (a)], [**54**, Th. 70]. Hint: [§ 17, Exer. 12, (vi)].

Exercise 18. (i) Same as [§6, Exer. 8].
(ii) See Exercise 1 and [§3, Exer. 18, (iv)].

Exercise 19. Cf. [**14**]. The parenthetical result on extremal points is due to Kadison [**43**] (cf. [**23**, Ch. I, §1, Exer. 2]).

Exercise 20. If there exists a counterexample A, then equivalence must fail to be additive in A (Proposition 4), the parallelogram law must fail to hold in A (Proposition 7), and A must have a nonzero finite direct summand (Exercise 4); in particular, if there exists a counterexample, then there exists a finite counterexample.

Exercise 21. If the answer is yes, then every Baer $*$-ring satisfying the parallelogram law has GC (Proposition 7). If the answer is no, a counterexample must fail to satisfy any of the conditions (1)–(3) of Exercise 9.

Exercise 22. If the answer is yes, then PC and GC are equivalent conditions on a Baer $*$-ring (Proposition 7).

§ 15

The principal references for structure theory (§§15–19) are [**67**], [**18**], [**19**], [**47**] and [**54**].

Definition 1. [**67**, Def. 7.1.1].

Proposition 3. [**67**, Lemma 7.1.1].

Definition 2. [**51**, p. 533], [**54**, pp. 10, 36]. For von Neumann algebras, see [**18**, Déf. 4.4] (the term used there is 'irréductible'); for AW^*-algebras, see [**47**, p. 241].

Proposition 6. Cf. [**19**, Lemme 3.3], [**47**, Lemma 4.7], [**54**, Th. 13].

Proposition 8. Cf. [**19**, Lemme 1.1], [**54**, pp. 12–14]. For a general formulation of such phenomena in lattices, see [**60**, Def. 1.3].

Theorem 1. [**19**], [**47**], [**51**, p. 533], [**54**, Ths. 10, 11].

Definition 4. For factorial von Neumann algebras, the nomenclature goes back to [**67**, Ch. VIII].

Exercise 1. (iii) In detail, see [§18, Exer. 13].

Exercise 2. Cf. [**52**, p. 6], [**54**, p. 37, Exer. 2].

Exercise 9. [**81**, Part II, Cor. 10.2], [**23**, Ch. I, §8, Th. 1; Ch. III, §2, No. 4, Cor. 3].

Exercise 10. [**60**, Th. 1.1].

Exercise 11. [**18**, Ths. 10, 14], [**23**, Ch. III, §8, Th. 1]. For finite AW^*-algebras, the question of the existence of trace is open.

Exercise 12. (i) Hint: [§17, Prop. 1] and [§3, Exer. 18, (ii)].
(ii) Hint: [§19, Lemma 1].

Exercise 13. (i), (ii) [**10**].
(iii) [**29**].

Exercise 14. Cf. Proposition 6 and [§4, Exer. 3].

§ 16

Proposition 2. [**47**, proof of Lemma 3.1], [**54**, Th. 29].

Exercise 1. [**54**, p. 40, Exer. 13].

Exercise 2. Cf. [**42**, p. 40, Prop. 1].

Exercise 4. Cf. [**67**, Ch. IV, Th. III], [**23**, Ch. I, § 2, Prop. 7]. For the case $n=1$, see [§ 6, Exer. 14].

Exercise 5. Cf. [**29**, p. 333], [**10**, Lemma 1].

§ 17

Proposition 1. Cf. [**67**, Lemma 7.1.3], [**18**, Lemme 4.2], [**47**, Lemma 4.3], [**54**, Th. 37].

Definition 1. [**18**, Déf. 3.2].

Proposition 2. [**47**, Lemma 3.5], [**52**, p. 41], [**54**, Th. 43].

Proposition 3. [**47**, Lemma 4.4], [**52**, p. 40, Th. 4], [**54**, Th. 44].

Theorem 1. Cf. [**67**, Lemma 7.2.3], [**19**, Lemme 1.3], [**47**, Lemma 4.5], [**52**, p. 41, Cor.], [**54**, p. 65, Exer. 1].

Corollary. Cf. [**54**, p. 66, Exer. 2].

Theorem 2. Cf. [**67**, Lemma 7.3.5], [**19**, Lemme 1.5], [**47**, Th. 6.2], [**52**, p. 43, Th. 6], [**54**, Th. 56].

Proposition 5. Cf. [**19**, Lemme 1.6], [**47**, Th. 5.7], [**54**, p. 88, Exer. 2].

Exercise 2. Used in [§ 30, Lemma 5].

Exercise 4. Hint: $f=xy$ is an idempotent, algebraically equivalent to 1 [cf. § 1, Exer. 6]. Write $fA=eA$, e a projection [§ 1, Exer. 7]; then e is algebraically equivalent to 1. Cf. [§ 1, Exer. 8].

Exercise 5. Hint: Writing $f=xy$, $e=1-\mathrm{RP}(1-f)$, one has $fA=R(\{1-f\})=eA$. Cf. Exercise 4.

Exercise 6. Worked out in [§ 46, Prop. 1, (5)]. Cf. [**23**, Ch. III, § 1, Exer. 3].

Exercise 7. Cf. [**54**, p. 71, Exer. 2]. Hint: [§ 16, Exer. 1].

Exercise 9. Hint: Drop down to $(e\cup f)A(e\cup f)$.

Exercise 10. (i) Same proof as for [§ 34, Prop. 1]. Cf. [**47**, Th. 6.3], [**54**, Th. 67]. (ii) Cf. Exercise 9.

Exercise 11. [**51**, p. 524, Th.].

Exercise 12. (iv) Cf. [**47**, Th. 6.6, (b)], [**54**, pp. 119–120].
(v) Hint: Proposition 4, criterion (2a).
(vi) Cf. [**63**, Lemma 2.4].
(vii) Cf. [**47**, Th. 6.6, (c)]. Hint: One can suppose that A is finite and $e\cup f=1$. Write $e=e'+e''$, $f=f'+f''$ as in [§ 13, Prop. 5]. Setting $h=e'\cup f'$, apply (iii) to e', f' in hAh; since $e''\sim f''$ by Exercise 3, part (vi) is applicable to e'', f'' in $(1-h)A(1-h)$.

(viii) Under hypothesis (2) or (3), cf. [**54**, Th. 71] (but note that the above sketch of (vii) avoids continuous geometry). Under hypothesis (4), see [**47**, Th. 6.6, (c)].
(ix) Hint: [§14, Exer. 12].
(x) A theorem of Fillmore [**27**].
(xi) The question is asked in [**54**, p. 120], [**55**, p. 11].

Exercise 13. (1) Hint: [§14, Exer. 16] and Theorem 3. Cf. the proof of [§58, Prop. 1].
(2) Hint: [§13, Th. 1], [§8, Lemma 3] and (1).

Exercise 14. Cf. [**18**, Lemme 6.3.β]. For the proof, cf. [§26, Prop. 5].

Exercise 15. Cf. [**63**, Th. 2.1, (ii), (2,ζ)], [**54**, p. 87, Exer. 1].

Exercise 16. Cf. [**67**, Lemma 7.2.1], [**19**, Lemme 1.8], [**80**, Lemma 3.5].

Exercise 17. (i) Cf. [**19**, Lemme 1.7], [**82**, p. 404, Remark 1.1], [**80**, Lemma 3.4]. Hint: Decompose according to [§15, Exer. 7], and apply Proposition 4 and Theorem 1 (with $m=2$) to the pieces.
(ii) See Exercise 12, (x).

Exercise 18. Cf. [**80**, Lemma 3.7]. Hint: Expand $\{g\}$ to a maximal homogeneous partition; cf. the proof of Proposition 2.

Exercise 19. (i) Hint: [§14, Exer. 15].
(iii) Hint: Cf. the proof of [§18, Lemma 2 to Prop. 3].

Exercise 20. (i) [**10**, Th. 3]. Hint: [§14, Exer. 12] and Exercise 3.
(ii) See [§15, Exer. 13].

Exercise 21. (i) Hint: [§14, Prop. 5], [§20, Prop. 3].
(ii) Hint: [§14, Prop. 4].

Exercise 22. (i) [**28**].

Exercise 23. Hint: Using criterion (2b) of Proposition 4, construct inductively an infinite sequence of orthogonal projections equivalent to e.

§ 18

Definition 1. Cf. [**19**, Déf. 3.3], [**48**, p. 469], [**52**, p. 32, Def. 1].

Proposition 1. Cf. [**19**, Lemme 3.4], [**48**, Lemma 19], [**54**, Th. 53].

Proposition 3. Cf. [**19**, Lemme 3.9], [**47**, Lemma 4.8], [**54**, Th. 46].

Lemma 2. Cf. [**35**, §61, Th. 3], [**52**, p. 34, Lemma 5].

Theorem 1. [**52**, p. 34, Th. 1].

Theorem 4. [**48**, Lemma 18]. For the case of von Neumann algebras, see [**81**, Part II, Th. 10], [**19**, Th. 2]; in this case, all the $h_\aleph$ are uniquely determined (see Exercise 10).

Proposition 4. Cf. [**47**, Lemma 4.10], [**54**, Th. 47]. In the presence of GC, the proposition is also immediate from the dimension theory of projections [§42, Lemma 2]; in particular, the von Neumann algebra case is already covered by [**18**, Th. 8].

Proposition 5. Cf. [**47**, Lemma 4.11], [**51**, p. 533], [**54**, Th. 48].

Proposition 6. Cf. [**52**, p. 50].

Exercise 2. (i) Hint: [§14, Exer. 15] or Lemma 1 to Proposition 3.
(ii) Hint: Lemma 2 to Proposition 3 and [§6, Cor. 3, (i) of Prop. 4].

Exercise 5. Cf. [**54**, p. 105]. Hint: Let $h=C(e)$ and write $h=h_1+h_2$, where $h_1=hC(f)$ and $h_2=h-h_1$. Note that $C(h_1 e)=C(h_1 f)=h_1$ and apply Propositions 1 and 6 to get $h_1 e \precsim h_1(1-f)$. On the other hand, $h_2 f=0$. Finally, $e=h_1 e+h_2 e$.

Exercise 7. A theorem of Kaplansky [**48**, Th. 2] (for an alternative exposition, see [**49**, Ths. 8 and 4]).

Exercise 8. A theorem of Widom [**90**, Th. 4.1]. K. Saitô has extended the result to the case that B is a semifinite AW^*-subalgebra with center Z [**79**, Th. 5.2].

Exercise 9. [**49**, Th. 8].

Exercise 10. (i) Cf. [**81**, Part II, Lemma 2.7], [**48**, p. 471, footnote].
(ii) [**48**, Th. 4].
(iii) [**19**, Lemma 3.5].
(iv) Kaplansky has conjectured that the answer is negative [**49**, p. 843, footnote].

Exercise 11. Cf. [**29**, p. 334].

Exercise 14. (i) Cf. [§4, Exer. 3].

Exercise 16. (ii) Cf. [§20, Exer. 12].

§ 19

Theorem 1. Cf. [**18**, Lemma 6.2], [**47**, Lemma 4.12], [**54**, Th. 49].

Exercise 1. Hint: [§26, Prop. 5].

Exercise 2. (i) Hint: By structure theory, reduce to the case that A is properly infinite [§17, Cor. of Th. 1] or continuous (Corollary of Theorem 1).
(ii) Hint: [§16, Exer. 5].

§ 20

Theorem 1. Cf. [**47**, Th. 5.5], [**48**, Lemma 20], [**51**, p. 534], [**54**, Ths. 54, 64].

Exercise 1. Same as [§11, Exer. 5].

Exercise 2. (a) implies (b): See Theorem 3.
(b) implies (c): [§13, Prop. 2].
(c) implies (a): A has PC [§14, Prop. 3] hence also GC [§14, Prop. 7].

Exercise 3. Hint: [§14, Prop. 5]. See also [§14, Exer. 4].

Exercise 4. [**54**, pp. 103–104, 109].

Exercise 6. See [§14, Exer. 5]. The more general version appears in [**54**, Ths. 54, 64 and 63].

Exercise 7. Better yet, see [§21, Exer. 11].

Exercise 8. Cf. [**62**, Lemma 1]. Hint: [§14, Exer. 15], [§17, Th. 1] and Theorem 1.

Exercise 9. Cf. [**34**, Lemma 2.1].
(i) Hint: Write $e=\sup e_n$ with e_n an orthogonal sequence such that $e_n \sim e$ for all n. Expand (e_n) to a (necessarily countable) maximal homogeneous partition (g_n).

Invoke [§17, Prop. 2] to get a central projection h and a partition (f_n) of h equivalent to (hg_n). Note that $he_n \sim f_n$ and apply Proposition 3.

(ii) Hint: Exhaustion, via (i) and Proposition 3.

Exercise 11. (i) The question is raised in [**47**, p. 249].

(ii) An affirmative answer to (ii) would imply an affirmative answer to (i), as follows. Write $x^*x=r^2$, $r\geq 0$, let $C=\{x^*x\}''$, let e_n be an orthogonal sequence of projections in C such that $\sup e_n=\mathrm{RP}(r)=\mathrm{RP}(x)$ and re_n has an inverse y_n in e_nC [§8, Prop. 1, (4)], and let $w_n=xy_n$; continuing as in the proof of [§21, Prop. 1], one would arrive at a factorization $x=wr$ with $w^*w=\mathrm{RP}(x)$, $ww^*=\mathrm{LP}(x)$.

(iii) An affirmative answer to (iii) would also imply an affirmative answer to (i); this is clear from the remarks for (ii) and the proof of Theorem 3.

(iv) Even assuming GC, the prognosis for (iii) seems unfavorable; note, for example, the lack of an analogue of structure theory (in the sense of Section 15) for Rickart C^*-algebras.

Exercise 12. (i) The equivalence of (a) and (b) is Theorem 2. The proof that (c) implies (b) is easy if A is properly infinite (Exercise 3) or if A is Type $\mathrm{I}_{\mathrm{fin}}$ [cf. §18, Prop. 5]; the residual case of Type $\mathrm{II}_{\mathrm{fin}}$ is stubborn [**54**, Th. 52].

(ii) Cf. [§11, Exer. 3].

§ 21

Proposition 2. See also Exercise 1. The von Neumann algebra case appears in [**67**, Lemma 4.4.1]; for the case of AW^*-algebras, see [**80**, Lemma 4.2], [**94**, Lemma 2.1].

Exercise 1. [**54**, Th. 65]. Hint: Partial isometries are addable in A [§20, Exer. 6].

Exercise 3. [**54**, p. 109].

Exercise 4. Hint: Observe that all elements of A_1 have countable spectrum, and construct an element x of A_1 for which the element w of Proposition 3 could only have uncountable spectrum. For example, let $\xi_1, \xi_2, \xi_3, \ldots$ be an orthonormal basis of $\mathscr{H}$ and let x be the compact operator defined by $x\xi_n=(1/n)\xi_{n+1}$. Thus $x=ur$, where $r\xi_n=(1/n)\xi_n$ and $u\xi_n=\xi_{n+1}$. If A_1 had WPD, there would exist a factorization $x=ws$ with s in the bicommutant of r^2 in A_1, w in A_1, and $s^*s=r^2$. Argue that w is a weighted shift $w\xi_n=\mu_n\xi_{n+1}$ with weights μ_n of absolute value 1, therefore w has uncountable spectrum (equal to the closed unit disc [cf. **37**, Problem 78]), contrary to $w\in A_1$.

Exercise 7. Better yet, see [§17, Exer. 6].

Exercise 8. Cf. [**14**, Th. 1].

Exercise 9. Cf. [**14**, Th. 3].

Exercise 11. [**39**].

§ 22

Proposition 1. [**92**, Lemma 2.1].

Definition 1. [**56**, Def. 2], [**92**, Def. 2.1].

Proposition 2. Cf. [**92**, Lemma 2.3], [**3**, Lemma 3.6].

Definition 2. [**21**, Déf. 3.3].

Theorem 1. Cf. [**21**, Lemme 3.9], [**3**, p. 502], [**63**, Lemma 3.3].

Theorem 2. Proved for AW^*-algebras by F. B. Wright [**92**, Th. 2.4].

Exercise 1. [**64**, Th. 6.4].

Exercise 3. Hint: I is the left ideal generated by $\mathfrak{p}$ (Proposition 2) and is closed [§ 8, Lemma 3]; cf. Proposition 4.

Exercise 4. (i) Cf. [**47**, proof of Th. 2.3].
(ii) [**23**, Ch. I, § 3, Cor. 3 of Th. 2]. Hint: If (e_ι) is any orthogonal family of projections in L, then the net of finite sums converges to $\sup e_\iota$ in each of the indicated topologies.

Exercise 5. (i) Example: $A = \mathscr{L}(\mathscr{H})$, $\mathscr{H}$ infinite-dimensional, I the ideal of compact operators. Cf. [§ 15, Exer. 13, (i)].
(iii) For the application to AW^*-algebras, see [§ 15, Exer. 13, (ii)].

Exercise 6. Hint: Let e be a nonzero idempotent in $\bar{I}$; since eAe is a Banach algebra with unity, it can have no dense proper ideal [**75**, Cor. 2.1.4].

Exercise 7. Proved for AW^*-algebras by Yen [**95**, Lemma 1].

Exercise 8. Cf. [**80**, Cor. 4.2], [**93**, Th. 4.1].

Exercise 9. (i) Hint: If $g \sim e \leq e_1 \cup e_2$, then there exists a unitary element u such that $g \leq u(e_1 \cup e_2)u^*$.

§ 23

Proposition 1. Cf. [**3**, Th. 3.7], [**63**, Lemma 3.4].

Proposition 2. Cf. [**92**, Lemma 3.4], [**3**, Lemma 3.10].

Exercise 2. Cf. [**92**, Th. 3.2]. The argument for (ii) is given in [§ 45, Lemma 2].

Exercise 3. [**3**, Lemma 2.1].

Exercise 4. See Proposition 5.

Exercise 8. (i) Cf. [§ 7, Exer. 8].
(ii) Cf. [§ 22, Exer. 4, (i)].
(iii) Follows easily from (i) and (ii).

Exercise 9. (ii) and (iii) are covered by Exercise 8.
(i), (iv) See [**23**, Ch. I, § 4, Exer. 9 and Cor. 2 of Th. 2].
(v) Consider the family of finite subsups [cf. § 3, Prop. 7].

§ 24

Proposition 3. Cf. [**59**, Ch. V, Satz 3.1], [**63**, p. 86].

Exercise 1. Hint: [§ 23, Exer. 4].

Exercise 2. This is the usual form of 'weak centrality', proved for finite von Neumann algebras by Godement [**32**, Lemme 15], for general von Neumann algebras by Misonou [**66**, Th. 3] and for AW^*-algebras by F. B. Wright [**92**, Th. 2.5]. Incidentally, Dixmier's central approximation theorem [**18**, Th. 7], on which Misonou's proof is based, is also valid for AW^*-algebras, and somewhat more generally [**33**, Th. 1].

Exercise 3. (i) Cf. [**75**, Cor. 2.1.4].

(ii) For the one-dimensionality of the center of a primitive Banach algebra with unity, see [**50**, Lemma 9], [**75**, Cor. 2.4.5].

(iii) Cf. [**42**, p. 4, Th. 1].

(iv) [**24**, p. 49, Th. 2.9.7], [**75**, Th. 4.9.6].

Exercise 4. Cf. [**92**, Th. 2.6], [**3**, Lemma 1.5].

Exercise 5. (iii) See [§ 23, Exer. 1, (viii)].

Exercise 6. (i) Cf. [§ 6, Exer. 20].

§ 25

Definition 1. Dimension functions may also be defined in certain infinite situations; without going into details, the penalty for dropping finiteness of the ring is to admit infinite-valued functions in the definition of a dimension function.

For factorial von Neumann algebras, dimension goes back to Murray and von Neumann, who used it as the basis for classifying factors according to type [**67**, Ch. VIII]. Dimension was defined for an irreducible (i. e., 'centerless') continuous geometry [§ 34, Def. 1] by von Neumann [**71**, p. 52, Th. 6.9]; for any continuous geometry by Iwamura [**41**]; for a finite von Neumann algebra by Dixmier [**18**]; for a finite AW^*-algebra by Kaplansky [**47**, pp. 247–248]; for any von Neumann algebra by Segal [**82**, Th. 1]; for any AW^*-algebra by Sasaki [**80**, Th. 5.1] and Feldman [**106**]; for any Baer $*$-ring satisfying the parallelogram law, by S. Maeda ([**63**, Th. 2.1], [**65**, § 10]). Without attempting to cover the literature of dimension in lattices, we mention also the memoir of Loomis [**58**], the paper of Ramsay [**73**] and the survey of Holland [**110**]. {Warning: The term 'center' is used in a variety of ways in lattice theory.}

The exposition in the present chapter (especially for Type II) leans most heavily on the construction of dimension in [**18**], as transcribed for finite AW^*-algebras in [**2**, Appendix III].

§ 26

Propositions 1–13. Cf. [**18**, Déf. 4.3, Lemme 6.3].

Proposition 14. Cf. [**18**, Lemmes 4.10, 4.11].

Proposition 16. Cf. [**18**, Lemme 6.4].

Exercise 1. Hint: By hypothesis, there exists a simple abelian projection with $(1:e)=n$; cf. [§ 18, Cor. of Prop. 1], [§ 15, Prop. 6].

§ 27

Exercise 2. Same as [§ 20, Exer. 9].

§ 29

Exercise 1. This is worked out in [§ 42, Lemma 2].

§ 30

Proposition 4. Cf. [**18**, Lemme 6.10].

Proposition 5. Cf. [**18**, Lemme 6.13].

§ 33

Theorem 1. See the notes for [§ 25, Def. 1].

Theorem 2. Cf. [**59**, Ch. V, p. 115, Hilfsatz 1.5].

Theorem 3. Cf. [**41**, Th. 4], [**59**, Ch. V, Satz 1.6].

Theorem 4. For the case of a finite von Neumann algebra, see [**82**, Lemma 1.1].

Exercise 1. This is an easy application of Theorem 2. For a direct (dimension-free) argument, see [**47**, Lemma 6.4], [**54**, Th. 68].

Exercise 4. Cf. [**59**, Appendix II].

§ 34

Proposition 1. [**47**, Th. 6.3], [**54**, Th. 67].

Definition 1. The concept is due to von Neumann [**69**], [**71**, pp. 1–2, Axioms I–V]. See also the treatise of F. Maeda [**59**, Ch. V, Def. 2.1; p. 237, Appendix II] and the brief survey in Birkhoff's book [**11**, Ch. XI, §§ 8–11].

Theorem 1. Cf. [**63**, Th. 2.2], [**54**, Th. 69].

Exercise 2. [**51**, p. 524, Th.].

Exercise 3. The result is due to U. Sasaki (see Apendix II of F. Maeda's book [**59**]).

§ 36

Proposition 1. Since $\tilde{A}$ is a continuous geometry [§ 34, Th. 1], this is a special case of a theorem of Kawada, Higuti and Matusima [**56**, Satz 4].

Corollary. Cf. [**54**, Th. 66].

Exercise 1. The result, which is immediate from the Corollary, is due to F. B. Wright [**92**, Th. 2.7]. For finite von Neumann algebras, it was proved by Godement [**32**, Lemme 15, Cor.].

Exercise 2. Cf. [**75**, Th. 2.3.11].

§ 38

Proposition 1. Cf. [**71**, pp. 273–274, Ths. 2.14, 2.15], [**59**, Ch. IV, Satz 4.5].

Proposition 2. Cf. [**71**, p. 274, Th. 2.16], [**59**, Ch. IV, Satz 4.6].

§ 39

Theorem 1. Cf. [**56**, Satz 3], [**59**, Ch. V, Satz 3.1].

Corollary. [**63**, p. 86].

Exercise 2. (iii) Hint: See the proof of [§ 40, Lemma].

Exercise 3. Cf. [**3**, Lemma 1.2].

§ 40

Lemma. Cf. [**3**, Cor. 1.7].

§ 41

Theorem 1. See the references for [§ 44, Th. 1].

§ 43

Theorem 1. See the references for [§ 44, Th. 1].

§ 44

Theorem 1. Since the projection lattice of A is a continuous geometry [§ 34, Th. 1], the present theorem is essentially an application of the reduction theory of continuous geometries. The latter is due to Iwamura [**41**] (see also [**56**], [**59**, Ch. V, Satz 3.2]). For the case that A is a finite AW^*-algebra, see [**3**, Ths. 5.3, 5.7]; for A a finite Baer $*$-ring satisfying the (EP)-axiom and the (SR)-axiom, see [**63**, p. 87, Cor. 1]; for A a regular Baer $*$-ring, see [**63**, p. 87, Cor. 2] (also [**87**, Th. 6]).

The key proofs in Sections 41–43 are modeled on arguments in Wright's paper on the reduction of finite AW^*-algebras into finite AW^*-factors [**92**, Ths. 4.1, 4.2, 5.1]. The latter theory (as generalized by Yen [**95**]) is exposed in Section 45.

§ 45

Lemma 2. [**92**, Th. 3.2]. See also [§ 23, Exer. 2].

Theorem 1. For a finite AW^*-algebra with trace [cf. § 15, Exer. 11], the theorem is due to F. B. Wright [**92**]; the assumption of a trace (whose existence is an open problem) was removed by Yen [**95**]. The exposition in this section is strongly influenced by [**3**]. {But the order of events here is the reverse of that in [**3**], where the A/I theorem is deduced from the A/M theorem.}

Exercise 1. Cf. [**70**], [**81**, Part I], [**19**, Section IV], [**23**, Ch. II].

Exercise 2. This is a special case of a result of Feldman [**26**, Th. 1]. Sketch of argument: If $x \mapsto x^\natural$ is the center-valued trace function of A [cf. § 15, Exer. 11], then $x \in M$ implies $x^\natural \in M$ [**18**, Th. 12], therefore a positive linear form f on A/M may be defined by the formula $f(\bar{x}) = x^\natural(\sigma)$ $(x \in A)$. Since $x \in M$ iff $(x^* x)^\natural \in M$ [**32**, Lemme 15], f is faithful. By uniqueness of dimension, $D(e) = e^\natural$ $(e \in \tilde{A})$, therefore $\bar{e} \mapsto f(\bar{e}) = D(e)(\sigma)$ is the dimension function of A/M; in particular, f is completely additive on projections. By the proof of the cited theorem of Feldman, A/M is represented as a von Neumann algebra on the Hilbert space derived canonically from f.

Exercise 3. Strong semisimplicity is noted in [§ 36, Exer. 1].

Exercise 4. (i) Sketch: $\tilde{M}$ is a maximal p-ideal of A; write $\tilde{M}=\mathscr{I}_\sigma=\{e\in\tilde{A}: D(e)(\sigma)=0\}$ for suitable $\sigma\in\mathscr{X}$. Then $N=\{c\in Z: c(\sigma)=0\}$ is a maximal ideal of Z with $\tilde{M}=\{e\in\tilde{A}: D(e)\in N\}$. Evidently $\tilde{N}\subset\mathscr{I}_\sigma\cap Z=\tilde{M}\cap Z$; since N is the closed linear span of its projections, $N\subset M\cap Z$ (hence $N=M\cap Z$ by the maximality of N). Then $e\in\tilde{M}$ implies $D(e)\in N\subset M\cap Z$.
(ii) See [§ 24, Exer. 4, (iv)] and (i).
(iii) See [§ 24, Exer. 3, (iv)].
(iv) Cf. the proof of Lemma 3. The result is worked out in [**3**, Cor. 1.7].

Exercise 5. Hints: (i) [§ 27, Exer. 3] and Exercise 4.
(ii) [§ 8, proof of Lemma 2].
(iii) Exhaustion on (ii). Cf. [§ 22, Exer. 4, (i)].

Exercise 6. Hint: Exercise 5, (iii).

Exercise 7. Hint: Exercise 6.

Exercise 8. Hints: (i) [§ 8, proof of Lemma 2].
(ii) [§ 33, Th. 3] and Exercise 5, (i).

Exercise 9. Hints: (i) [§ 24, Exer. 3, (iii)].
(ii) [§ 36, Exer. 2].
(iii) [§ 36, Cor. of Prop. 1].
(iv) A C^*-algebra—or any $*$-algebra of operators on a Hilbert space—is semisimple [cf. **75**, Th. 4.1.19].

Exercise 10. Hints: (i) Note that $J=\{c\in Z: \mathrm{RP}(c)(\sigma)=0\}$. In the notation of the book of Gillman and Jerison [**31**, p. 62], $J=\mathbf{O}_\sigma$.
(ii) [§ 4, Exer. 3].
(iv) [§ 61, Th. 1].
(v) [§ 42, Exer. 1, (i)].
(vi) Let e_1 be a faithful abelian projection in A. Then e_1Ae_1 is $*$-isomorphic with Z [§ 6, Cor. 2 of Prop. 4], thus J coincides with the ideal described in [§ 42, Exer. 1, (ii)].
(vii) [§ 16, Exer. 2].
(viii) [**54**, p. 17, Exer. 3].
(x) If B is a ring with radical R, then B_n has radical R_n [**42**, Ch. I, § 7, Th. 3].

Exercise 11. Sketch: The crux of the matter is to prove that (a) implies (b). Assuming Z infinite-dimensional, it is to be shown that there exists a nonclosed maximal-restricted ideal I. One easily reduces to the case of Type I or Type II. The latter is covered by Exercise 8. Assuming A is of Type I, write $A=A_1\oplus A_2\oplus A_3\oplus\cdots$ with A_i of Type I_{n_i}, $n_1<n_2<n_3<\cdots$. If every A_i has finite-dimensional center, then there are infinitely many A_i and we are in the situation of Exercise 7 (cf. [§ 33, Th. 4] for orthoseparability). Otherwise, some A_i has infinite-dimensional center; dropping down, we can suppose $A=Z_n$, with $Z=C(\mathscr{X})$ infinite-dimensional. Then Z is not regular (in the sense of von Neumann [§ 51, Def. 1]), therefore there exists $\sigma\in\mathscr{X}$ such that, in the notation of Exercise 10, $J\neq N$ [**31**, p. 63, (2) $\Leftrightarrow$ (8)].

Exercise 12. For the Stone-Čech compactification, see [**31**].

§ 46

Exercise 3. Cf. [§ 7, Exer. 5], [§ 13, Exer. 2], [§ 20, Exer. 6].

§ 47

Definition 1. Cf. [**67**, Def. 16.2.1], [**82**, Def. 2.1], [**4**, Def. 1.1].

Definition 5. Cf. [**67**, Def. 4.2.1, Lemma 16.2.3], [**82**, Def. 2.1], [**4**, Def. 2.3].

Theorem 1. For the case of a finite von Neumann factor, the theorem (slightly reformulated) is due to Murray and von Neumann [**67**, Ch. XVI, Th. XV]; their construction was subsequently generalized to arbitrary von Neumann algebras by Segal [**82**, Cor. 5.2]. The case of a finite AW^*-algebra is treated in [**4**] (see also the paper of Roos [**76**, Cor. of Th. 2]), subsequently generalized to arbitrary AW^*-algebras by K. Saitô [**77**], [**78**]. The present result appears in [**8**].

Exercise 1. (ii) The only nontrivial point is that $e_n \uparrow 1$ implies $e \cap e_n \uparrow e$ [§ 34, Prop. 2].

Exercise 4. See the proof of [§ 48, Prop. 2].

§ 48

Theorem 2. A considerable improvement on [**8**], where this was proved only under heavy additional hypotheses [**8**, Prop. 8.7].

Exercise 3. Hint: Proposition 1.

Exercise 7. See [§ 20, Exer. 6].

Exercise 8. Hint: [§ 53, Lemma to Prop. 4], [§ 36, Cor. of Prop. 1], [§ 16, Exer. 2].

§ 49

Exercise 1. The main change is that in the proof of Theorem 1 one defines $v_\rho = w_\rho\left(e_\rho - \sup_{\sigma<\rho} e_\sigma\right)$.

Exercise 2. Cf. [§ 48, Exer. 7].

§ 50

Exercise 2. Hint: [§ 48, Exer. 6].

Exercise 3. A counterexample may be constructed as follows.

(i) Let F be a field, equipped with the identity involution. It is trivial that F is a Baer $*$-ring satisfying the (EP)-axiom and the (SR)-axiom. An element a of F is positive in the sense of [§ 13, Def. 8] if and only if it is a sum of squares; we say that a has *length* n if it is the sum of n squares but no fewer. If F has an element of length ≥ 2, it cannot satisfy the (UPSR)-axiom.

(ii) For every positive integer n, let F_n be a field having an element a_n of length $\geq n$. {It is a theorem of J. W. S. Cassels [**99**] that if $t_1, \ldots, t_n$ are indeterminates, then $t_1^2 + \cdots + t_n^2$ is not the sum of $n-1$ squares in the rational function field $\mathbb{R}(t_1, \ldots, t_n)$.} Let $A = \prod_{n=1}^{\infty} F_n$ be the complete direct product of the F_n [§ 1, Exer. 13]. Then A is a commutative, regular Baer $*$-ring satisfying the (EP)-axiom and the (SR)-axiom; since the partial isometries in A are the elements $w = (w_n)$ with $w_n = 0, 1$ or -1, it is clear that partial isometries in A are addable.

(iii) With notation as in (ii), let $x=(a_n)$. For $m=1,2,3,\ldots$ let $h_m=(\delta_{mn})$, $e_m=h_1+\cdots+h_m$. Then $e_m \uparrow 1$, $e_m x=(a_1,\ldots,a_m,0,0,0,\ldots)$ is ≥ 0 for all m, but x is not ≥ 0.

Exercise 4. The problem is easily reduced to the case that $e=1$. From $e_\rho \leq 1$ it is immediate that $x^* e_\rho x \leq x^* x$, thus $x^* x$ is an upper bound for the family $(x^* e_\rho x)$. On the other hand, if y is a self-adjoint element of B such that $x^* e_\rho x \leq y$ for all ρ, it is to be shown that $x^* x \leq y$. By Theorem 1 (generalized to well-directed families), it suffices to find a well-directed family (f_ρ) of projections with $f_\rho \uparrow 1$ and $f_\rho(y-x^* x) f_\rho \geq 0$ for all ρ. Set $f_\rho = x^{-1}(e_\rho)$. Then $f_\rho \uparrow 1$ [cf. § 47, Lemma 5]; moreover, $(1-e_\rho)x f_\rho = 0$, that is, $x f_\rho = e_\rho x f_\rho$, thus $f_\rho x^* x f_\rho = f_\rho(x^* e_\rho x) f_\rho \leq f_\rho y f_\rho$.

§ 51

Regrettably, the numbering of the conditions (3°), (4°), etc. deviates from that in [**8**].

Lemma. Note that the (SR)-axiom may be substituted for the (UPSR)-axiom here.

Definition 1. The concept of regularity is due to von Neumann [cf. **71**, p. 70, Def. 2.2; p. 114, Th. 4.5].

Proposition 3. Cf. [**71**, p. 114, Th. 4.5].

Corollaries 1, 2. The possibility of combining sequences of elements along an orthogonal sequence of projections in a regular Baer $*$-ring is a theme in [**51**, Section 3].

Corollary 3. In particular, if A is regular and satisfies (1), then A is the direct sum of finitely many factors of Type I (Exercise 11, (i)). A more general result of Vidav [**89**]: If A is a regular Baer $*$-ring in which $\sum_1^n x_i^* x_i = 0$ implies $x_1 = \cdots = x_n = 0$ (n any positive integer), and if A satisfies (1), then A is of Type I.

Exercise 1. Cf. [**54**, p. 130, Th. A].

Exercise 5. See [**71**, p. 70, Th. 2.2].

Exercise 6. Hint: [§ 5, Prop. 3].

Exercise 8. Hint: [§ 23, Exer. 1].

Exercise 9. Hint: [§ 50, Lemma].

Exercise 10. Cf. [**7**].

Exercise 11. (i) See the notes for Corollary 3.

Exercise 12. See [**71**, p. 81, Th. 2.13], [**13**, esp. p. 167].

Exercise 14. Hint: Exercise 7.

Exercise 15. See [**72**, Th. 2].

Exercise 16. Cf. [§ 50, Exer. 1].

Exercise 17. Theorems of Kaplansky [**51**, Ths. 2, 3].

Exercise 18. Cf. [**89**, Lemma 6].

Exercise 19. Cf. [§ 49, Exer. 2].

§ 52

Theorem 1. For A a finite von Neumann factor, the theorem (slightly reformulated) is due to von Neumann [cf. **71**, p. 89, (VI)]; the case of a finite von Neumann algebra is covered by the construction of Segal [**82**, Cor. 5.2] (cf. [**105**, Th. 2]). The case of a finite AW^*-algebra is treated in [**4**]; the present result appears in [**8**].

Roos [**76**] has observed that any Baer ring A may be enlarged to a regular ring R—the maximal ring of right quotients of A in the sense of Utumi [**88**]; when A is a finite Baer $*$-ring satisfying the (EP)-axiom and the (SR)-axiom, R coincides with the regular Baer $*$-ring that accrues to the projection geometry of A (which is a continuous geometry; cf. the remarks at the end of Section 46) [**76**, Th. 2]; R coincides with $\mathbf{C}$ when A is a finite AW^*-algebra [**76**, Cor. of Th. 2], but in general the connection between R and $\mathbf{C}$ (for the various extensions of [§ 47, Th. 1]) remains to be explored.

Theorem 2. Proved in [**9**]. For the case that A is a finite AW^*-algebra, see [**6**, p. 177, Lemma]. Cf. the discussion in [**51**, Section 3].

Exercise 4. Cf. [§ 51, Exer. 19].

Exercise 5. Hint: Partial isometries in A are addable [§ 20, Exer. 6]; see [§ 49, Exer. 2], [§ 51, Exer. 1] and Exercise 4.

Exercise 7. Hint: [§ 51, Th. 1] (as generalized in [§ 51, Exer. 19]); see the proof of Theorem 2.

Exercise 8. (i) Sketch: A also satisfies (2°) [§ 20, Exer. 6], thus $\mathbf{C}$ is regular by Exercise 4. Since $\mathbf{C}$ satisfies LP$\sim$RP [§ 48, Exer. 7] and has no abelian central projections, $\mathbf{C}_n$ is $*$-regular for all n [§ 51, Exer. 15]. {In effect, [§ 51, Exer. 15] is a substitute for [§ 50, Prop. 1]—at the cost of excluding abelian summands.}

Incidentally, if, in place of (2°′), A is assumed only to satisfy (2°), then at least $\mathbf{C}_2$ is $*$-regular. {Proof: The relations $x, y \in A$, $x^*x + y^*y = 0$ imply $x = y = 0$ [§ 51, proof of Lemma], therefore the relations $\mathbf{x}, \mathbf{y} \in \mathbf{C}$, $\mathbf{x}^*\mathbf{x} + \mathbf{y}^*\mathbf{y} = 0$ imply $\mathbf{x} = \mathbf{y} = 0$ [§ 50, proof of Prop. 1]. Since $\mathbf{C}$ is regular (Exercise 4), so is $\mathbf{C}_2$ [§ 51, Exer. 13].}

§ 53

Theorem 1. [**8**, Prop. 8.6]. There is an intriguing analogue for the maximal ring of right quotients [**88**, Th. 4] (cf. the notes for [§ 52, Th. 1]).

Exercise 1. [**8**, Lemma 8.10].

Exercise 2. [**8**, Prop. 8.12].

Exercise 3. (i), (ii) [**8**, Prop. 8.13].
(iii) [**8**, Prop. 9.3].

Exercise 5. Hint: Exercise 4.

Exercise 6. Same proof as for the case that A is a finite AW^*-algebra [**6**, Th. 8].

§ 54

Definition 1. The concept appears in an unpublished manuscript of von Neumann [**53**, L4]. See also [**3**, p. 501]. For a discussion of boundedness in $*$-regular rings, see [**89**].

Proposition 1. Cf. [**3**, Lemma 3.11], [**89**].

Theorem 1. [**8**, Cor. 9.5].

Theorem 2. An analogous result is proved in [**3**, Th. 3.12] for certain Rickart $*$-rings.

Exercise 1. Hint: [§ 51, Exer. 11, (i)].

Exercise 2. (i) Cf. [§ 51, Exer. 18].
(ii) Cf. [§ 53, Exer. 3].

Exercise 4. (i) Hint: [§ 52, Exer. 6], [§ 51, Cor. 2 of Th. 1].
(ii) Hint: [§ 53, Th. 1, (i)].

Exercise 6. (ii) Hint: Exercise 4, (ii).
(iv) For the case that A is a finite AW^*-algebra, see [**3**, Th. 4.2].

§ 55

Exercise 2. Hint: [§ 8, Lemma 2].

Exercise 5. Same as [§ 11, Exer. 5].

Exercise 6. No. Sketch of example: Let A be the $*$-ring of all ultimately real sequences of complex numbers [§ 54, Exer. 9]. Since the relations $x_1, \ldots, x_n \in A$, $x_1^* x_1 + \cdots + x_n^* x_n = 0$ imply $x_1 = \cdots = x_n = 0$, A_n is a Rickart $*$-ring for all n [§ 51, Exer. 13]. But partial isometries in A are not addable [cf. § 20, Exer. 4], thus A_n is not a Baer $*$-ring for any $n \geq 2$ (Exercise 5).

§ 56

Exercise 1. (ii) Cf. [**5**, p. 44], [**91**, Th. 9].
(iii) Cf. [**54**, p. 38, Exer. 8], [**46**, p. 3].

Exercise 2. Hint: [§ 51, Exer. 9, 18] and Proposition 3.

Exercise 3. See [§ 51, Prop. 4].

Exercise 5. (a) implies (b): See the proof of Proposition 3.
The implications (b)$\Rightarrow$(c)$\Rightarrow$(d)$\Rightarrow$(e) are trivial.
(e) implies (a): Suppose $x \in C$, $x^* x \leq 1$. Write $x = wr$ with $x \in A$ and $r \geq 0$ [§ 53, Prop. 5]. Since $0 \leq r \leq 1$ [§ 53, Exer. 3, (i)], one can form $s = (r(1-r))^{\frac{1}{2}}$ [§ 53, Def. 1]; the matrix

$$\begin{pmatrix} r & s \\ s & 1-r \end{pmatrix}$$

is a projection in C_2 [cf. **37**, Solution 177, p. 327], therefore $r \in A$ by hypothesis.

§ 57

Proposition 2. Cf. [**5**, Lemma 3.3].

Exercise 1. Hint: [§ 54, Exer. 4, (ii)].

Exercise 2. Hint: [§ 54, Th. 1].

Exercise 3. Condition (C) is discussed in a paper of Prijatelj and Vidav [**72**, esp. Th. 1]. (iii) Hint: [§ 51, Exer. 6], [§ 21, Prop. 3].

§ 58

Proposition 1. Cf. [**5**, Lemma 4.1].

Exercise 1. Hint: [§ 17, Exer. 6].

Exercise 2. Sketch: By a theorem of Kaplansky [**51**, Th. 2], every regular Baer $*$-ring—in particular B—has the desired property ($yx=1$ implies $xy=1$) and is therefore finite. Let I be a maximal ideal of B. Since I is strict [§ 51, Exer. 8], by reduction theory B/I is a regular Baer $*$-factor satisfying LP$\sim$RP [§ 44, Th. 1]. Note that $B_n/I_n \cong (B/I)_n$. If B/I is of Type II, then $(B/I)_n$ is also a regular Baer $*$-factor of Type II [§ 62, Exer. 8], therefore B_n/I_n has the desired property by Kaplansky's theorem. If B/I is of Type I_r, then $(B/I)_n$ is the ring of linear mappings on an nr-dimensional vector space over a division ring ([§ 18, Exer. 17], [§ 56, Exer. 1, (ii)]), therefore B_n/I_n has the desired property by linear algebra. Since the intersection of the I—hence the I_n—is $\{0\}$ [§ 36, Prop. 1], it follows that B_n has the desired property too. (Incidentally, if B has no abelian summand, then B_n is $*$-regular [§ 51, Exer. 15].)

Exercise 3. Since C is a regular Baer $*$-ring satisfying LP$\sim$RP [§ 48, Exer. 7], Exercise 2 is applicable. (Incidentally, C_n is $*$-regular when A has no abelian summand or when A satisfies (1°).)

§ 60

Lemma 3. Cf. [**5**, Lemma 5.4].

Lemma 4. Cf. [**5**, Lemma 5.5].

§ 61

Theorem 1. The AW^* case, as proved in [**5**, p. 37], was based on the theory of AW^*-modules [**49**].

§ 62

Corollary 1. This is proved in [**5**].

Exercise 1. See [§ 51, Exer. 17]. Better yet, see [§ 58, Exer. 3].

Exercise 3. A theorem of Deckard and Pearcy [**15**, Th. 2].

Exercise 4. See [§ 48, Exer. 7, 8].

Exercise 5. See [§ 45, Exer. 10].

Exercise 6. A theorem of Yohe [**96**].

Exercise 8. For the proof that B_n is a regular Baer $*$-factor, view the results of von Neumann [**71**, p. 230, Th. 17.4; p. 236, Lemma 18.6], Kaplansky [**51**, Th. 3], and Halperin [**135**, Th. 1] in the light of [§ 51, Exer. 15]. The fact that B_n also satisfies LP$\sim$RP is a recent result of J.L. Burke [**136**, Th. 1].

Exercise 9. Cf. [§ 52, Exer. 5], Exercise 8, [§ 56, Prop. 3], and [§ 17, Exer. 17]. A substitute for the (PS)-axiom is criterion (a) of [§ 56, Exer. 5].

Bibliography

[1] Banach, S.: *Théorie des opérations linéaires.* Warsaw 1932. Reprinted New York: Chelsea 1955.

[2] Berberian, S. K.: *The regular ring of a finite AW*-algebra.* Ph. D. Thesis, University of Chicago, Chicago, Ill. 1955.

[3] — *On the projection geometry of a finite AW*-algebra.* Trans. Amer. Math. Soc. **83**, 493–509 (1956).

[4] — *The regular ring of a finite AW*-algebra.* Ann. of Math. (2) **65**, 224–240 (1957).

[5] — *$N \times N$ matrices over an AW*-algebra.* Amer. J. Math. **80**, 37–44 (1958).

[6] — *Note on a theorem of Fuglede and Putnam.* Proc. Amer. Math. Soc. **10**, 175–182 (1959).

[7] — *A note on the algebra of measurable operators of an AW*-algebra.* Tôhoku Math. J. (2) **22**, 613–618 (1970).

[8] — *The regular ring of a finite Baer $*$-ring.* J. Algebra (to appear).

[9] — *Matrix rings over Baer $*$-rings.* Unpublished manuscript, February 1971.

[10] — *Equivalence of projections.* Proc. Amer. Math. Soc. (to appear).

[11] Birkhoff, G.: *Lattice theory.* Third edition, Amer. Math. Soc. Colloq. Publ., Vol. 25. Providence, R. I.: American Mathematical Society 1967.

[12] Brown, A.: *On the absolute equivalence of projections.* Bul. Inst. Politehn. Iaşi **4**, Fasc. 3–4, pp. 5–6 (1958).

[13] Brown, B., McCoy, N. H.: *The maximal regular ideal of a ring.* Proc. Amer. Math. Soc. **1**, 165–171 (1950).

[14] Choda, H.: *An extremal property of the polar decomposition in von Neumann algebras.* Proc. Japan Acad. **46**, 341–344 (1970).

[15] Deckard, D., Pearcy, C.: *On matrices over the ring of continuous complex valued functions on a Stonian space.* Proc. Amer. Math. Soc. **14**, 322–328 (1963).

[16] Dixmier, J.: *Position relative de deux variétés linéaires fermées dans un espace de Hilbert.* Revue Scientifique **86**, 387–399 (1948).

[17] — *Étude sur les variétés et les opérateurs de Julia, avec quelques applications.* Thèse, Université de Paris, Paris: Gauthier-Villars 1949.

[18] — *Les anneaux d'opérateurs de classe finie.* Ann. Sci. École Norm. Sup. **66**, 209–261 (1949).

[19] — *Sur la réduction des anneaux d'opérateurs.* Ann. Sci. École Norm. Sup. **68**, 185–202 (1951).

[20] — *Sur certains espaces considérés par M. H. Stone.* Summa Brasil. Math. **2**, 151–182 (1951).

[21] — *Applications ♮ dans les anneaux d'opérateurs.* Compositio Math. **10**, 1–55 (1952).

[22] — *Formes linéaires sur un anneau d'opérateurs.* Bull. Soc. Math. France **81**, 9–39 (1953).

[23] — *Les algèbres d'opérateurs dans l'espace hilbertien (Algèbres de von Neumann).* Deuxième édition, Paris, Gauthier-Villars 1969.

[24] — *Les C*-algèbres et leurs représentations.* Deuxième edition, Paris, Gauthier-Villars 1969.

[25] Dyer, J.: *Concerning AW*-algebras.* Notices Amer. Math. Soc. **17**, 788 (1970), Abstract 677-47-5.

[26] Feldman, J.: *Embedding of AW*-algebras.* Duke Math. J. **23**, 303–307 (1956).

[27] Fillmore, P. A.: *Perspectivity in projections lattices.* Proc. Amer. Math. Soc. **16**, 383–387 (1965).

[28] — *On products of symmetries.* Canad. J. Math. **18**, 897–900 (1966).

[29] — Topping, D. M.: *Operator algebras generated by projections.* Duke Math. J. **34**, 333–336 (1967).

[30] Gel'fand, I. M., Naĭmark, M. A.: *On the embedding of normed rings into the ring of operators in Hilbert space* (Russian). Mat. Sb. (54) **12**, 197–213 (1943).

[31] Gillman, L., Jerison, M.: *Rings of continuous functions.* Princeton, N. J.: Van Nostrand 1960.

[32] Godement, R.: *Mémoire sur la théorie des caractères dans les groupes localement compacts unimodulaires.* J. Math. Pures Appl. **30**, 1–110 (1951).

[33] Goldman, M.: *Structure of AW*-algebras. I.* Duke Math. J. **23**, 23–34 (1956).

[34] Griffin, E. L., Jr.: *Some contributions to the theory of rings of operators. II.* Trans. Amer. Math. Soc. **79**, 389–400 (1955).

[35] Halmos, P. R.: *Introduction to Hilbert space and the theory of spectral multiplicity.* New York: Chelsea 1951.

[36] — *Lectures on Boolean algebras.* Princeton, N. J.: Van Nostrand 1963.

[37] — *A Hilbert space problem book.* Princeton, N. J.: Van Nostrand 1967.

[38] Halpern, H.: *Embedding as a double commutator in a Type I AW*-algebra.* Trans. Amer. Math. Soc. **148**, 85–98 (1970).

[39] Herman, L.: *A Loomis *-ring satisfies the polar decomposition axiom.* Preprint.

[40] Herstein, I. N., Montgomery, S.: *A note on division rings with involutions.* Michigan Math. J. **18**, 75–79 (1971).

[41] Iwamura, T.: *On continuous geometries. I.* Japan. J. Math. **19**, 57–71 (1944).

[42] Jacobson, N.: *Structure of rings.* Revised edition, Amer. Math. Soc. Colloq. Publ., Vol. 37. Providence, R. I.: American Mathematical Society 1964.

[43] Kadison, R. V.: *Isometries of operator algebras.* Ann. of Math. (2) **54**, 325–338 (1951).

[44] Kaplansky, I.: *Regular Banach algebras.* J. Indian Math. Soc. **12**, 57–62 (1948).

[45] — *Normed algebras.* Duke Math. J. **16**, 399–418 (1949).

[46] – *Forms in infinite-dimensional spaces.* An. Acad. Brasil. Ci. **22**, 1–17 (1950).

[47] — *Projections in Banach algebras.* Ann. of Math. (2) **53**, 235–249 (1951).

[48] — *Algebras of type I.* Ann. of. Math. (2) **56**, 460–472 (1952).

[49] — *Modules over operator algebras.* Amer. J. Math. **75**, 839–858 (1953).

[50] — *Ring isomorphisms of Banach algebras.* Canad. J. Math. **6**, 374–381 (1954).

[51] — *Any orthocomplemented complete modular lattice is a continuous geometry.* Ann. of Math. (2) **61**, 524–541 (1955).

[52] — *Rings of operators.* University of Chicago Mimeographed Lecture Notes (Notes by S. Berberian, with an Appendix by R. Blattner). Chicago, Ill.: University of Chicago 1955.

[53] — *Von Neumann's characterization of factors of Type II_1.* Published in *John von Neumann: Collected works.* Vol. III, pp. 562–563. New York: Pergamon 1961.

[54] — *Rings of operators.* New York: Benjamin 1968.

[55] — *Algebraic and analytic aspects of operator algebras.* CBMS Regional Con-

ference Series in Mathematics, No. 1. Providence, R. I.: American Mathematical Society 1970.
[56] Kawada, Y., Higuti, K., Matusima, Y.: *Bemerkungen zur vorangehenden Arbeit von Herrn T. Iwamura.* Japan. J. Math. **19**, 73–79 (1944).
[57] Lebow, A.: *A Schroeder-Bernstein theorem for projections.* Proc. Amer. Math. Soc. **19**, 144–145 (1968).
[58] Loomis, L. H.: *The lattice theoretic background of the dimension theory of operator algebras.* Mem. Amer. Math. Soc., No. 18. Providence, R. I.: American Mathematical Society 1955.
[59] Maeda, F.: *Kontinuierliche Geometrien.* Berlin-Göttingen-Heidelberg: Springer 1958.
[60] — *Decomposition of general lattices into direct summands of types I, II and III.* J. Sci. Hiroshima Univ. Ser. A Math. **23**, 151–170 (1959).
[61] Maeda, S.: *Dimension functions on certain general lattices.* J. Sci. Hiroshima Univ. Ser. A Math. **19**, 211–237 (1955).
[62] — *Lengths of projections in rings of operators.* J. Sci. Hiroshima Univ. Ser. A Math. **20**, 5–11 (1956).
[63] — *On the lattice of projections of a Baer ∗-ring.* J. Sci. Hiroshima Univ. Ser. A Math. **22**, 75–88 (1958).
[64] — *On a ring whose principal right ideals generated by idempotents form a lattice.* J. Sci. Hiroshima Univ. Ser. A Math. **24**, 509–525 (1960).
[65] — *Dimension theory on relatively semi-orthocomplemented complete lattices.* J. Sci. Hiroshima Univ. Ser. A Math **25**, 369–404 (1961).
[66] Misonou, Y.: *On a weakly central operator algebra.* Tôhoku Math. J. **4**, 194–202 (1952).
[67] Murray, F. J., Neumann, J. von: *On rings of operators.* Ann. of Math. (2) **37**, 116–229 (1936).
[68] Neumann, J. von: *Zur Algebra der Funktionaloperatoren und Theorie der normalen Operatoren.* Math. Ann. **102**, 370–427 (1929).
[69] — *Continuous geometry.* Proc. Nat. Acad. Sci. U.S.A. **22**, 92–100 (1936).
[70] — *On rings of operators. Reduction theory.* Ann. of. Math. (2) **50**, 401–485 (1949).
[71] — *Continuous geometry.* Princeton, N. J.: Princeton University Press 1960.
[72] Prijatelj, N., Vidav, I.: *On special ∗-regular rings.* Michigan Math. J. **18**, 213–221 (1971).
[73] Ramsay, A.: *Dimension theory in complete orthocomplemented weakly modular lattices.* Trans. Amer. Math. Soc. **116**, 9–31 (1965).
[74] Rickart, C. E.: *Banach algebras with an adjoint operation.* Ann. of Math. (2) **47**, 528–550 (1946).
[75] — *General theory of Banach algebras.* Princeton, N. J.: Van Nostrand 1960.
[76] Roos, J.-E.: *Sur l'anneau maximal de fractions des AW*-algèbres et des anneaux de Baer.* C. R. Acad. Sci. Paris Sér. A–B **266**, A120–A123 (1968).
[77] Saitô, K.: *On the algebra of measurable operators for a general AW*-algebra.* Tôhoku Math. J. (2) **21**, 249–270 (1969).
[78] — *A non-commutative theory of integration for a semi-finite AW*-algebra and a problem of Feldman.* Tôhoku Math. J. (2) **22**, 420–461 (1970).
[79] — *On the embedding as a double commutator in a type I AW*-algebra.* Tôhoku Math. J. (2) **23**, 541–557 (1971).
[80] Sasaki, U.: *Lattices of projections in AW*-algebras.* J. Sci. Hiroshima Univ. Ser. A Math. **19**, 1–30 (1955).
[81] Segal, I. E.: *Decompositions of operator algebras. I, II.* Mem. Amer. Math. Soc., No. 9. New York: American Mathematical Society 1951.

[82] — *A non-commutative extension of abstract integration.* Ann. of Math. (2) **57**, 401–457 (1953); correction, ibid. **58**, 595–596 (1953).

[83] Skornyakov, L.A.: *Complemented modular lattices and regular rings.* Edinburgh: Oliver & Boyd, 1964.

[84] Steen, S.W.P.: *An introduction to the theory of operators.* Proc. London Math. Soc. (2) **41**, 361–392 (1936); Part II, ibid. (2) **43**, 529–543 (1937); Part III, ibid. (2) **44**, 398–411 (1938); Part IV, Proc. Cambridge Philos. Soc. **35**, 562–578 (1939); Part V, ibid. **36**, 139–149 (1940).

[85] Stone, M.H.: *Boundedness properties in function-lattices.* Canad. J. Math. **1**, 176–186 (1949).

[86] Takeda, Z., Turumaru, T.: *On the property "Position p′".* Math. Japon. **2**, 195–197 (1952).

[87] Teleman, S.: *On the regular rings of John von Neumann.* Rev. Roumaine Math. Pures Appl. **15**, 735–742 (1970).

[88] Utumi, Y.: *On quotient rings.* Osaka J. Math. **8**, 1–18 (1956).

[89] Vidav, I.: *On some $*$-regular rings.* Acad. Serbe Sci. Arts. Glas, Publ. Inst. Math. (Beograd) **13**, 73–80 (1959).

[90] Widom, H.: *Embedding in algebras of type I.* Duke Math. J. **23**, 309–324 (1956).

[91] Wolfson, K.G.: *Baer rings of endomorphisms.* Math. Ann. **143**, 19–28 (1961).

[92] Wright, F.B.: *A reduction for algebras of finite type.* Ann. of Math. (2) **60**, 560–570 (1954).

[93] — *The ideals in a factor.* Ann. of Math. (2) **68**, 475–483 (1958).

[94] Yen, T.: *Trace on finite AW^*-algebras.* Duke Math. J. **22**, 207–222 (1955).

[95] — *Quotient algebra of a finite AW^*-algebra.* Pacific J. Math. **6**, 389–395 (1956).

[96] Yohe, C.R.: *Commutative rings whose matrix rings are Baer rings.* Proc. Amer. Math. Soc. **22**, 189–191 (1969).

Supplementary Bibliography

[97] Akemann, C.A.: *Left ideal structure of C*-algebras*. J. Functional Analysis **6**, 305–317 (1970).

[98] Birkhoff, G.: *What can lattices do for you?*. Published in *Trends in lattice theory* (Abbott, J.C., Editor), pp. 1–40. New York: Van Nostrand Reinhold 1970.

[99] Cassels, J.W.S.: *On the representation of rational functions as sums of squares*. Acta Arith. **9**, 79–82 (1964).

[100] Clark, W.E.: *Baer rings which arise from certain transitive graphs*. Duke Math. J. **33**, 647–656 (1966).

[101] Davies, E.B.: *The structure of $\sum$*-algebras*. Quart. J. Math. Oxford Ser. (2) **20**, 351–366 (1969).

[102] Dye, H.A.: *On the geometry of projections in certain operator algebras*. Ann. of Math. (2) **61**, 73–89 (1955).

[103] Džanseitov, K.K.: *Some properties of special *-regular rings* (Russian). Izv. Akad. Nauk SSSR Ser. Math. **27**, 279–286 (1963).

[104] Feldman, J.: *Nonseparability of certain finite factors*. Proc. Amer. Math. Soc. **7**, 23–26 (1956).

[105] — *Isomorphisms of finite type II rings of operators*. Ann. of Math. (2) **63**, 565–571 (1956).

[106] — *Some connections between topological and algebraic properties in rings of operators*. Duke Math. J. **23**, 365–370 (1956).

[107] — Fell, J.M.G.: *Separable representations of rings of operators*. Ann. of Math. (2) **65**, 241–249 (1957).

[108] Fell, J.M.G.: *Representations of weakly closed algebras*. Math. Ann. **133**, 118–126 (1957).

[109] Halpern, H.: *The maximal GCR ideal in an AW*-algebra*. Proc. Amer. Math. Soc. **17**, 906–914 (1966).

[110] Holland, S.S., Jr.: *The current interest in orthomodular lattices*. Published in *Trends in lattice theory* (Abbott, J.C., Editor), pp. 41–126. New York: Van Nostrand Reinhold 1970.

[111] Johnson, B.E.: *AW*-algebras are QW*-algebras*. Pacific J. Math. **23**, 97–99 (1967).

[112] Kadison, R.V.: *Operator algebras with a faithful weakly-closed representation*. Ann. of Math. (2) **64**, 175–181 (1956).

[113] — Pedersen, G.K.: *Equivalence in operator algebras*. Math. Scand. **27**, 205–222 (1970).

[114] Kehlet, E.T.: *On the monotone sequential closure of a C*-algebra*. Math. Scand. **25**, 59–70 (1969).

[115] Maeda, F., Maeda, S.: *Theory of symmetric lattices*. Berlin-Heidelberg-New York: Springer 1970.

[116] Maeda, S.: *On relatively semi-orthocomplemented lattices*. J. Sci. Hiroshima Univ. Ser. A Math. **24**, 155–161 (1960).

[117] Mewborn, A.C.: *Regular rings and Baer rings*. Math. Z. **121**, 211–219 (1971).
[118] Miles, P.: *B*-algebra unit ball extremal points*. Pacific J. Math. **14**, 627–637 (1964).
[119] Nakai, M.: *Some expectations in AW*-algebras*. Proc. Japan Acad. **34**, 411–416 (1958).
[120] Ono, T.: *Local theory of rings of operators. I, II*. J. Math. Soc. Japan **10**, 184–216, 438–458 (1958).
[121] Pedersen, G.K.: *Measure theory for C*-algebras*. Math. Scand. **19**, 131–145 (1966); Part II, ibid. **22**, 63–74 (1968); Part III, ibid. **25**, 71–93 (1969); Part IV, ibid. **25**, 121–127 (1969).
[122] — *Operator algebras with weakly closed abelian subalgebras*, preprint.
[123] Pollingher, A., Zaks, A.: *On Baer and quasi-Baer rings*. Duke Math. J. **37**, 127–138 (1970).
[124] Raphael, R.M.: *Algebraic extensions of commutative regular rings*. Canad. J. Math. **22**, 1133–1155 (1970).
[125] Reid, G.A.: *A generalisation of W*-algebras*. Pacific J. Math. **15**, 1019–1026 (1965).
[126] Sakai, S.: *On the group isomorphism of unitary groups in AW*-algebras*. Tôhoku Math. J. **7**, 87–95 (1955).
[127] — *A characterization of W*-algebras*. Pacific J. Math. **6**, 763–773 (1956).
[128] — *C*-algebras and W*-algebras*. Berlin-Heidelberg-New York: Springer 1971.
[129] Schwartz, J.T.: *W*-algebras*. New York: Gordon and Breach 1967.
[130] Widom, H.: *Approximately finite algebras*. Trans. Amer. Math. Soc. **83**, 170–178 (1956).
[131] Wolfson, K.G.: *Baer subrings of the ring of linear transformations*. Math. Z. **75**, 328–332 (1961).
[132] Wright, J.D.M.: *A spectral theorem for normal operators on a Kaplansky-Hilbert module*. Proc. London Math. Soc. (3) **19**, 258–268 (1969).
[133] Yen, T.: *Isomorphism of unitary groups in AW*-algebras*. Tôhoku Math. J. **8**, 275–280 (1956).
[134] — *Isomorphisms of AW*-algebras*. Proc. Amer. Math. Soc. **8**, 345–349 (1957).
[135] Halperin, I.: *Extension of the rank function*. Studia Math. **27**, 325–335 (1966).
[136] Burke, J.L.: *On the property* (PU) *for *-regular rank rings*. Preprint.

Index

Die Grundlehren der mathematischen Wissenschaften in Einzeldarstellungen mit besonderer Berücksichtigung der Anwendungsgebiete

Eine Auswahl

---

67. Byrd/Friedman: Handbook of Elliptic Integrals for Engineers and Scientists
68. Aumann: Reelle Funktionen
74. Boerner: Darstellungen von Gruppen
76. Tricomi: Vorlesungen über Orthogonalreihen
77. Behnke/Sommer: Theorie der analytischen Funktionen einer komplexen Veränderlichen
78. Lorenzen: Einführung in die operative Logik und Mathematik
86. Richter: Wahrscheinlichkeitstheorie
87. van der Waerden: Mathematische Statistik
94. Funk: Variationsrechnung und ihre Anwendung in Physik und Technik
97. Greub: Linear Algebra
99. Cassels: An Introduction to the Geometry of Numbers
104. Chung: Markov Chains with Stationary Transition Probabilities
107. Köthe: Topologische lineare Räume
114. MacLane: Homology
116. Hörmander: Linear Partial Differential Operators
117. O'Meara: Introduction to Quadratic Forms
120. Collatz: Funktionalanalysis und numerische Mathematik
121./122. Dynkin: Markov Processes
123. Yosida: Functional Analysis
124. Morgenstern: Einführung in die Wahrscheinlichkeitsrechnung und mathematische Statistik
125. Itô/McKean jr.: Diffusion Processes and their sample Paths
126. Lehto/Virtanen: Quasikonforme Abbildungen
127. Hermes: Enumerability, Decidability, Computability
128. Braun/Koecher: Jordan-Algebren
129. Nikodým: The Mathematical Apparatus for Quantum Theories
130. Morrey jr.: Multiple Integrals in the Calculus of Variations
131. Hirzebruch: Topological Methods in Algebraic Geometry
132. Kato: Perturbation Theory for Linear Operators
133. Haupt/Künneth: Geometrische Ordnungen
134. Huppert: Endliche Gruppen I
135. Handbook for Automatic Computation. Vol. 1/Part a: Rutishauser: Description of ALGOL 60
136. Greub: Multilinear Algebra
137. Handbook for Automatic Computation. Vol. 1/Part b: Grau/Hill/Langmaack: Translation of ALGOL 60
138. Hahn: Stability of Motion
139. Mathematische Hilfsmittel des Ingenieurs. 1. Teil
140. Mathematische Hilfsmittel des Ingenieurs. 2. Teil
141. Mathematische Hilfsmittel des Ingenieurs. 3. Teil
142. Mathematische Hilfsmittel des Ingenieurs. 4. Teil
143. Schur/Grunsky: Vorlesungen über Invariantentheorie
144. Weil: Basic Number Theory
145. Butzer/Berens: Semi-Groups of Operators and Approximation

146. Treves: Locally Convex Spaces and Linear Partial Differential Equations
147. Lamotke: Semisimpliziale algebraische Topologie.
148. Chandrasekharan: Introduction to Analytic Number Theory
149. Sario/Oikawa: Capacity Functions
150. Iosifescu/Theodorescu: Random Processes and Learning
151. Mandl: Analytical Treatment of One-dimensional Markov Processes
152. Hewitt/Ross: Abstract Harmonic Analysis. Vol. 2: Structure and Analysis for Compact Groups. Analysis on Locally Compact Abelian Groups
153. Federer: Geometric Measure Theory
154. Singer: Bases in Banach Spaces I
155. Müller: Foundations of the Mathematical Theory of Electromagnetic Waves
156. van der Waerden: Mathematical Statistics
157. Prohorov/Rozanov: Probability Theory
159. Köthe: Topological Vector Spaces I
160. Agrest/Maksimov: Theory of Incomplete Cylindrical Functions and Their Applications
161. Bhatia/Szegö: Stability Theory of Dynamical Systems
162. Nevanlinna: Analytic Functions
163. Stoer/Witzgall: Convexity and Optimization in Finite Dimensions I
164. Sario/Nakai: Classification Theory of Riemann Surfaces
165. Mitrinović/Vasić: Analytic Inequalities
166. Grothendieck/Dieudonné: Eléments de Géométrie Algébrique I
167. Chandrasekharan: Arithmetical Functions
168. Palamodov: Linear Differential Operators with Constant Coefficients
170. Lions: Optimal Control Systems Governed by Partial Differential Equations
171. Singer: Best Approximation in Normed Linear Spaces by Elements of Linear Subspaces
172. Bühlmann: Mathematical Methods in Risk Theory
173. F. Maeda/S. Maeda: Theory of Symmetric Lattices
174. Stiefel/Scheifele: Linear and Regular Celestial Mechanics. Perturbed Two-body Motion—Numerical Methods—Canonical Theory
175. Larsen: An Introduction of the Theory of Multipliers
176. Grauert/Remmert: Analytische Stellenalgebren
177. Flügge: Practical Quantum Mechanics I
178. Flügge: Practical Quantum Mechanics II
179. Giraud: Cohomologie non abélienne
180. Landkoff: Foundations of Modern Potential Theory
181. Lions/Magenes: Non-Homogeneous Boundary Value Problems and Applications I
182. Lions/Magenes: Non-Homogeneous Boundary Value Problems and Applications II
183. Lions/Magenes: Non-Homogeneous Boundary Value Problems and Applications III. In preparation
184. Rosenblatt: Markov Processes. Structure and Asymptotic Behavior
185. Rubinowicz: Sommerfeldsche Polynommethode
186. Wilkinson/Reinsch: Handbook for Automatic Computation II, Linear Algebra
187. Siegel/Moser: Lectures on Celestial Mechanics
188. Warner: Harmonic Analysis on Semi-Simple Lie Groups I
189. Warner: Harmonic Analysis on Semi-Simple Lie Groups II
190. Faith: Algebra: Rings, Modules, and Categories I. In preparation
191. Faith: Algebra: Rings, Modules, and Categories II. In preparation
192. Maltsev: Algebraic Systems. In preparation
193. Pólya/Szegö: Problems and Theorems in Analysis. Vol. 1. In preparation
194. Igusa: Theta Functions